VOLUME SEVEN HUNDRED AND TWO

METHODS IN
ENZYMOLOGY

Siderophore and Related Natural
Products Biosynthesis

METHODS IN ENZYMOLOGY

Editors-in-Chief

ANNA MARIE PYLE

*Departments of Molecular, Cellular and Developmental
Biology and Department of Chemistry
Investigator, Howard Hughes Medical Institute
Yale University*

DAVID W. CHRISTIANSON

*Roy and Diana Vagelos Laboratories
Department of Chemistry
University of Pennsylvania
Philadelphia, PA*

Founding Editors

SIDNEY P. COLOWICK and NATHAN O. KAPLAN

VOLUME SEVEN HUNDRED AND TWO

METHODS IN ENZYMOLOGY

Siderophore and Related Natural Products Biosynthesis

Edited by

TIMOTHY WENCEWICZ
*Department of Chemistry,
Washington University in St. Louis,
St. Louis, MO, United States*

Academic Press is an imprint of Elsevier
50 Hampshire Street, 5th Floor, Cambridge, MA 02139, United States
525 B Street, Suite 1650, San Diego, CA 92101, United States
125 London Wall, London, EC2Y 5AS, United Kingdom

First edition 2024

Copyright © 2024 Elsevier Inc. All rights are reserved, including those for text and data mining, AI training, and similar technologies.

Publisher's note: Elsevier takes a neutral position with respect to territorial disputes or jurisdictional claims in its published content, including in maps and institutional affiliations.

No part of this publication may be reproduced or transmitted in any form or by any means, electronic or mechanical, including photocopying, recording, or any information storage and retrieval system, without permission in writing from the publisher. Details on how to seek permission, further information about the Publisher's permissions policies and our arrangements with organizations such as the Copyright Clearance Center and the Copyright Licensing Agency, can be found at our website: www.elsevier.com/permissions.

This book and the individual contributions contained in it are protected under copyright by the Publisher (other than as may be noted herein).

Notices
Knowledge and best practice in this field are constantly changing. As new research and experience broaden our understanding, changes in research methods, professional practices, or medical treatment may become necessary.

Practitioners and researchers must always rely on their own experience and knowledge in evaluating and using any information, methods, compounds, or experiments described herein. In using such information or methods they should be mindful of their own safety and the safety of others, including parties for whom they have a professional responsibility.

To the fullest extent of the law, neither the Publisher nor the authors, contributors, or editors, assume any liability for any injury and/or damage to persons or property as a matter of products liability, negligence or otherwise, or from any use or operation of any methods, products, instructions, or ideas contained in the material herein.

ISBN: 978-0-443-29678-9
ISSN: 0076-6879

For information on all Academic Press publications
visit our website at https://www.elsevier.com/books-and-journals

Publisher: Zoe Kruze
Editorial Project Manager: Saloni Vohra
Production Project Manager: James Selvam
Cover Designer: Gopalakrishnan Venkatraman

Typeset by MPS Limited, India

Contents

Contributors   *xvii*

1. Kinetic analysis of the three-substrate reaction mechanism of an NRPS-independent siderophore (NIS) synthetase   **1**

Andrew M. Gulick, Lisa S. Mydy, and Ketan D. Patel

1. Introduction	2
2. Protein production and purification	5
2.1 Equipment	5
2.2 Buffers, strains, and reagents	5
2.3 Procedure	6
2.4 Notes	7
3. NADH coupled adenylation assay	8
3.1 Equipment	8
3.2 Reagents	8
3.3 Procedure	9
3.4 Notes	11
4. Kinetic analysis to distinguish among potential mechanisms	11
4.1 Possible mechanisms	11
4.2 Initial double reciprocal plots	13
4.3 Slope and intercept replots	13
4.4 Notes	15
5. Alternate approaches	15
6. Summary and conclusions	15
Acknowledgment	17
References	17

2. Experimental methods for evaluating siderophore–antibiotic conjugates   **21**

Rachel N. Motz, Ghazal Kamyabi, and Elizabeth M. Nolan

1. Introduction	22
2. Methods	25
2.1 General considerations for evaluation of SACs	25
2.2 Safety precautions and general practices	26
2.3 Antimicrobial activity (AMA) assay	26

2.4 Time-kill kinetics	33
2.5 Competition with unmodified Ent	36
2.6 ^{57}Fe uptake assay	40
3. Summary	46
References	46

3. A continuous fluorescence assay to measure nicotianamine synthase activity

51

Thiago M. Pasin, Kathleen M. Meneely, Deegan M. Ruiz, and Audrey L. Lamb

1. Introduction	52
2. Assay design	54
3. Before you begin	55
4. Key resources table	55
5. Methods and equipment	56
5.1 Equipment	56
5.2 Reagents	56
6. Preparation of reagents	57
6.1 Purification of methylthioadenosine nucleosidase (MTAN)	57
6.2 Purification of adenine deaminase (AD)	60
6.3 Check for activity of MTAN and AD	64
7. Step-by-step method details	66
7.1 Prepare standard curve	66
7.2 Run the assay	68
8. Quantification and statistical analysis	69
9. Summary	72
References	73

4. ITC-based kinetics assay for NIS synthetases

75

Katherine M. Hoffmann, Jocelin D. Hernandez, Eliana G. Goncuian, and Nathan L. March

1. Introduction	76
2. General overview of the method	79
3. Before you begin	80
3.1 Protein preparation	80
4. Materials and equipment	81
5. Step-by-step method details	81
5.1 Optional recovery of siderophores	82
6. Expected outcomes	83
7. Quantification and statistical analysis	84

Contents

vii

8. Optimization and troubleshooting — 84
   8.1 Problem: enzyme solubility — 84
9. Summary — 84
References — 85

5. An in vitro assay to explore condensation domain specificity from non-ribosomal peptide synthesis

89

Minuri Ratnayake, Y.T. Candace Ho, Xinyun Jian, and Max J. Cryle

1. Introduction — 90
2. General method and statistical analysis — 95
3. Molecular design of PCP_2-C_3 SpyCatcher and PCP_3 SpyTag constructs — 96
   3.1 Equipment — 97
   3.2 Reagents — 97
   3.3 Procedure — 97
   3.4 Notes — 101
4. Protein expression and purification — 102
   4.1 Reagents — 102
   4.2 Procedure — 102
   4.3 Notes — 105
5. Synthesis of chemical reagents — 105
   5.1 Equipment — 106
   5.2 Reagents — 106
   5.3 Synthesis — 107
   5.4 Notes — 108
6. *In vitro* reconstitution of NRPS C-domain — 108
   6.1 Equipment — 108
   6.2 Reagents — 108
   6.3 Procedure — 109
   6.4 Notes — 110
7. LC-HRMS/MS analysis of methylamine cleaved peptide products — 111
   7.1 Material and equipment — 111
   7.2 Buffer and reagents — 111
   7.3 Procedures — 111
   7.4 Notes — 114
8. Intact protein PPant ejection LC-ESI-Q-TOF-MS analysis of chemically stabilised peptide products — 114
   8.1 Equipment — 115
   8.2 Buffers and reagents — 115
   8.3 Procedure — 115
   8.4 Notes — 116

viii Contents

9. Conclusions 117
Acknowledgements 117
References 118

6. The production of siderophore analogues using precursor-directed biosynthesis 121

Tomas Richardson-Sanchez, Thomas J. Telfer, Cho Z. Soe,
Kate P. Nolan, Michael P. Gotsbacher, and Rachel Codd

1. Introduction 122
  1.1 The clinical natural product desferrioxamine B 122
  1.2 Structural diversification of desferrioxamine B using precursor-directed biosynthesis 124
  1.3 Generation of constitutional isomers of desferrioxamine B analogues 125
  1.4 Mass spectrometry to identify desferrioxamine B analogues and constitutional isomers 126
  1.5 Scope of precursor-directed biosynthesis in producing desferrioxamine B analogues 129
  1.6 Theoretical maxima of desferrioxamine B analogues produced using precursor-directed biosynthesis 132
2. Materials and equipment 133
  2.1 Bacteria and chemicals 133
  2.2 Consumables 135
  2.3 General equipment 135
  2.4 LC-MS-Q instrumentation 135
  2.5 LC-MS/MS-QQQ instrumentation 135
3. Protocol 136
  3.1 Before you begin 136
  3.2 Preparing solutions and use of high grade reagents 136
  3.3 Preparing frozen stocks of *Streptomyces pilosus* 136
  3.4 Precursor-directed biosynthesis using non-native diamine substrates 137
  3.5 Measuring siderophore production 139
  3.6 Siderophore purification 139
  3.7 Analysing desferrioxamine B analogues and constitutional isomers 141
4. Summary 142
Acknowledgements 142
References 142

Contents ix

7. Preparation of coenzyme F430 biosynthetic enzymes and intermediates   **147**

Prosenjit Ray, Chelsea R. Rand-Fleming, and
Steven O. Mansoorabadi

1. Introduction   148
2. Expression and purification of the coenzyme F430 biosynthesis enzymes   151
   2.1 HemC   152
   2.2 HemD   153
   2.3 SirA   154
   2.4 SirC   155
   2.5 CfbA   157
   2.6 CfbB   158
   2.7 CfbCD   159
   2.8 CfbE   160
   2.9 McrD   161
3. Synthesis and purification of coenzyme F430 biosynthetic intermediates   162
   3.1 Sirohydrochlorin   163
   3.2 Ni-sirohydrochlorin   163
   3.3 Ni-sirohydrochlorin a,c-diamide   164
   3.4 $15,17^3$-seco-F430-17^3-acid   165
   3.5 Coenzyme F430   166
4. Concluding remarks   167
Acknowledgments   167
References   168

8. Purification and biochemical characterization of methanobactin biosynthetic enzymes   **171**

Reyvin M. Reyes and Amy C. Rosenzweig

1. Introduction   172
2. Expression and purification of *M. trichosporium* OB3b MbnBC complexes   174
   2.1 Plasmid construction and transformation into *Escherichia coli*   174
   2.2 Large-scale growth and induction of MbnBC expression   175
   2.3 Purification of MbnBC for biochemical studies   176
   2.4 Purification of MbnBC for crystallization   177
3. *In vitro* modification of MbnA using purified MbnBC   179
4. Crystallization and structure determination of *M. trichosporium* OB3b MbnBC   181

5. Expression and purification of *M. trichosporium* OB3b MbnN	182
6. Conclusions	185
Acknowledgments	185
References	185

9. Discovery, isolation, and characterization of diazeniumdiolate siderophores — **189**

Melanie Susman, Jin Yan, Christina Makris, and Alison Butler

1. Introduction	190
2. Bioinformatics and genome mining to predict bacteria producing graminine-containing siderophores	192
2.1 Materials	192
2.2 Constructing an SSN	192
2.3 Identifying bacterial strains of interest	194
3. Bacterial growth conditions for production of siderophores	195
3.1 Materials	195
3.2 Protocols for bacterial culturing	196
4. Detection and isolation of the siderophores	198
4.1 Materials	198
4.2 CAS assay for general siderophore detection	199
4.3 Extraction of siderophores	200
4.4 Purification of siderophores	202
5. Identification and characterization of graminine-containing siderophores	203
5.1 Materials	204
5.2 Mass spectrometry and stable isotope labeling	205
5.3 Marfey's amino acid analysis	206
5.4 Photoreactivity of the diazeniumdiolate group	208
5.5 NMR spectroscopy characterization of the *C*-diazeniumdiolate and the photoproducts	211
6. Conclusions and future outlook	212
Acknowledgements	212
References	213

10. Linking biosynthetic genes to natural products using inverse stable isotopic labeling (InverSIL) — **215**

Tashi C.E. Liebergesell and Aaron W. Puri

1. Introduction	216
1.1 The gene-to-molecule approach for natural product discovery	216

Contents

xi

1.2 Inverse stable isotopic labeling (InverSIL)	216
1.3 Using InverSIL to link quorum sensing signal synthase genes to their products	217
2. Key resources	218
3. Equipment and reagents	219
3.1 Inverse labeling of microbial culture	219
3.2 Natural product extraction	219
3.3 Analysis of labeled microbial extracts by untargeted mass spectrometry	219
3.4 Data analysis to identify inverse labeled natural products	220
4. Method	220
4.1 Inverse labeling of microbial culture	220
4.2 Natural product extraction	221
4.3 Analysis of labeled microbial extracts by untargeted mass spectrometry	222
4.4 Data analysis to identify inverse labeled natural products	222
5. Notes	225
Acknowledgments	226
References	226

11. 4-Aldrithiol-based photometric assay for detection of methylthioalkylmalate synthase activity

229

Vivian Kitainda and Joseph Jez

1. Introduction	230
2. Using continuous enzyme assays for MAMS activity	232
2.1 Materials and equipment	232
2.2 Reagents	233
2.3 MAMS spectrophotometric enzyme assay protocol	233
3. Analysis and interpretation of spectrophotometric enzyme assay for *Brassica juncea* MAMS isoforms	236
3.1 pH screen	236
3.2 Buffer screen	236
3.3 Metal cofactor screen	238
3.4 Michaelis-Menten kinetic analysis	238
3.5 Leucine inhibition assays	242
4. Spectrophotometric alternatives to the 4,4'-DTP assay	242
5. Conclusion and outlook	243
References	244

12. Methods for biochemical characterization of flavin-dependent *N*-monooxygenases involved in siderophore biosynthesis

Noah S. Lyons, Sydney B. Johnson, and Pablo Sobrado

1. Introduction	248
2. NADPH oxidation	252
2.1 Equipment	253
2.2 Reagents	253
2.3 Procedure	254
2.4 Notes	254
3. Oxygen consumption assays	254
3.1 Equipment	256
3.2 Reagents	256
3.3 Procedure	256
3.4 Notes	256
4. Iodine oxidation assay	257
4.1 Equipment	258
4.2 Reagents	258
4.3 Procedure	259
4.4 Notes	260
5. Determination of hydrogen peroxide formation and reaction uncoupling	260
5.1 Colorimetric measurement of H_2O_2	261
5.2 Continuous measurement of H_2O_2 using catalase	262
5.3 Reagents	262
5.4 Procedure	262
5.5 Notes	262
6. Fmoc-Cl derivatization	263
6.1 Equipment	263
6.2 Reagents	263
6.3 Procedure	264
6.4 Notes	265
7. Determination of reaction stereospecificity	265
7.1 Equipment	267
7.2 Expression and purification of TbADH	267
7.3 Synthesis of (R)-[4-^{2}H]-NADPH	269
7.4 Expression and purification of LmG6PDH	270
7.5 Synthesis of (S)-[4-^{2}H]-NADPH	271
8. Fluorescence anisotropy	272
8.1 Equipment	274

8.2	Reagents	274
8.3	Procedure	274
9.	Targeting flavin dynamics for drug discovery	274
10.	Summary and conclusions	276
Acknowledgments		276
References		277

13. Siderophore-dependent ferrichelatases

C.E. Merrick, N.M. Gulati, and T.A. Wencewicz

281

1.	Introduction	282
2.	General methods and statistical analysis	291
3.	Overexpression and purification of N-His6-FhuD2	292
	3.1 Equipment	292
	3.2 Buffers, strains, and reagents	292
	3.3 Procedure	293
	3.4 Optional steps	295
	3.5 Notes	296
4.	FhuD2 binding assay: intrinsic tryptophan fluorescence quenching	297
	4.1 Equipment	297
	4.2 Buffers and reagents	297
	4.3 Procedure	297
	4.4 Notes	298
5.	FhuD2 binding assay: siderophore affinity chromatography	300
	5.1 Equipment	300
	5.2 Buffers and reagents	300
	5.3 Procedure	300
	5.4 Optional steps	301
	5.5 Notes	302
6.	Synthesis of DFO-NBD and Fe(III)-siderophores	302
	6.1 Equipment	302
	6.2 Reagents	303
	6.3 Procedure: synthesis of DFO-NBD	303
	6.4 Procedure: synthesis of iron bound siderophores	303
	6.5 Notes	304
7.	Ferrichelatase assay	304
	7.1 Equipment	304
	7.2 Buffers and reagents	305
	7.3 Procedure: fluorescence based FhuD2 iron exchange assay	305
	7.4 Procedure: validation by LC–MS	306
	7.5 Notes	307

8. Siderophore EDTA competition assay	307
8.1 Equipment	307
8.2 Buffers and reagents	307
8.3 Procedure	307
8.4 Optional steps	309
8.5 Notes	309
9. *S. aureus* growth studies	309
9.1 Equipment	309
9.2 Strains and reagents	309
9.3 Procedure	310
9.4 Notes	310
10. Summary and conclusions	311
Acknowledgments	312
References	312

14. Native metabolomics for mass spectrometry-based siderophore discovery
317

Marquis T. Yazzie, Zachary L. Reitz, Robin Schmid, Daniel Petras, and Allegra T. Aron

1. Introduction	318
2. General methods and assembly of instrumentation	322
3. Sample preparation for native metabolomics	325
3.1 Materials and equipment	326
3.2 Step-by-step method details	326
3.3 Alternative methods/procedures	327
4. Mass spectrometry acquisition parameters	328
4.1 Materials and equipment	328
4.2 LC-MS^2 acquisition parameters	329
4.3 Post-column pH adjustment	330
4.4 Post-column metal infusion	332
5. Data analysis using ion identity molecular networking	333
5.1 Materials and equipment	334
5.2 Feature finding in mzmine: step-by-step method details	334
5.3 Metal adduct specific modules: step-by-step method details	336
5.4 Mzmine compound annotation, molecular networking, data export for GNPS2	342
5.5 Visualization in GNPS, GNPS2 & Cytoscape	343
5.6 Layering with other sources of info	345
6. Limitations and considerations	345
7. Summary and conclusions	347

Contents xv

Acknowledgments 347
References 347

15. Anaerobic heme recycling by gut microbes: Important methods for monitoring porphyrin production 353

Ronivaldo Rodrigues da Silva, Arnab Kumar Nath,
Victoria Adedoyin, Emmanuel Akpoto, and Jennifer L. DuBois

1. Introduction 354
2. Methods 355
   2.1 Preparing heme and porphyrin-containing extracts from biological materials 355
3. Procedure 355
   3.1 Discontinuous quantification of hemes and porphyrins by HPLC 358
   3.2 Quantification of heme and porphyrins by UV–visible (UV–vis) and fluorescence emission spectroscopy 360
   3.3 Distinguishing porphyrins with catabolic and anabolic origins using stable isotopes and LC-MS 364
References 370

16. Predicting metallophore structure and function through genome mining 371

Zachary L. Reitz

1. Introduction 372
2. Materials and methods 374
   2.1 Retrieving records from the antiSMASH database 374
   2.2 Running antiSMASH jobs on the web server 374
   2.3 Retrieving records from MIBiG 375
   2.4 Running BLAST 376
3. Metallophore genomics: from sequence to molecule to function 376
   3.1 Biosynthesis of metal-chelating small molecules 377
   3.2 Active transport through the cell membrane and subsequent metal release 378
   3.3 Metal-responsive regulatory elements 379
4. Predicting metallophore pathways in antiSMASH 381
   4.1 Rule-based BGC detection 381
   4.2 Navigating antiSMASH results 382
   4.3 Identifying putative metallophore BGC regions among antiSMASH results 384
5. Predicting metallophore structure through comparative analyses 385

	5.1	Strategies for comparing BGCs	385
	5.2	Predicting the structure of a peptide metallophore from *Dickeya dadantii* 3937	387
	5.3	Predicting the structure of a putatively novel metallophore in *Acinetobacter pittii* PHEA-2	390
6.	Next steps for the reader		396
Acknowledgments			398
References			398

Contributors

Victoria Adedoyin
Department of Chemistry and Biochemistry, Montana State University, Bozeman, MT, United States

Emmanuel Akpoto
Department of Chemistry and Biochemistry, Montana State University, Bozeman, MT, United States

Allegra T. Aron
Department of Chemistry and Biochemistry, University of Denver, Denver, CO, United States

Alison Butler
Department of Chemistry & Biochemistry, University of California, Santa Barbara, CA, United States

Rachel Codd
School of Medical Sciences, The University of Sydney, Sydney, NSW, Australia

Max J. Cryle
Department of Biochemistry and Molecular Biology, The Monash Biomedicine Discovery Institute, Monash University; EMBL Australia, Monash University, Clayton, VIC, Australia; ARC Centre of Excellence for Innovations in Peptide and Protein Science

Jennifer L. DuBois
Department of Chemistry and Biochemistry, Montana State University, Bozeman, MT, United States

Eliana G. Goncuian
Department of Chemistry, California Lutheran University, Thousand Oaks, CA, United States

Michael P. Gotsbacher
School of Medical Sciences, The University of Sydney, Sydney, NSW, Australia

N.M. Gulati
Department of Chemistry, Washington University in St. Louis, St. Louis, MO, United States

Andrew M. Gulick
Department of Structural Biology, University at Buffalo, SUNY, Buffalo, NY, United States

Jocelin D. Hernandez
Department of Chemistry, California Lutheran University, Thousand Oaks, CA, United States

Y.T. Candace Ho
Department of Chemistry, University of Warwick, Coventry, United Kingdom

Katherine M. Hoffmann
Department of Chemistry, California Lutheran University, Thousand Oaks, CA, United States

Joseph Jez
Department of Biology, Washington University in St. Louis, St. Louis, MO, United States

Xinyun Jian
Department of Biochemistry and Molecular Biology, The Monash Biomedicine Discovery Institute, Monash University; EMBL Australia, Monash University, Clayton, VIC, Australia; ARC Centre of Excellence for Innovations in Peptide and Protein Science

Sydney B. Johnson
Department of Biochemistry, Virginia Tech, Blacksburg, VA, United States

Ghazal Kamyabi
Department of Chemistry, Massachusetts Institute of Technology, Cambridge, MA, United States

Vivian Kitainda
Department of Biology, Washington University in St. Louis, St. Louis, MO, United States

Audrey L. Lamb
Department of Chemistry, University of Texas at San Antonio, San Antonio, TX, United States

Tashi C.E. Liebergesell
Department of Chemistry and the Henry Eyring Center for Cell and Genome Science, University of Utah, Salt Lake City, UT, United States

Noah S. Lyons
Department of Biochemistry, Virginia Tech, Blacksburg, VA, United States

Christina Makris
Department of Chemistry & Biochemistry, University of California, Santa Barbara, CA, United States

Steven O. Mansoorabadi
Department of Chemistry and Biochemistry, Auburn University, Auburn, AL, United States

Nathan L. March
Department of Chemistry, California Lutheran University, Thousand Oaks, CA, United States

Kathleen M. Meneely
Department of Chemistry, University of Texas at San Antonio, San Antonio, TX, United States

C.E. Merrick
Department of Chemistry, Washington University in St. Louis, St. Louis, MO, United States

Rachel N. Motz
Department of Chemistry, Massachusetts Institute of Technology, Cambridge, MA, United States

Lisa S. Mydy
Department of Medicinal Chemistry, University of Michigan, Ann Arbor, MI, United States

Arnab Kumar Nath
Department of Chemistry and Biochemistry, Montana State University, Bozeman, MT, United States

Elizabeth M. Nolan
Department of Chemistry, Massachusetts Institute of Technology, Cambridge, MA, United States

Kate P. Nolan
School of Medical Sciences, The University of Sydney, Sydney, NSW, Australia

Thiago M. Pasin
Department of Chemistry, University of Texas at San Antonio, San Antonio, TX, United States

Ketan D. Patel
Department of Structural Biology, University at Buffalo, SUNY, Buffalo, NY, United States; Interfaculty of Microbiology and Infection Medicine, University of Tubingen, Tubingen, Germany

Daniel Petras
Department of Biochemistry, University of California Riverside, Riverside, CA, United States

Aaron W. Puri
Department of Chemistry and the Henry Eyring Center for Cell and Genome Science, University of Utah, Salt Lake City, UT, United States

Chelsea R. Rand-Fleming
Department of Chemistry and Biochemistry, Auburn University, Auburn, AL, United States

Minuri Ratnayake
Department of Biochemistry and Molecular Biology, The Monash Biomedicine Discovery Institute, Monash University; EMBL Australia, Monash University, Clayton, VIC, Australia; ARC Centre of Excellence for Innovations in Peptide and Protein Science

Prosenjit Ray
Department of Chemistry and Biochemistry, Auburn University, Auburn, AL, United States

Zachary L. Reitz
Department of Ecology, Evolution and Marine Biology, University of California, Santa Barbara, CA, United States

Reyvin M. Reyes
Departments of Molecular Biosciences and of Chemistry, Northwestern University, Evanston, IL, United States

Tomas Richardson-Sanchez
School of Medical Sciences, The University of Sydney, Sydney, NSW, Australia

Amy C. Rosenzweig
Departments of Molecular Biosciences and of Chemistry, Northwestern University, Evanston, IL, United States

Deegan M. Ruiz
Department of Chemistry, University of Texas at San Antonio, San Antonio, TX, United States

Robin Schmid
Institute of Organic Chemistry and Biochemistry, Czech Academy of Sciences, Prague, Czechia

Pablo Sobrado
Department of Biochemistry; Center for Drug Discovery, Virginia Tech, Blacksburg, VA; Department of Chemistry, Missouri University of Science and Technology, Rolla, MO, United States

Cho Z. Soe
School of Medical Sciences, The University of Sydney, Sydney, NSW, Australia

Melanie Susman
Department of Chemistry & Biochemistry, University of California, Santa Barbara, CA, United States

Thomas J. Telfer
School of Medical Sciences, The University of Sydney, Sydney, NSW, Australia

T.A. Wencewicz
Department of Chemistry, Washington University in St. Louis, St. Louis, MO, United States

Jin Yan
Department of Chemistry & Biochemistry, University of California, Santa Barbara, CA, United States

Marquis T. Yazzie
Department of Chemistry and Biochemistry, University of Denver, Denver, CO, United States

Ronivaldo Rodrigues da Silva
Department of Chemistry and Biochemistry, Montana State University, Bozeman, MT, United States

CHAPTER ONE

Kinetic analysis of the three-substrate reaction mechanism of an NRPS-independent siderophore (NIS) synthetase

Andrew M. Gulick[a,*], Lisa S. Mydy[b], and Ketan D. Patel[a]

[a]Department of Structural Biology, University at Buffalo, SUNY, Buffalo, NY, United States
[b]Department of Medicinal Chemistry, University of Michigan, Ann Arbor, MI, United States
*Corresponding author. e-mail address: amgulick@buffalo.edu

Contents

1. Introduction	2
2. Protein production and purification	5
2.1 Equipment	5
2.2 Buffers, strains, and reagents	5
2.3 Procedure	6
2.4 Notes	7
3. NADH coupled adenylation assay	8
3.1 Equipment	8
3.2 Reagents	8
3.3 Procedure	9
3.4 Notes	11
4. Kinetic analysis to distinguish among potential mechanisms	11
4.1 Possible mechanisms	11
4.2 Initial double reciprocal plots	13
4.3 Slope and intercept replots	13
4.4 Notes	15
5. Alternate approaches	15
6. Summary and conclusions	15
Acknowledgment	17
References	17

Abstract

The biosynthesis of many bacterial siderophores employs a member of a family of ligases that have been defined as NRPS-independent siderophore (NIS) synthetases. These NIS synthetases use a molecule of ATP to produce an amide linkage between a carboxylate and an amine. Commonly used carboxylate substrates include citrate or α-ketoglutarate, or derivatives thereof, while the amines are often hydroxamate

Methods in Enzymology, Volume 702
ISSN 0076-6879, https://doi.org/10.1016/bs.mie.2024.06.012
Copyright © 2024 Elsevier Inc. All rights are reserved, including those for text and data mining, AI training, and similar technologies.

derivatives of lysine or ornithine, or their decarboxylated forms cadaverine and putrescine. Enzymes that employ three substrates to catalyze a reaction may proceed through alternate mechanisms. Some enzymes use sequential mechanisms in which all three substrates bind prior to any chemical steps. In such mechanisms, substrates can bind in a random, ordered, or mixed fashion. Alternately, other enzymes employ a ping-pong mechanism in which a chemical step occurs prior to the binding of all three substrates. Here we describe an enzyme assay that will distinguish among these different mechanisms for the NIS synthetase, using lucA, an enzyme involved in the production of aerobactin, as the model system.

1. Introduction

When faced with a low-iron environment, many microorganisms produce the biosynthetic and uptake machinery for siderophores, small molecule iron chelators, that are secreted into the environment to acquire iron (Barry & Challis, 2009; Hider & Kong, 2010; Lamb, 2015). Siderophores are often classified chemically into different families on the basis of the biosynthetic pathway that is used for their production and for the chemical functional groups that are involved in the coordination of iron. A common mechanism for production of peptide siderophores employs the nonribosomal peptide synthetases (NRPSs), a family of modular, assembly line proteins that convert amino acids into small peptide chains. Independent of the ribosomes, NRPS products often contain non-proteinogenic amino acids and chemical modifications that are incorporated into the final product (Miethke & Marahiel, 2007). Many NRPS siderophores additionally contain aryl caps composed of salicylic acid (2-hydroxybenzoic acid) or 2,3-dihydroxybenzoic acid (Kudo, Miyanaga, & Eguchi, 2019; Quadri, 2000) as well as oxazoline and thiazoline heterocycles that are formed by the cyclodehydration of serine, threonine, and cysteine residues.

In contrast to the NRPS derived siderophores, a second family of compounds have been termed NRPS-independent siderophores (NISs) (Challis, 2005; Kadi, Oves-Costales, Barona-Gomez, & Challis, 2007; Oves-Costales, Kadi, & Challis, 2009). NISs generally contain one or more amine compounds that are converted to a hydroxamate through the activity of a hydroxylase and an acyl transferase. Often, the pathway also contains a decarboxylase resulting in a three-protein cascade of that converts lysine or ornithine to cadaverine or putrescine, respectively, followed by the N-hydroxylation and N-acylation to produce the final hydroxamate. The hydroxamates are then installed on a carboxylate substrate, often citrate, α-ketoglutarate, or succinate.

These combined hydroxamate, hydroxyls, and carboxylate groups on different NISs facilitate iron binding. Common NISs include desferrioxamine (Ronan et al., 2018; Yang et al., 2022; Yang, Banas, Rivera, & Wencewicz, 2023), petrobactin (Nusca et al., 2012), staphyloferrin (Tang et al., 2020), and aerobactin (Bailey, Drake, Grant, & Gulick, 2016; Bailey et al., 2018; Carbonetti & Williams, 1984).

Also shared within the NIS biosynthetic gene cluster is an NIS synthetase, a ligase that uses ATP to catalyze the formation of an amide bond between an amine substrate bearing the hydroxamate and the carboxylate substrate (Challis, 2005; Kadi et al., 2007; Oves-Costales et al., 2009). These enzymes catalyze a two-step reaction involving the formation of an acyl-adenylate intermediate and inorganic pyrophosphate (PPi). The amine then attacks the adenylate to displace AMP and form the amide linkage. The NIS synthetase enzymes therefore consume three substrates, ATP, the carboxylate, and the amine, and produce three products, AMP, PPi, and the final amide product. Aerobactin, an NIS that is produced by many Gram-negative bacteria (Carbonetti & Williams, 1984; Ford, Cooper, & Williams, 1986), including hypervirulent strains of *Klebsiella pneumoniae* (Nassif & Sansonetti, 1986), is a well-characterized NIS biosynthetic pathway that contains two NIS synthetases, IucA and IucC. Aerobactin is produced from primary metabolites lysine and citrate through the actions of the hydroxylase IucD, the acetyltransferase IucB, and the two NIS synthetases (Fig. 1). Combined, the four-step pathway results in the production of aerobactin that is secreted from the cell. The ability to produce aerobactin is a critical virulence factor for hypervirulent strains of *K. pneumoniae* (Russo & Gulick, 2019; Russo et al., 2014; Russo, Olson, MacDonald, Beanan, & Davidson, 2015).

Enzymes that use three substrates can employ different mechanisms that can broadly be divided into two general classes. Enzymes with sequential mechanisms bind all three substrates prior to any chemical steps. The binding order for the substrates may be random, ordered, or partially ordered, with one substrate binding in a fixed position (first, second, last) and the order of binding of the remaining two being unimportant. In contrast, some three-substrate enzymes employ a ping-pong mechanism in which one or two substrates fail to bind prior to the first chemical step. Some ligases, such as the adenylate-forming family to which the NRPS adenylation domains belong, use a *bi-uni-uni-bi* reaction mechanism (Gulick, 2009). For example, AMP-forming acyl-CoA synthetases catalyze the formation of acetyl-CoA by first binding the carboxylate and ATP, which are converted to an acyl-AMP intermediate and PPi. Following release of PPi,

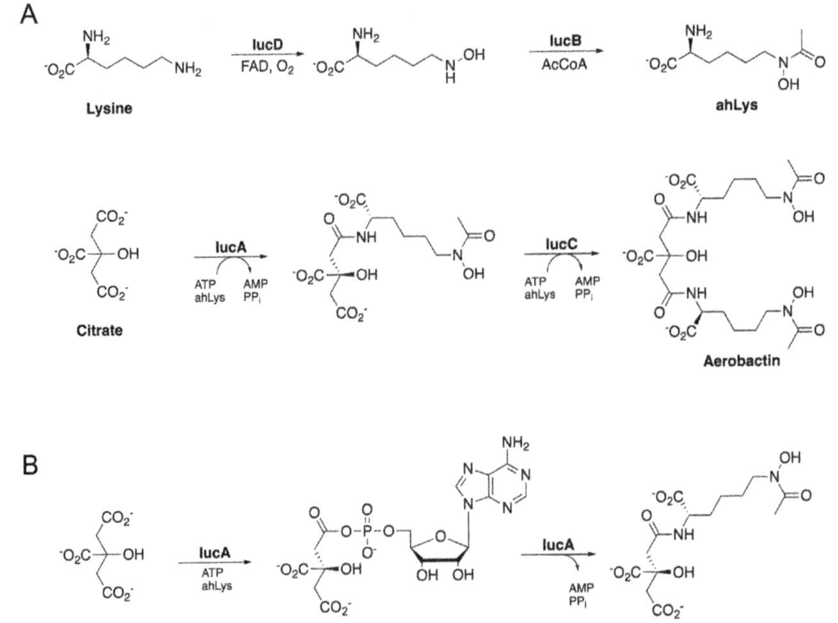

Fig. 1 Aerobactin biosynthesis. (A) Biosynthesis of aerobactin from lysine and citrate. The hydroxylase IucD and acetyl transferase IucB produce *N6*-acetyl-*N6*-hydroxylysine (ahLys), which is then installed in two reactions catalyzed by the NIS synthetases IucA and IucC. (B) The NIS synthetase IucA catalyzes formation of citryl-ahLys through a two-step ATP-dependent reaction.

CoA binds and the enzyme produces acetyl- or propionyl-CoA and AMP, which are both released from the enzyme (Horswill & Escalante-Semerena, 2002; Reger, Carney, & Gulick, 2007; Wu et al., 2008).

Here, we present a kinetic approach to distinguish among the different catalytic mechanisms of the three substrate NIS synthetase enzymes. The approach involves initial velocity measurements at multiple substrate concentrations, which are combined to distinguish among the different approaches. Methods are described for protein purification, kinetic measurements using an assay that couples AMP formation to NADH oxidation through three coupling enzymes (Wu & Hill, 1993), and the approach for plotting the initial rates to determine the mechanism of substrate binding. This technique has been used with the NIS synthetase IucA to identify an ordered binding mechanism, with ATP binding first, followed by citrate, and ahLys (Mydy, Bailey, Patel, Rice, & Gulick, 2020). The use of this technique to characterize other NIS synthetases should confirm a consistent ordered mechanism among different members of this family.

2. Protein production and purification

For enzyme assays, proteins should be purified to homogeneity, allowing accurate determination of concentration and ensuring consistent results can be obtained across different experiments. While the general strategy adopted will be specific to a particular protein, an overall strategy that has proven successful for numerous enzymes, including multiple NIS synthetases, employs a pET plasmid that allows for the production of a protein harboring a 6× histidine purification tag and a protease cleavage site that allows for the removal of the tag. In our laboratory, a TEV protease site has been incorporated into the pET15 plasmid. After lysis of the cell paste, the lysate is clarified via high-speed ultracentrifugation, passed through an immobilized metal ion affinity chromatography (IMAC) column charged with Ni^{2+}, and eluted with increasing concentrations of imidazole. The eluted protein is dialyzed to remove imidazole, and simultaneously treated with TEV protease to cleave the tag. The cleaved, partially purified protein is then passed over the IMAC column a second time, removing contaminants, uncleaved enzyme, cleaved His–tag, and the TEV protease. Finally, the protein is polished through a size exclusion chromatography step to remove aggregates. The protein is dialyzed into final storage buffer and frozen at −80 °C in small aliquots until needed for kinetic experiments.

2.1 Equipment
- New Brunswick Innova 44R shaker/incubators
- BioRad NGC Quest Chromatograph Systems chromatography system

2.2 Buffers, strains, and reagents
- *Escherichia coli* BL21(DE3) cell line containing pET15TEV-*iucA* expression plasmid
- LB Media, autoclaved (10 g tryptone, 10 g NaCl, 5 g yeast extract, per L)
- Lysis Buffer (50 mM HEPES, pH 7.5 at 4 °C, 250 mM NaCl, 20 mM imidazole, 0.2 mM triscarboxyethylphosphine (TCEP), and 10% glycerol)
- Elution Buffer (50 mM HEPES, pH 7.5 at 4 °C, 250 mM NaCl, 300 mM imidazole, 0.2 mM TCEP, and 10% glycerol)
- Storage buffer (50 mM HEPES, pH 7.5 at 4 °C, 150 mM NaCl, 0.2 mM TCEP)
- Isopropyl-β-D-thiogalactoside (IPTG), 100 mM stock in water, filter sterilized
- 5 mL HisTrap HP chromatography IMAC column, Cytiva Life Sciences

- Superdex-200 16/600 Size Exclusion Chromatography Column, Cytiva Life Sciences
- TEV protease, isolated as described (Raran-Kurussi, Cherry, Zhang, & Waugh, 2017; Tropea, Cherry, & Waugh, 2009)

2.3 Procedure

2.3.1 Protein expression

1. Grow an overnight starter culture of expression cells in 5 mL sterile LB media containing 100 μg/mL ampicillin in a sterile 15 mL culture tube at 37 °C in a shaker.
2. Inoculate 1 L of sterile LB media containing 100 μg/mL ampicillin in a 2 L baffled shaker flask. Shake cells at 37 °C for 3-5 h, monitor growth at OD_{600}.
3. At an OD_{600} of 0.6 – 0.8, chill cells on ice for 10 min, cool incubator to 16 °C. Prior to returning cells to the incubator, induce protein expression with 500 μM sterile IPTG. Continue protein expression overnight.
4. After 16–18 h of growth and expression, harvest cells by centrifugation at 6000g for 15 min. The cell paste can be frozen and stored at −80 °C until use, or directly continued into purification.

Purification

5. Cell paste is resuspended in a lysis buffer at a concentration of ~5–10 mL buffer per 1 g paste. Cells are lysed by sonication at 4 °C at 50% amplitude, using 20 cycles of 30 s on, 45 s off. The cell lysate is clarified by ultracentrifugation at 185×10^3 g for 45 min. The supernatant is then filtered over a 0.45 μm polysulfone membrane to prepare for the chromatographic purification.
6. Clarified lysate is passed over a pre-equilibrated 5 mL HisTrap HP column. Upon loading the full lysate, the column is washed with lysis buffer until absorbance returns to baseline, approximately 10 column volumes.
7. The column is then washed with lysis buffer containing 50 mM imidazole. If an FPLC system is used in which the A line contains lysis buffer and the B line contains the elution buffer, this wash step can be performed using 90% A and 10% B, which combines to 50 mM imidazole. The wash will result in the elution of several weakly bound contaminating proteins and should be monitored for return of the absorbance to baseline.

8. Elute the protein with 100% B, resulting in the elution of the bound tagged protein.
9. Fractions containing the eluted protein should be identified through SDS-PAGE, the chromatographic trace, or UV absorbance. The fractions should be combined, concentrated to ~5–10 mL, and dialyzed against Lysis Buffer at 4 °C to remove imidazole. The protein is then dialyzed for 2–4 h against 500 mL of buffer. At this stage, the dialysis bag is carefully opened, and TEV protease added at a ratio of approximately 1:100 mg:mg with target protein. The dialysis bag is then be transferred to a fresh 500 mL of buffer for dialysis overnight.
10. The following morning, harvest the protein from the dialysis bag, clarify the sample via centrifugation in a 4 °C microcentrifuge or filter through 0.45 μm polysulfone membrane, and load the protein onto a 5 mL HisTrap HP column that has been equilibrated with lysis buffer. Untagged protein will pass through the column to be collected. The column can be washed and eluted with 10% and 100% buffer B; fractions from these steps can be monitored by SDS-PAGE to assess cleavage efficiency.
11. Fractions containing the untagged protein can be combined and concentrated.
12. The protein should be passed over a size exclusion chromatography column, such as Superdex-S200 16/600. Protein should be monitored for homogeneity and elution as a single peak. Comparison with standards of known molecular weight and oligomeric status can inform understanding of the oligomeric state of the enzyme. The column should be preequilibrated with final storage buffer. Eluted protein can be concentrated to desired final concentration and frozen in small aliquots by directly pipetting into liquid nitrogen, and storage at −80 °C.

2.4 Notes

1. The expression system should be tailored to the protein target and may involve alternate expression plasmids or host strains. Similarly, the strategy for cell growth and protein induction may not be uniform for all NIS synthetases. The strategy described here has worked for many bacterial enzymes involved in siderophore biosynthesis that have been studied in our lab.
2. Purification and storage buffers will be protein-specific but should be designed to maintain protein solubility. The inclusion of additives such as reducing agents (dithiothreitol (DTT) and TCEP) or glycerol can be tested empirically for improvement in protein yields and stability.

3. Purification can be done manually in the absence of a FPLC system, using gravity flow of lysate and protein samples over the column. Flow-through and elution fractions should be analyzed via SDS-PAGE to identify protein.

3. NADH coupled adenylation assay

To rapidly characterize enzyme turnover in a generalizable format that can be employed for all NIS synthetases, an AMP-detection is assay is used that combines three enzymes, myokinase, pyruvate kinase, and lactate dehydrogenase, to couple the production of a single molecule of AMP to the conversion of two molecules of NADH to NAD^+. The disappearance of NADH is monitored spectrophotometrically because of its absorbance at 340 nm. The coupled reaction scheme involves the reaction of the AMP product with ATP to form two molecules of ADP that is catalyzed by myokinase. The ADP is a substrate for pyruvate kinase, which uses a molecule of phosphoenolpyruvate to form pyruvate and ATP. Finally, the pyruvate molecule is a substrate for lactate dehydrogenase that couples reduction of pyruvate to lactate with oxidation of NADH, to form NAD^+. The reaction is monitored in real-time and is readily adaptable to a 96-well plate reader with UV absorbance capabilities. The reaction is monitored for linear initial rate readings, that can be readily converted via an extinction coefficient of $\varepsilon = 6220\,M^{-1}\,cm^{-1}$ to initial rate in units of $\mu M/min$.

3.1 Equipment
- Biotek Synergy 4 Plate reader
- 96-well flat bottom UV transparent microplates (Caplugs)

3.2 Reagents
Common reagents.
- 1 U/L Pyruvate Kinase, Sigma-Aldrich
- 1 U/L Lactate Dehydrogenase, Sigma-Aldrich
- 10 U/L Myokinase, Sigma-Aldrich
- 300 mM Phosphoenolpyruvate, Sigma-Aldrich
- 10 mM NADH, Sigma-Aldrich
- 1 M MOPS, pH 7.5
- 1 M $MgCl_2$
- 25 mM ATP

Reagents specific to the NIS Synthetase under investigation
- 50 mM Citrate
- 50 mM ahLys (Bailey et al., 2018)
- Purified NIS synthetase from step 2, diluted to 10× concentration, here 10 μM.

3.3 Procedure

Substrate concentrations are then combined such that the varied substrate is provided in a dilution series and combined with the two remaining substrates that are held at a constant ratio in relation to their K_M values. It is therefore necessary to determine preliminary apparent K_M values through conventional kinetic approaches. An appropriate volume of master mix is produced that will be combined with enzyme and substrates in the multiwell-plate. The following protocol employs six substrate concentrations for the varied substrate, for which double reciprocal plots are generated at five concentrations of the fixed substrates. As each data point is measured in triplicate, for each line that is generated, it is necessary to generate a master mix for 18 reactions. For each reaction, 80 μL of master mix will be provided to initiate the reaction, combining to 1.44 mL of master mix. To ensure adequate material, 1.5 mL of master mix is created for each line that is generated.

For IucA, apparent K_M values were determined to be 50 μM for ATP, 540 μM for citrate, and 790 μM for ahLys. A representative recipe is provided (Table 1) to generate the master mix and perform the reaction for varying ATP and the lowest concentrations of citrate and ahLys. Four more master mixes will be generated to be used to produce the first series of plots for varied citrate and constant values of ATP and ahLys at a constant ratio relative to their apparent K_M value.

1. Create the master mix solution that will be combined with the varied substrate. This recipe produces a master mix for a final reaction concentration of 54 μM citrate and 79 μM ahLys, roughly 0.1× apparent K_M value.
2. Thaw enzyme rapidly then store on ice. Dilute enzyme stock at 10× working concentration, here 10 μM.
3. Create varied substrate dilution series at six 10× concentrations that bracket the apparent K_M value. For ATP, with K_M value of 50 μM, create ATP stock concentrations at 250, 375, 500, 800, and 2000 μM.
4. Arrange three columns of the 96 well plate to perform one series of replicates for the double reciprocal plot. The first three wells in the top row receive 10 μL of enzyme added to one side of the well, and 10 μL of

Table 1 Master mix recipe.

Component	Stock concentration	Volume stock	Master mix final concentration[a]
Pyruvate kinase	1000 U/mL	18.8 μL	12.5 U/mL
Myokinase	10,000 U/mL	1.9 μL	12.5 U/mL
Lactate dehydrogenase	1000 U/mL	18.8 μL	12.5 U/mL
Phosphoenolpyruvate	300 mM	18.8 μL	3.75 mM
NADH	10 mM	150 μL	1000 μM
MOPS	1 M	375 μL	250 mM
$MgCl_2$	1 M	28.1 μL	18.74 mM
Citrate $(0.1\times K_M)$	25 mM	4.1 μL	67.5 μM
ahLys $(0.1\times K_M)$	25 mM	5.9 μL	98.75 μM
Water		878.6 μL	
Total volume		1500 μL	

[a]Final concentration in master mix is 1.25× concentration in the final reaction.

water to the other. These three wells will serve as a negative control, or blank. To the remainder of the wells in the column add 10 μL of enzyme and 10 μL of the increasing concentrations of ATP, such that 10 μL of the 25 μM ATP is added to row 2, 10 μL of 37.5 μM ATP is added to row 3, and so on. Be careful not to allow enzyme and substrate to mix so that the reaction is initiated by addition of 80 μL of the master mix. In a separate column add 250 μL of master mix to be added later for triplicate wells of reaction. Incubate plate in plate reader for 5 min to allow reaction set up to achieve 37 °C.

5. To each well, as rapidly as possible, add 80 μL of master mix with a multichannel pipettor, allowing the substrate, enzyme, and master mix components to mix.

6. Initiate the reaction and monitor the reaction for 10 min. When the identity of the blank wells are provided, the plate reader provides corrected values for the $\Delta OD/min$. These values can be converted into an initial rate that can be used for generation of the double reciprocal plot.

3.4 Notes

1. While modern kinetic analysis uses least squares fitting for the determination of kinetic constants, the approach here with double reciprocal, Lineweaver Burk plots, allows for facile graphical representation of the data that is indicative of different mechanisms, as described in the next section.
2. Initial values for apparent K_M values should be determined via standard techniques by saturating two of the three substrates and varying the third. Initial velocity rates are plotted against the concentration of the varied substrate. These values can be used for defining the fixed concentrations of non-varied substrates at a constant ratio relative to their K_M value.
3. For reactions lacking a nucleophile like ahLys, a surrogate nucleophile like 50–150 mM hydroxylamine could be also used.
4. For reactions requiring a specific buffer, MOPS could be replaced with a buffer like 100–250 mM HEPES pH 7.5.

4. Kinetic analysis to distinguish among potential mechanisms

The initial rate plots generated in Section 3 are then used to identify the kinetic mechanism, which are distinguished by the linearity and non-linearity of the different plots, as well as the position of the intercepts of different plots. These features have been described in detail for three-substrate reaction mechanisms (Rudolph & Fromm, 1979; Segel, 1975). Because of the errors in measurements, particularly at low substrate concentration(s), and minor deviations from linearity, it can be difficult to fully distinguish among potential mechanisms. Nonetheless, careful analysis of the plots and potentially additional biochemical and biophysical methods can be employed to distinguish among the possible catalytic mechanisms.

4.1 Possible mechanisms

Three-substrate reaction mechanisms can be divided broadly into ordered and ping-pong mechanisms (Table 2). With each, there are five mechanisms that relate to the nature of substrate binding within each mechanism. The five ordered mechanisms encompass the first five mechanisms and describe reactions in which (1) substrate binding is completely random, (2) substrates are required to bind in an ordered fashion, and (3–5) partially ordered reactions in which the second, third, and first substrate binds in a requisite position while the remaining two substrates can bind randomly.

Table 2 Graphical analysis of kinetic data for ter-reactant enzymes.[a]

Mechanism	Initial rate plots[b]			Slope and intercept replots[c]					
	1/A	1/B	1/C	A_{Slope}	A_{Int}	B_{Slope}	B_{Int}	C_{Slope}	C_{Int}
1. Random	I	I	I	N	N	N	N	N	N
2. Ordered	I	I	I	N	N	N	L	N	N
3. Partially Random (AC Random)	I	I	I	N	N	N0	L	N	N
4. Partially Random (AB Random)	I	I	0	N0	N	N0	N	N	*
5. Partially Random (BC Random)	I	I	I	N0	N	N	L	N	L
6. Hexa Uni Ping Pong	P	P	P	L	L	L	L	L	L
7. Ordered Bi Uni Uni Bi Ping Pong	I	I	P	L	L	L	L	L	N
8. Ordered Uni Uni Bi Bi Ping Pong	P	I	I	L	N	L	L	L	L
9. Random Bi Uni Uni Bi Ping Pong	I	I	P	L	L	L	L	L	N
10. Random Uni Uni Bi Bi Ping Pong	P	I	I	L	N	L	L	L	L

[a]Reprinted with permission from Mydy, L. S., Bailey, D. C., Patel, K. D., Rice, M. R., & Gulick, A. M. (2020). The siderophore synthetase IucA of the aerobactin biosynthetic pathway uses an ordered mechanism. *Biochemistry, 59*, 2143–2153. Copyright 2020 American Chemical Society.
[b]*I*, lines intersect to the left of the $1/v$ axis. *0*, lines intersect on the $1/v$ axis. *P*, parallel lines.
[c]*N*, non-linear. *L*, linear. *N0*, non-linear, passes through the origin. *, all intercepts are the same.

The five ping–pong reaction mechanisms are defined by the number of substrates that bind and are released between the different chemical steps. In the Hexa Uni mechanism (mechanism 6), there is a chemical step that occurs between the binding of each substrate and the release of a product. In the Bi Uni Uni Bi (mechanism 7) and Uni Uni Bi Bi (mechanism 8) mechanisms, two or one substrate, respectively, bind prior to the first chemical step. Mechanisms 9 and 10 are similar, although involve a random binding of substrates. We note that often, although not always, ping pong reaction mechanisms can involve covalent acyl-enzyme intermediates. Although none is expected for NIS synthetases, the possibility of bound but not covalent intermediates, as seen with the acyl-CoA synthetase adenylate intermediates (Gulick, 2009), raises the possibility of a ping-pong enzyme mechanism.

4.2 Initial double reciprocal plots

The initial rate plots obtained with different substrate concentrations are now replotted to distinguish among the ten mechanisms. Three sets of plots for IucA were created, using varied citrate at fixed concentrations of ATP and ahLys, varied ATP at fixed concentrations of citrate and ahLys, and varied ahLys at fixed concentrations of ATP and citrate. These plots (Fig. 2) clearly show a series of intercepting lines, which clearly rules out any of the ping pong mechanisms. That none of the points of intersection lies on the Y-axis, allows us to also rule out mechanism 4.

4.3 Slope and intercept replots

To distinguish among the remaining mechanisms, the slopes and intercepts of the lines fit in the initial rate plots are replotted against the reciprocal of the concentration of one of the fixed substrates (Fig. 3). Analyzing these plots in relation to the table from above provides insight into the final reaction mechanism. All of the replots show better fits to non-linear than linear. However, the citrate intercept replot (in Fig. 3D) approximates a straight line and may indeed represent a straight line. We interpreted this in our earlier work (Mydy et al., 2020) as supporting an ordered mechanism in which citrate binds second (Mechanism 2). We ultimately concluded that the best fit to all of the data, including additional kinetic studies using saturating amounts of the two fixed substrates, the impact of substrate binding on stabilizing the protein as indicated by a melting temperature analysis, and the crystal structure of IucA bound to ATP, that the catalytic mechanism is ordered with ATP binding preceding citrate, and finally ahLys.

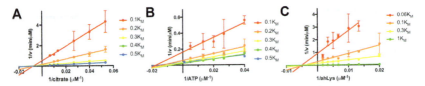

Fig. 2 Initial Rate plots for IucA, varying concentrations of one substrate at fixed concentrations of the remaining two substrates at a constant ratio relative to their respective K_M values. (A) Citrate concentration was varied at multiple concentrations of ATP and ahLys. (B) ATP concentration was varied at multiple concentrations of citrate and ahLys. (C) ahLys concentration was varied at multiple concentrations of ATP and citrate. Double reciprocal plots were created illustrating intersecting patterns for all three substrates. *Reprinted with permission from Mydy, L. S., Bailey, D. C., Patel, K. D., Rice, M. R., & Gulick, A. M. (2020). The siderophore synthetase IucA of the aerobactin biosynthetic pathway uses an ordered mechanism. Biochemistry, 59, 2143–2153. Copyright 2020 American Chemical Society.*

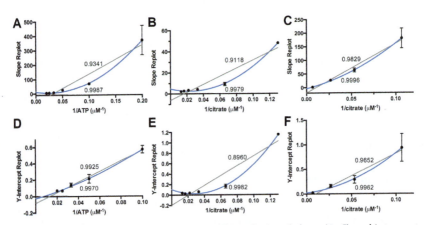

Fig. 3 Slope and Intercept re-plots for IucA. The calculated slope (A–C) and intercepts (D–F) from Fig. 2 are plotted against the concentration of one of the fixed, non-varied substrates. Fits of linear and non-linear plots are provided, along with the R^2 value of the fit. The data from Fig. 2A, varying citrate, are replotted in panels (A) slope and (D) intercept against the reciprocal concentration of ATP. The data from Fig. 2B, varying ATP, are replotted in panels (B) slope and (E) intercept against the reciprocal concentration of citrate. The data from Fig. 2A, varying ahLys, are replotted in panels (C) slope and (F) intercept against the reciprocal concentration of citrate. *Reprinted with permission from Mydy, L. S., Bailey, D. C., Patel, K. D., Rice, M. R., & Gulick, A. M. (2020). The siderophore synthetase IucA of the aerobactin biosynthetic pathway uses an ordered mechanism. Biochemistry, 59, 2143–2153. Copyright 2020 American Chemical Society.*

4.4 Notes

1. We note that some studies of ter-reactant enzymes include additional potential mechanisms that include equilibrium or steady-state binding of certain substrates, or Theorell-Chance approaches that are adopted by enzymes in which the reaction and product release occur very quickly upon biding the third substrate (Viola & Cleland, 1982). For simplicity, we have reduced the number of possible mechanisms to the ten shown here.

2. It should be noted that errors in curves and minor deviations from linear can in some instances be difficult to distinguish. Many of the replots of Fig. 3, for example, fit reasonably well to both linear and non-linear curves. Ultimately, investigators should consider multiple techniques and look for consistent observations that match all known features of the enzyme, structure, and chemical mechanism.

5. Alternate approaches

Alternate methods are available to analyze ter-reactant enzyme mechanisms, some of which were also performed and supported our data with the NIS synthetase IucA (Mydy et al., 2020). One approach is to saturate the enzyme with two, nonvarying substrates at > 100× apparent K_M value (Rudolph & Fromm, 1979). This strategy uses similar tables that associate with different ordered and sequential methods. One limitation of this approach can result from solubility or availability of custom substrates that may prevent the full analysis or suitably high concentrations.

6. Summary and conclusions

Bacterial siderophores play important roles in the adaptation to iron limiting environments, including the sites of infections for many human pathogens. Because of the diverse nature of the siderophore structures, understanding their biosynthesis may enable the development of novel inhibitors that block siderophore production and block growth in iron limiting media (Lamb, 2015). Many siderophores are derived from the activity of the modular, assembly-line NRPS proteins, and the development of specific inhibitors of NRPS enzymes (Ferreras, Ryu, Di Lello, Tan, & Quadri, 2005; Miethke et al., 2006; Neres et al., 2008; Neres et al., 2013; Shelton et al., 2022), associated peptide modification enzymes on the siderophore pathway (Drake & Gulick, 2011; Theriault et al., 2013; Wurst et al., 2014), and enzymes involved in the generation of NRPS building blocks (Meneely et al., 2014; Payne, Kerbarh, Miguel, Abell, & Abell, 2005; Vasan et al., 2010) have all been described.

In contrast, mechanistic and inhibition studies of NIS synthetases have been comparatively under-explored. The development of inhibitors requires the elucidation of the mechanism and understanding of the substrate binding properties. The discovery of an ordered mechanism for IucA and the formation of a quarternary complex with ATP, citrate, and ahLys suggests that the active site that must accommodate all three substrates prior to any chemical steps. This observation should be repeated with additional NIS synthetases. It is notable that some NIS synthetases are iterative, catalyzing the polymerization and often cyclization of multiple copies of a single molecule that contains a carboxylate on one end and an amine on the other. The dimerization, trimerization, and cyclization of N-hydroxy-N-succinylcadaverine or putrescine molecules result in a family of hydroxamate-containing siderophores avaroferrin, putrebactin, desferrioxamine, and others (Ronan et al., 2018; Rütschlin, Gunesch, & Böttcher, 2017; Yang et al., 2023). The mechanistic investigation of extension and cyclization, while likely employing a similar mechanism as IucA, should be interrogated experimentally.

The structural and mechanistic investigation additionally identifies a distinction between the NIS synthetases and NRPS adenylation domains, both of which activate a carboxylate for amide bond formation through an adenylation step. Many NRPS enzymes have been studied with isosteric acyl sulfamoyl adenylate inhibitors that mimic the acyl adenylate intermediate (Ferreras et al., 2005; Miethke et al., 2006; Somu et al., 2006). These high potency inhibitors were recently tested for DesD, the NIS synthetase involved in the biosynthesis of desferrioxamine (Yang et al., 2022). The inhibitor was surprising much less efficient, with an IC50 value that was $\sim 1000\times$ greater than observed for the best NRPS inhibitors. Structural examination provided a reasonable explanation. NIS synthetases bind ATP with the β- and γ-phosphates buried deep within the protein core, making multiple interactions with several conserved arginine residues. The lack of binding energy provided in the adenylate mimic reduces the potency of the inhibitors. Indeed, addition of inorganic pyrophosphate to the inhibition experiment increased the potency of the acyl sulfamate inhibitor (Yang et al., 2022).

The investigation of catalytic mechanisms of enzymes remains an important determinant to the understanding of the protein function in the context of a broader biosynthetic pathway. Probing the function of individual enzymes informs our understanding of enzymes from uncharacterized biosynthetic pathways that may produce important natural products that have yet to be discovered.

Acknowledgment

Work in our lab is supported by NIH grant GM136235 (to AMG). The authors thank Daniel C. Bailey, Eric J. Drake, and Matthew R. Rice for early assistance and studies on IucA.

References

Bailey, D. C., Alexander, E., Rice, M. R., Drake, E. J., Mydy, L. S., Aldrich, C. C., & Gulick, A. M. (2018). Structural and functional delineation of aerobactin biosynthesis in hypervirulent *Klebsiella pneumoniae*. *The Journal of Biological Chemistry, 293*(20), 7841–7852.

Bailey, D. C., Drake, E. J., Grant, T. D., & Gulick, A. M. (2016). Structural and functional characterization of aerobactin synthetase IucA from a hypervirulent pathotype of *Klebsiella pneumoniae*. *Biochemistry, 55*(25), 3559–3570.

Barry, S. M., & Challis, G. L. (2009). Recent advances in siderophore biosynthesis. *Current Opinion in Chemical Biology, 13*(2), 205–215.

Carbonetti, N. H., & Williams, P. H. (1984). A cluster of five genes specifying the aerobactin iron uptake system of plasmid ColV-K30. *Infection and Immunity, 46*(1), 7–12.

Challis, G. L. (2005). A widely distributed bacterial pathway for siderophore biosynthesis independent of nonribosomal peptide synthetases. *Chembiochem: A European Journal of Chemical Biology, 6*(4), 601–611.

Drake, E. J., & Gulick, A. M. (2011). Structural characterization and high-throughput screening of inhibitors of PvdQ, an NTN hydrolase involved in pyoverdine synthesis. *ACS Chemical Biology, 6*(11), 1277–1286.

Ferreras, J. A., Ryu, J. S., Di Lello, F., Tan, D. S., & Quadri, L. E. (2005). Small-molecule inhibition of siderophore biosynthesis in *Mycobacterium tuberculosis* and *Yersinia pestis*. *Nature Chemical Biology, 1*(1), 29–32.

Ford, S., Cooper, R. A., & Williams, P. H. (1986). Biochemical genetics of aerobactin biosynthesis in *Escherichia coli*. *FEMS Microbiology Letters, 36*(2-3), 281–285.

Gulick, A. M. (2009). Conformational dynamics in the acyl-CoA synthetases, adenylation domains of non-ribosomal peptide synthetases, and firefly luciferase. *ACS Chemical Biology, 4*, 811–827.

Hider, R. C., & Kong, X. (2010). Chemistry and biology of siderophores. *Natural Product Reports, 27*(5), 637–657.

Horswill, A. R., & Escalante-Semerena, J. C. (2002). Characterization of the propionyl-CoA synthetase (PrpE) enzyme of *Salmonella enterica*: Residue Lys592 is required for propionyl-AMP synthesis. *Biochemistry, 41*(7), 2379–2387.

Kadi, N., Oves-Costales, D., Barona-Gomez, F., & Challis, G. L. (2007). A new family of ATP-dependent oligomerization-macrocyclization biocatalysts. *Nature Chemical Biology, 3*(10), 652–656.

Kudo, F., Miyanaga, A., & Eguchi, T. (2019). Structural basis of the nonribosomal codes for nonproteinogenic amino acid selective adenylation enzymes in the biosynthesis of natural products. *Journal of Industrial Microbiology & Biotechnology, 46*(3-4), 515–536.

Lamb, A. L. (2015). Breaking a pathogen's iron will: Inhibiting siderophore production as an antimicrobial strategy. *Biochimica et Biophysica Acta, 1854*(8), 1054–1070.

Meneely, K. M., Luo, Q., Riley, A. P., Taylor, B., Roy, A., Stein, R. L., ... Lamb, A. L. (2014). Expanding the results of a high throughput screen against an isochorismate-pyruvate lyase to enzymes of a similar scaffold or mechanism. *Bioorganic & Medicinal Chemistry*.

Miethke, M., Bisseret, P., Beckering, C. L., Vignard, D., Eustache, J., & Marahiel, M. A. (2006). Inhibition of aryl acid adenylation domains involved in bacterial siderophore synthesis. *The FEBS Journal, 273*(2), 409–419.

Miethke, M., & Marahiel, M. A. (2007). Siderophore-based iron acquisition and pathogen control. *Microbiology and Molecular Biology Reviews: MMBR, 71*(3), 413–451.

Mydy, L. S., Bailey, D. C., Patel, K. D., Rice, M. R., & Gulick, A. M. (2020). The siderophore synthetase IucA of the aerobactin biosynthetic pathway uses an ordered mechanism. *Biochemistry, 59*(23), 2143–2153.

Nassif, X., & Sansonetti, P. J. (1986). Correlation of the virulence of *Klebsiella pneumoniae* K1 and K2 with the presence of a plasmid encoding aerobactin. *Infection and Immunity, 54*(3), 603–608.

Neres, J., Engelhart, C. A., Drake, E. J., Wilson, D. J., Fu, P., Boshoff, H. I., ... Aldrich, C. C. (2013). Non-nucleoside inhibitors of BasE, an adenylating enzyme in the siderophore biosynthetic pathway of the opportunistic pathogen *Acinetobacter baumannii. Journal of Medicinal Chemistry, 56*(6), 2385–2405.

Neres, J., Labello, N. P., Somu, R. V., Boshoff, H. I., Wilson, D. J., Vannada, J., ... Aldrich, C. C. (2008). Inhibition of siderophore biosynthesis in *Mycobacterium tuberculosis* with nucleoside bisubstrate analogues: Structure-activity relationships of the nucleobase domain of 5′-O-[N-(salicyl)sulfamoyl]adenosine. *Journal of Medicinal Chemistry, 51*(17), 5349–5370.

Nusca, T. D., Kim, Y., Maltseva, N., Lee, J. Y., Eschenfeldt, W., Stols, L., ... Sherman, D. H. (2012). Functional and structural analysis of the siderophore synthetase AsbB through reconstitution of the petrobactin biosynthetic pathway from *Bacillus anthracis. The Journal of Biological Chemistry, 287*(19), 16058–16072.

Oves-Costales, D., Kadi, N., & Challis, G. L. (2009). The long-overlooked enzymology of a nonribosomal peptide synthetase-independent pathway for virulence-conferring siderophore biosynthesis. *Chemical Communications (Camb), 43*, 6530–6541.

Payne, R. J., Kerbarh, O., Miguel, R. N., Abell, A. D., & Abell, C. (2005). Inhibition studies on salicylate synthase. *Organic & Biomolecular Chemistry, 3*(10), 1825–1827.

Quadri, L. E. (2000). Assembly of aryl-capped siderophores by modular peptide synthetases and polyketide synthases. *Molecular Microbiology, 37*(1), 1–12.

Raran-Kurussi, S., Cherry, S., Zhang, D., & Waugh, D. S. (2017). Removal of affinity tags with TEV protease. *Methods in Molecular Biology, 1586*, 221–230.

Reger, A. S., Carney, J. M., & Gulick, A. M. (2007). Biochemical and crystallographic analysis of substrate binding and conformational changes in acetyl-CoA synthetase. *Biochemistry, 46*(22), 6536–6546.

Ronan, J. L., Kadi, N., McMahon, S. A., Naismith, J. H., Alkhalaf, L. M., & Challis, G. L. (2018). Desferrioxamine biosynthesis: Diverse hydroxamate assembly by substrate-tolerant acyl transferase DesC. *Philosophical Transactions of the Royal Society of London. Series B, Biological Sciences, 373*(1748).

Rudolph, F. B., & Fromm, H. J. (1979). Plotting methods for analyzing enzyme rate data. *Methods in Enzymology, 63*, 138–159.

Russo, T. A., & Gulick, A. M. (2019). Aerobactin synthesis proteins as antivirulence targets in hypervirulent *Klebsiella pneumoniae. ACS Infectious Diseases*.

Russo, T. A., Olson, R., MacDonald, U., Beanan, J., & Davidson, B. A. (2015). Aerobactin, but not yersiniabactin, salmochelin and enterobactin, enables the growth/survival of hypervirulent (hypermucoviscous) *Klebsiella pneumoniae* ex vivo and in vivo. *Infection and Immunity, 83*, 3325–3333.

Russo, T. A., Olson, R., MacDonald, U., Metzger, D., Maltese, L. M., Drake, E. J., & Gulick, A. M. (2014). Aerobactin mediates virulence and accounts for the increased siderophore production under iron limiting conditions by hypervirulent (hypermucoviscous) *Klebsiella pneumoniae*. *Infection and Immunity, 82*, 2356–2367.

Rütschlin, S., Gunesch, S., & Böttcher, T. (2017). One enzyme, three metabolites: Shewanella algae controls siderophore production via the cellular substrate pool. *Cell Chemical Biology, 24*(5), 598–604.e510.

Segel, I. H. (1975). *Enzyme kinetics: Behavior and analysis of rapid equilibrium and steady-state enzyme systems.* New York: John Wiley & Sons.

Shelton, C. L., Meneely, K. M., Ronnebaum, T. A., Chilton, A. S., Riley, A. P., Prisinzano, T. E., & Lamb, A. L. (2022). Rational inhibitor design for *Pseudomonas aeruginosa* salicylate adenylation enzyme PchD. *Journal of Biological Inorganic Chemistry: JBIC: A Publication of the Society of Biological Inorganic Chemistry, 27*(6), 541–551.

Somu, R. V., Boshoff, H., Qiao, C., Bennett, E. M., Barry, C. E., 3rd, & Aldrich, C. C. (2006). Rationally designed nucleoside antibiotics that inhibit siderophore biosynthesis of *Mycobacterium tuberculosis*. *Journal of Medicinal Chemistry, 49*(1), 31–34.

Tang, J., Ju, Y., Zhou, J., Guo, J., Gu, Q., Xu, J., & Zhou, H. (2020). Structural and biochemical characterization of SbnC as a representative Type B siderophore synthetase. *ACS Chemical Biology, 15*(10), 2731–2740.

Theriault, J. R., Wurst, J., Jewett, I., Verplank, L., Perez, J. R., Gulick, A. M., & Schreiber, S. (2013). Identification of a small molecule inhibitor of *Pseudomonas aeruginosa* PvdQ acylase, an enzyme involved in siderophore pyoverdine synthesis. Probe Reports from the NIH Molecular Libraries Program [Internet]. Bethesda, MD: National Center for Biotchnology Information (US).

Tropea, J. E., Cherry, S., & Waugh, D. S. (2009). Expression and purification of soluble His (6)-tagged TEV protease. *Methods in Molecular Biology, 498*, 297–307.

Vasan, M., Neres, J., Williams, J., Wilson, D. J., Teitelbaum, A. M., Remmel, R. P., & Aldrich, C. C. (2010). Inhibitors of the salicylate synthase (MbtI) from *Mycobacterium tuberculosis* discovered by high-throughput screening. *ChemMedChem, 5*(12), 2079–2087.

Viola, R. E., & Cleland, W. W. (1982). Initial velocity analysis for terreactant mechanisms. *Methods in Enzymology, 87*, 353–366.

Wurst, J. M., Drake, E. J., Theriault, J. R., Jewett, I. T., VerPlank, L., Perez, J. R., ... Gulick, A. M. (2014). Identification of inhibitors of PvdQ, an enzyme involved in the synthesis of the siderophore pyoverdine. *ACS Chemical Biology, 9*(7), 1536–1544.

Wu, R., Cao, J., Lu, X., Reger, A. S., Gulick, A. M., & Dunaway-Mariano, D. (2008). Mechanism of 4-chlorobenzoate:coenzyme a ligase catalysis. *Biochemistry, 47*(31), 8026–8039.

Wu, M. X., & Hill, K. A. (1993). A continuous spectrophotometric assay for the aminoacylation of transfer RNA by alanyl-transfer RNA synthetase. *Analytical Biochemistry, 211*(2), 320–323.

Yang, J., Banas, V. S., Patel, K. D., Rivera, G. S. M., Mydy, L. S., Gulick, A. M., & Wencewicz, T. A. (2022). An acyl-adenylate mimic reveals the structural basis for substrate recognition by the iterative siderophore synthetase DesD. *The Journal of Biological Chemistry, 298*(8), 102166.

Yang, J., Banas, V. S., Rivera, G. S. M., & Wencewicz, T. A. (2023). Siderophore synthetase DesD catalyzes N-to-C condensation in desferrioxamine biosynthesis. *ACS Chemical Biology, 18*(6), 1266–1270.

CHAPTER TWO

Experimental methods for evaluating siderophore–antibiotic conjugates

Rachel N. Motz[1], Ghazal Kamyabi[1], and Elizabeth M. Nolan*

Department of Chemistry, Massachusetts Institute of Technology, Cambridge, MA, United States
*Corresponding author. e-mail address: lnolan@mit.edu

Contents

1. Introduction	22
2. Methods	25
2.1 General considerations for evaluation of SACs	25
2.2 Safety precautions and general practices	26
2.3 Antimicrobial activity (AMA) assay	26
2.4 Time-kill kinetics	33
2.5 Competition with unmodified Ent	36
2.6 ^{57}Fe uptake assay	40
3. Summary	46
References	46

Abstract

Siderophore–antibiotic conjugates (SACs) are of past and current interest for delivering antibacterials into Gram-negative bacterial pathogens that express siderophore receptors. Studies of SACs are often multifaceted and involve chemical and biological approaches. Major goals are to evaluate the antimicrobial activity and uptake of novel SACs and use the resulting data to inform further mode-of-action studies and molecular design strategies. In this chapter, we describe four key methods that we apply when investigating the antimicrobial activity and uptake of novel SACs based on the siderophore enterobactin (Ent). These methods are based on approaches from the siderophore literature as well as established protocols for antimicrobial activity testing, and include assays for evaluating SAC antimicrobial activity, time-kill kinetics, siderophore competition, and bacterial cell uptake using ^{57}Fe. These assays have served us well in characterizing our Ent-based conjugates and can be applied to study SACs that use other siderophores as targeting vectors.

---

[1] Co-first author.

Methods in Enzymology, Volume 702
ISSN 0076-6879, https://doi.org/10.1016/bs.mie.2024.06.004
Copyright © 2024 Elsevier Inc. All rights are reserved, including those for text and data mining, AI training, and similar technologies.

1. Introduction

Investigations of new antibacterial agents and drug delivery strategies are important to achieve therapeutics that can bypass the permeability barrier of the Gram-negative outer membrane (OM) and combat the growing prevalence of antibiotic resistance (Darby et al., 2023; Zgurskaya, López, & Gnanakaran, 2015). Siderophores are bacterial secondary metabolites that bind tightly to Fe (III), and can be employed as delivery vectors for antibacterial warheads. Siderophore–antibiotic conjugates (SACs) aim to hijack bacterial Fe uptake machinery and the active uptake of ferric siderophores for the "Trojan horse" delivery of their drug cargoes (Almeida, da Costa, Sousa, & Resende, 2023; Rayner, Verderosa, Ferro, & Blaskovich, 2023). Siderophore-modified toxic cargos are naturally employed by some bacteria for interbacterial competition, including class IIb microcins (Thomas et al., 2004; Vassiliadis, Destoumieux-Garzón, Lombard, Rebuffat, & Peduzzi, 2010), and "sideromycins", such as albomycin (Gause, 1955). Inspired by Nature's strategy, a variety of synthetic SACs have been studied over the decades for their therapeutic potential (Rayner et al., 2023). Such efforts are highlighted in the clinical implementation of cefiderocol, a monocatechol-modified cephalosporin that is transported by OM receptors for catechol siderophores (Ito et al., 2016). Among drug cargos commonly employed within SACs, β-lactams have been popular for probing transport across the OM due to their periplasmic target (Ji, Miller, & Miller, 2012; Möllmann, Heinisch, Bauernfeind, Köhler, & Ankel-Fuchs, 2009; Pinkert et al., 2021; Zheng & Nolan, 2014). Fluoroquinolones, namely ciprofloxacin (Cipro), have been explored for siderophore-mediated cytoplasmic delivery, although it has been challenging to retain antibacterial activity following attachment to siderophores (Fardeau et al., 2014; Neumann, Sassone-Corsi, Raffatelli, & Nolan, 2018; Pandey et al., 2019; Wencewicz, Long, Möllmann, & Miller, 2013). Recently, applying siderophore conjugation to achieve drug repurposing, especially of anticancer agents, has gained attention (Guo & Nolan, 2022; Zhao et al., 2022).

Our contributions to the design and evaluation of SACs have included an initiative employing the siderophore enterobactin (Ent, Fig. 1A) as the delivery vector (Guo & Nolan, 2024). Ent is a native triscatecholate siderophore that was discovered approximately 50 years ago and has been extensively studied in both chemical and biological contexts. We have developed syntheses of monofunctionalized Ent and its conjugates with a variety of cargos (Fig. 1B) (Guo & Nolan, 2022; Sargun, Johnstone, Zhi, Raffatelli, & Nolan, 2021; Zheng & Nolan, 2014; Zheng, Bullock, & Nolan, 2012). Consequently, we

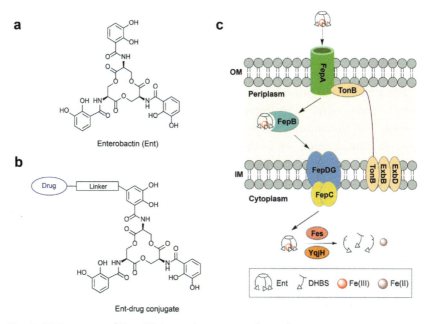

Fig. 1 (A) Structure of Ent; (B) general structure of Ent–drug conjugates; (C) cartoon overview of Ent transport machinery in *E. coli* K12.

have found Ent to be a valuable case study in the evaluation of SACs and, in this chapter, we focus on Ent as a model siderophore moiety. Ent is produced and utilized by members of Enterobacteriaceae including *Escherichia coli*, *Salmonella* species and *Klebsiella pneumoniae*. It scavenges Fe(III) in the extracellular environment and Fe(III)-bound Ent is transported through the OM by the TonB-dependent receptor FepA. Next, the periplasmic binding protein FepB delivers Fe(III)-bound Ent to the inner membrane (IM) permease FepCDG. Upon arrival in the cytoplasm, Fe(III)-bound Ent is hydrolyzed to three 2,3-dihydroxybezoyl serine (DHBS) monomers by the esterase Fes and the Fe(III) ion is reduced, presumably by the reductase YqjH, to release the Fe(II) ion for metabolic use (Fig. 1C) (Miethke & Marahiel, 2007; Miethke, Hou, & Marahiel, 2011). While it will not be a primary consideration in this chapter, additional strain-specific Ent uptake and processing machinery may also be involved (Hagan & Mobley, 2007; Hantke, Nicholson, Rabsch, & Winkelmann, 2003; Lin, Fischbach, Liu, & Walsh, 2005; Zhu, Valdebenito, Winkelmann, & Hantke, 2005).

Our research investigates the antibacterial activity of Ent-based conjugates and their cellular fates. Unlike the parent antibiotics, these conjugates benefit from an endogenous nutrient uptake system that transports the siderophore vector (e.g., FepABCDG). Moreover, the activity and behavior of SACs cannot be fully described by standard minimal inhibitory concentration (MIC) values. For SACs, it is crucial to consider and investigate the effect that conjugation has on the interaction of the siderophore moiety with the components of its uptake and processing machinery, as well as on the activity of the antibacterial warhead and its engagement with the target. Through consideration of both functional elements of SACs, we have incorporated the concepts and methods for studying siderophores with approaches to investigating antibiotics. This strategy has allowed us to apply and modify well-established assays from microbiology to better understand the efficacy and fate of our conjugates.

In this chapter, we provide protocols for four complementary and informative assays that are routinely used by our laboratory for evaluating the activity and uptake of SACs and related molecules: (i) antimicrobial activity (AMA) assays, (ii) time-kill kinetics assays, (iii) competition assays with unmodified siderophores, and (iv) [57]Fe uptake assays. We focus on these techniques because we consider them to be highly informative, broadly applicable, and well-developed. A repertoire of additional techniques exists that can provide insight into the function of siderophores and their conjugates, several of which have been reviewed recently (Southwell, Black, & Duhme-Klair, 2021). For instance, growth promotion assays are an approach for examining the ability of the siderophore moiety to deliver nutrient iron to cells (Zheng et al., 2012; Zscherp et al., 2021). Methods such as radioactive or fluorescent labeling are valuable for tracking the localization of siderophores and drug conjugates within bacterial cells (Ferreira et al., 2017; Hannauer, Barda, Mislin, Shanzer, & Schalk, 2010; Lee et al., 2016; Weizman et al., 1996). The delivery of drug cargo can often be studied using microscopic imaging, which reveals aberrant bacterial morphologies characteristic of drug-target engagement upon uptake of the conjugate (Guo & Nolan, 2022; Lai, Franke, Pinkert, Overwin, & Brönstrup, 2023; Sargun et al., 2021). In addition, studying bacterial mutants lacking components of siderophore biosynthesis, export, uptake, and processing machinery allows for a deeper understanding of SAC transport and activity (McPherson et al., 2012; Motz et al., 2024; Neumann et al., 2018; Nikaido & Rosenberg, 1990; Zheng et al., 2012). When appropriate, we describe the possibility of using bacterial mutant strains in assays highlighted in this chapter. Although our focus is on Ent-based SACs, we introduce the following assays in a generalized manner such that they can be applied to a variety of SACs with minimal modifications.

2. Methods
2.1 General considerations for evaluation of SACs

The modes of uptake and activity of SACs require careful consideration when designing experiments and selecting growth conditions. Here we focus on the amount of Fe available to bacterial cells and the siderophore moiety of the SAC, because this parameter can markedly affect the outcome of the following assays. To induce the expression of siderophore receptors, which is required for SAC uptake and subsequent activity, bacteria must be cultured under conditions of Fe limitation. Complex growth media like Luria-Bertani (LB) and Mueller-Hinton Broth (MHB) contain significant quantities of Fe, which can vary depending on the supplier and batch. Since early studies, metal binding proteins, such as transferrin or fetal bovine serum, or synthetic chelators, such as 2,2′-dipyridyl (DP) or ethylenediamine-N,N′-bis(2-hydroxyphenylacetic acid) (EDDHA), have been commonly employed to sequester Fe in such complex media, affording conditions of Fe limitation (Aoki et al., 2018; Caradec et al., 2023; Diarra et al., 1996; Ferreira et al., 2017; Ji et al., 2012; Möllmann et al., 2009; Pandey et al., 2019; Sassone-Corsi et al., 2016; Wencewicz et al., 2013; Zhao et al., 2022; Zscherp et al., 2021). More recently, Fe limitation has been achieved by employing chemically defined medium that lacks added iron (Fardeau et al., 2014; Lee et al., 2016; Neumann et al., 2018; Pinkert et al., 2021), or by pretreatment of a complex medium with Chelex® to remove Fe (Ito et al., 2016; Lai et al., 2023). Our lab initially employed 50% MHB (~2–4 μM Fe), a standard antibiotic susceptibility testing medium, supplemented with DP for studies of SACs (Zheng et al., 2012). Subsequently, we evaluated a chemically defined medium based on M9 minimal medium that provides Fe limitation (<1 μM Fe) (Neumann et al., 2018). We note that Fe-depleted cation-adjusted Mueller-Hinton Broth (ID-CAMHB) has recently emerged as a standard medium for evaluation of SACs such as cefiderocol (Ito et al., 2016). It is possible for SACs to exhibit varying levels of activity in different media even when Fe is limiting (Neumann et al., 2018).

While Fe limitation is key for induction of receptor expression, it will also induce production of native siderophores, which may compete with the SAC for Fe(III) and receptor binding. In addition, since Fe(III) binding is required for the siderophore/SAC to be recognized and transported by the receptors, the Fe(III)-bound SAC must be formed. This complex can be pre-formed and administered to the bacterial culture. Alternatively, the apo SAC can be administered and the siderophore moiety can scavenge Fe(III) from the growth medium. In some cases, such as when the MIC of the

SAC is greater than the concentration of Fe in the medium, preloading the SAC with Fe(III) may enhance its uptake and antimicrobial activity. We also note that administration of an apo SAC can result in Fe starvation. For instance, if an apo SAC scavenges Fe(III) from the growth medium and cannot be utilized by the bacterial cell, it acts as an extracellular Fe(III) chelator that prevents bacterial Fe acquisition and, at sufficient concentrations, causes growth inhibition. This possibility should be considered when initially evaluating the AMA of novel SACs.

2.2 Safety precautions and general practices

Abide by standard safety precautions when handling acids (hydrochloric acid and nitric acid) and organic solvents (DMSO, methanol). Be sure to employ appropriate sterile conditions, biosafety controls, and waste treatment/disposal for all bacterial pathogens. We note that in these procedures SAC stock solutions are not sterilized to avoid loss of material. These solutions are prepared in DMSO (which does not permit bacterial growth) and then diluted in filter- or autoclave-sterilized Milli-Q water. Microbiology assays are typically performed with two technical replicates (same bacterial colony) and at least three biological replicates (different bacterial colonies), ideally performed on at least two different days with at least two different synthetic batches of SAC. Using two different synthetic preparations of SAC increases the experimental rigor, ensuring a set of results is not dependent on a given batch of compound.

2.3 Antimicrobial activity (AMA) assay

2.3.1 Introduction

Following the successful synthesis and purification of a SAC, one of the first and most important steps of its evaluation is determining its AMA against the bacterial strain(s) of interest. The potency of a SAC is typically compared to that of the parent antibiotic. It is often hypothesized that successful siderophore-mediated drug delivery will increase cellular accumulation of the warhead and that SACs will achieve equivalent or enhanced AMA of the cargo against siderophore-utilizing bacteria. Nevertheless, it is possible that chemical modification to the antibiotic moiety may attenuate its growth inhibitory activity and that chemical modification of the siderophore may impede uptake of the SAC. To evaluate the AMA of SACs, the standard antibiotic activity testing procedures described by EUCAST and CLSI have been modified (CLSI, 2023; EUCAST, 2020). In these assays, bacteria are treated with varying concentrations of SAC and the

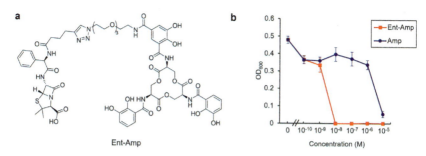

Fig. 2 (A) Structure of Ent-Amp. (B) AMA plot of Ent-Amp and Amp against *E. coli* CFT073. Performed in modified M9 medium, 30 °C, 20 h, mean ± sd (n = 3). *Data taken and replotted from Sargun, A., Johnstone, T. C., Zhi, H., Raffatellu, M., & Nolan, E. M. (2021). Enterobactin- and salmochelin-β-lactam conjugates induce cell morphologies consistent with inhibition of penicillin-binding proteins in uropathogenic Escherichia coli CFT073. Chemical Science, 12, 4041–4056.*

AMA is quantified as the MIC defined as the lowest concentration of an antibacterial agent which completely prevents visible growth (EUCAST, 1998). Modifications to the standard guidelines include the choice of growth medium, Fe supplementation, and the use of spectrometry to measure cell density. For example, our study of Ent-ampicillin (Ent-Amp, Fig. 2A) against *E. coli* CFT073 revealed a ~1000-fold decrease in MIC compared to ampicillin (Amp, a β-lactam antibiotic) in Fe-limited medium, where the optical density at 600 nm (OD_{600}) was reported as a measure of cell growth (Fig. 2B) (Sargun et al., 2021; Zheng & Nolan, 2014).

Beyond MIC determination, AMA assays can provide us with additional key information regarding the uptake and cellular fates of SACs. For instance, performing AMA assays under conditions of high and low Fe availability highlights the dependence of SAC activity on the expression of siderophore uptake systems (McPherson et al., 2012; Nikaido & Rosenberg, 1990; Zheng & Nolan, 2014). Employing bacterial mutant strains can establish the involvement of OM receptors, such as the role of FepA in transporting SACs containing catecholate siderophores (Möllmann et al., 2009; Neumann et al., 2018; Nikaido & Rosenberg, 1990; Zheng et al., 2012). In addition, mutants of the IM transporters can be employed to interrogate cytoplasmic delivery, including the role of FepCDG in transporting Ent-based SACs that target DNA and DNA replication (Guo & Nolan, 2022; Neumann et al., 2018). As siderophore conjugation generally narrows the AMA spectrum of an antibiotic towards bacteria that utilize the siderophore, AMA assays with co-cultures can inform the species or strain selectivity of a SAC. These assays often involve the use of selective or differentiating culture plates that allow colonies of each species or strain to be detected and quantified (Prinzi & Rohde, 2020). For example, Ent-Amp selectively inhibits the growth of *E. coli* CFT073 in co-cultures with Gram-positive organisms *Staphylococcus aureus* or *Lactobacillus rhamnosus* (Chairatana, Zheng, & Nolan, 2015; Sargun et al., 2021; Zheng & Nolan, 2014).

Here we describe the procedure for an AMA assay with Ent-Amp (Fig. 2A) against *E. coli* CFT073. The result of this experiment is summarized in Fig. 2B.

2.3.2 Materials
Equipment:
- Innova 4000 Shaker (able to reach 37 °C and 250 rpm)
- Beckman Coulter DU800 spectrophotometer
- BioTek Synergy HT plate reader (capable of measuring absorbance at 600 nm for 96-well plates)
- P10, P20, P200, P1000 Eppendorf pipettes, and multichannel P20 and P200 Eppendorf pipettes (if available)
- Pipette bulb; or electronic pipette controller

Consumables:
- Freezer stock of *E. coli* CFT073 (ATCC 700928) in LB and 25% glycerol
- Sterile 5 mL pipettes
- Sterile 15 mL culture tubes (polystyrene or polypropylene)
- Sterile inoculating plastic loop, or metal loop sterilized by flame
- Sterile modified M9 medium (Neumann et al., 2018):
  M9 minimal salts (REF248510, Becton Dickinson), 2 mM $MgSO_4$, 0.1 mM $CaCl_2$, 0.4% glucose, 0.2% casamino acids (Becton Dickinson), 16.5 mg/L thiamine hydrochloride, sterilize by filtering (0.22 μm)
- Sterile LB agar plates
- Disposable cuvettes (1.5 mL, d = 1 cm)
- Quartz cuvette with a lid (1.5 mL, d = 1 cm)
- DMSO
- MeOH
- Milli-Q water (autoclaved or filter sterilized)
- Sterile PCR strip tubes with lid
- Sterile transparent 96-well plates
- Sterile plastic 25 mL reservoir
- Eppendorf pipette tips (appropriate for pipettes used)
- Plastic wrap
- Ent-Amp and Amp (purified solids)

Alternatives:

- General laboratory equipment may be used in place of the models mentioned.
- This assay can also be performed in other growth media such as LB or MHB supplemented with DP. The concentration of the DP supplement will vary depending on the medium and bacterial strain being employed. For instance, 50% MHB supplemented with 200 μM DP is commonly used for studies of *E. coli* and a 600 μM DP supplement was required for monitoring Ent uptake by *Pseudomonas aeruginosa* in 50% MHB (Zheng et al., 2012). For some mutant strains with defects in Fe uptake that show attenuated growth under severe Fe limitation, lower DP concentrations may be required.

2.3.3 Procedure

Day 1:

1. Using a sterile inoculation loop, take a sample from the freezer stock of *E. coli* CFT073 and streak the surface of an LB agar plate.
2. Allow the plates to incubate at 37 °C for 10–16 h until the colonies are ~1 mm in diameter.

Day 2:

3. Using a 5 mL pipette, transfer 5 mL of modified M9 medium to a 15 mL culture tube.

   Note: We recommend working with the same growth medium for the starter culture, the diluted culture prepared on Day 3, and AMA assay. However, slow or attenuated growth in Fe-limited medium may necessitate a more nutrient-rich medium for the starter culture such as LB or MHB without DP. In such cases, it is necessary to consider factors such as impacts on the expression of siderophore uptake machinery and the availability of nutrient metals carried over from the initial medium.
4. To prepare the starter culture, pick one of the well-isolated colonies using an inoculation loop and inoculate in 5 mL growth medium. This culture will correspond to one biological replicate.
5. Place the tube tilted on a rack inside the shaker. Allow the culture to grow to the stationary phase at 37 °C for 14–16 h while shaking rapidly.

Day 3:

6. Using a 5 mL pipette, transfer 5 mL of modified M9 medium to a 15 mL culture tube.
7. Pipette 50 μL of the starter culture into the 5mL aliquot of fresh modified M9 medium (1:100 dilution).

8. Incubate the diluted culture at 37 °C for 2–4 h until it reaches the mid-log phase of bacterial growth. For *E. coli* CFT073, mid-log phase corresponds to OD_{600} ~0.6.

   Note: OD_{600} is a measure of turbidity and is the absorbance of the cell culture at $\lambda = 600$ nm. For most bacterial cells, a linear relationship exists between the cell concentration in the form of colony forming units (CFU/mL) and the OD_{600} of the culture. This relationship varies with the bacterial species and the spectrometer used and must be determined case-by-case (Van Alst, LeVeque, Martin, & DiRita, 2023). For *E. coli* CFT073 and our spectrometer, the OD_{600} of 0.6 is equivalent to $\sim 5 \times 10^{8}$ CFU/mL.

   Note: For the assay, the mid-log phase culture is used to test the AMA of the SAC against a consistent bacterial inoculum in a metabolically active state. It is recommended to obtain growth curves of each bacterial strain to identify the OD_{600} associated with this growth phase.

In the meantime, prepare the solutions of the SAC and parent antibiotic:

9. Prepare a stock solution of Ent-Amp by dissolving 1.4 mg of solid (1 μmol) in 100 μL DMSO. The resulting solution should have a concentration of ~10 mM.

   Note: the stock solution must be gently but thoroughly mixed, aliquoted, and stored at −20 °C. Storing the stock solution in small aliquots avoids cycles of freeze and thaw that will compromise the integrity of the SAC.

   Ent has an extinction coefficient of $9500\,M^{-1}\,cm^{-1}$ at 316 nm in MeOH resulting from absorption by the catechol moieties (Scarrow, Ecker, Ng, Liu, & Raymond, 1991). The concentration of the stock solution must be accurately determined using this extinction coefficient and a UV–vis spectrometer.

10. In a clean and dry quartz cuvette, pipette 800 μL of MeOH. Keep the lid closed to prevent evaporation.

11. Pipette 1 μL of the DMSO stock solution into the methanol, mix well, and monitor the absorbance at 316 nm using a UV–vis spectrometer. Obtain at least 3 data points by adding additional volumes of the stock solution to the same cuvette and measuring the absorbance.

    Note: Ideally independent solutions of the SAC (varying concentrations) in MeOH are made. The titration approach is used when the quantity of material is limited.

    Note: Some cargos may show absorbance at 316 nm, which contributes to the overall ε_{316} value of the conjugate. This potential feature of a SAC should be considered when determining a stock solution concentration.

Experimental methods for evaluating siderophore–antibiotic conjugates

12. Calculate the concentration of stock solution and dilute accordingly to prepare the working solution with 0.1 mM Ent-Amp in 10% DMSO in water.

   Note: freshly prepare the working solution in the volume needed for the assay. Any excess solution cannot be stored as the presence of water may hydrolyze the Ent trilactone ring over time.

13. Obtain 7 PCR tubes for serial dilution of the Ent-Amp working solution. Transfer 25 μL of the working solution to the first tube (highest concentration).

   Note: A 25 μL volume of SAC working solution is sufficient for 2 technical replicates. You may increase the volume according to your experimental design.

   Note: When treating the cells with Fe(III)-loaded conjugate, add 9 μL of 1 mM Fe(III) working solution to each 100 μL of the working solution of SAC. The 1 mM Fe(III) working solution can be prepared from stock solutions of $FeCl_3$ in 0.2 M HCl or $Fe(acac)_3$ in DMSO (Guo & Nolan, 2022; Neumann et al., 2018).

14. In the remaining 6 PCR tubes, pipette 22.5 μL of 10% DMSO (aq).

15. Serially dilute the working solution by transferring 2.5 μL of the working solution to the subsequent tube containing 10% DMSO (aq) (10-fold dilution). Mix well and repeat for the next 5 tubes. Ensure the last tube contains only the solvent (untreated control).

   Note: The concentration of the compound in each tube is 10 × the final concentration in the assay.

   Note: Alternatively, 2-fold serial dilutions can be carried out, but this dilution series may require a greater number of samples to span the relevant concentration range for observing the antibacterial activity of the SAC and parent antibiotic.

16. Next, prepare the working solution of the parent antibiotic, 0.1 mM of Amp in 10% DMSO (aq). Serially dilute this stock solution following the same procedure as Ent-Amp (steps 13–15).

Prepare bacterial cell cultures:

17. Transfer 700–800 μL of each culture to a disposable 1.5 mL cuvette and measure the OD_{600} using an UV–vis spectrometer. Ensure that the cells have grown to the mid-log phase as described in step 8.

   Note: use modified M9 medium as a blank on the spectrometer.

18. Dilute the cell culture to a cell density of $\sim 10^6$ CFU/mL using modified M9 medium in a sterile plastic reservoir. For *E. coli* CFT073,

this cell density is associated with OD_{600} of about 0.001, so a culture with the OD of 0.6 must be diluted 1:600.

Note: To ensure there is sufficient diluted culture and to facilitate pipetting from the reservoir, prepare at least 2 mL diluted culture per 2 technical replicates.

Prepare the assay plate:

19. Using a marker, label the lid of the 96-well plate with the compound (s), bacterial strain, date, and time. Mark the peripheral wells as blank as they will be holding the sterile growth medium only. You may include wells associated with appropriate controls, such as parent antibiotic, unconjugated siderophore, untreated control, etc. Be sure to dedicate at least two wells to each treatment condition (technical replicates).

20. Pipette 100 µL growth medium into all peripheral wells.

    Note: Fill every cell on the plate that does not hold samples with sterile growth medium. The presence of sterile medium limits evaporation in the plate, is used as baseline absorption, and serves as an indicator of contamination in the growth medium.

21. Plate 90 µL of diluted cell culture per appropriate well.

22. Pipette 10 µL of 10 × Ent-Amp or Amp in each appropriate well of the plate.

    Note: preparing the dilutions in PCR tubes allows for the facile use of multichannel pipette to efficiently add the treatment solution.

    Note: The total DMSO concentration in each well amounts to 1%. DMSO can inhibit bacterial growth, so it may be necessary to test its effect on the growth of the specific bacterial strain of interest.

23. Wrap plates with damp paper towel and plastic wrap to preserve the moisture and secure them into a shaker at 30 °C and 150 rpm for 20 h.

Day 4:

24. After 20 h, measure OD_{600} of the incubated plate with a plate reader.

25. The OD_{600} values from terminal wells, containing sterile medium, are associated with background absorption and the average OD_{600} of these wells constitutes the blank. Subtract this value from each data point and calculate the average OD_{600} values of wells with the same concentration of Ent-Amp or Amp (technical replicates). Plot mean OD_{600} values of 3 biological replicates as a function of SAC/antibiotic concentration. The standard deviation of the OD_{600} of multiple biological replicates is shown as error bars associated with each data point. Refer to Fig. 2B as an example.

2.4 Time-kill kinetics

2.4.1 Introduction

Time-kill kinetics assays evaluate the AMA of SACs by assessing the rate of cell killing by the warhead over time. Time-kill kinetics experiments are often employed to differentiate the activity of antibacterial agents as bacteriostatic or bactericidal, where a bacteriostatic agent stalls the growth without cell killing and a bactericidal agent kills 99.9% of the inoculum (Pankey & Sabath, 2004). For a SAC containing a bactericidal warhead, the time-kill kinetics assay reveals whether the conjugate retains the cell killing ability of the parent antibiotic, and the extent to which siderophore conjugation affects the rate of cell death. When treated with a high concentration of SAC or antibiotic greater than the MIC bacterial cultures are expected to exhibit a decrease in cell viability over the course of the incubation, measurable by OD_{600} or CFU enumeration. The active uptake of a SAC could potentially result in rapid intracellular accumulation of the warhead and hence accelerate the rate of cell killing relative to the parent antibiotic. Thus, time-kill kinetics assays are informative when comparing SACs to the corresponding parent antibiotic because an increased rate of cell death provides strong evidence for the active transport of the SAC into the cells. For example, treatment of *E. coli* CFT073 with 50 µM Ent-Amp results in a more rapid decrease of culture turbidity than for 100 µM Amp (Fig. 3) (Sargun et al., 2021). We note that rate enhancement is not observed for all SACs, namely cefiderocol and some other Ent-based SACs exhibit kinetics of cell killing similar to the parent antibiotics (Ito et al., 2018; Sargun et al., 2021). Different bacterial strains may also exhibit varying rates of cell killing when treated with the same SAC (Chairatana et al., 2015). Nevertheless, so long as the bactericidal warhead maintains its mode of action upon siderophore conjugation and the MIC value is similar or lower than that of the parent antibiotic, we expect to observe similar, if not accelerated, cell killing between the parent antibiotic and the SAC. An attenuated rate of cell killing may suggest impeded uptake of the antibiotic as a result of siderophore conjugation or that the activity of the antibiotic is compromised by conjugation to the siderophore.

Herein we describe the procedure for the time-kill kinetics assay with Ent-Amp (Fig. 2A) and the parent antibiotic Amp against *E. coli* CFT073. The result of this experiment is summarized in Fig. 3.

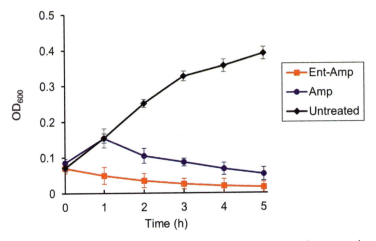

Fig. 3 Time-kill kinetics plot of Ent-Amp (50 µM), Amp (100 µM), and untreated control for *E. coli* CFT073. Performed in modified M9 medium, 37 °C, mean ± sd (n = 3). *Data were taken and replotted from Sargun, A., Johnstone, T. C., Zhi, H., Raffatellu, M., & Nolan, E. M. (2021). Enterobactin- and salmochelin-β-lactam conjugates induce cell morphologies consistent with inhibition of penicillin-binding proteins in uropathogenic Escherichia coli CFT073. Chemical Science, 12, 4041–4056.*

2.4.2 Materials

Equipment:
- Innova 4000 Shaker (able to reach 37 °C and 250 rpm)
- Beckman Coulter DU800 spectrophotometer
- P10, P20, P200, P1000 Eppendorf pipettes
- When monitoring by OD_{600}: BioTek Synergy HT plate reader (capable of measuring at 600 nm for 96-well plates)
- Pipette bulb; or electronic pipette controller
- Beckman Coulter Allegra X-15R Centrifuge

Consumables:
- Freezer stock of *E. coli* CFT073 (ATCC 700928) in LB and 25% glycerol
- Sterile 5 mL pipettes
- Sterile 15 mL polystyrene or polypropylene culture tubes
- Sterile inoculating plastic loop; or metal loop sterilized by flame
- Sterile modified M9 medium (as described in Section 2.3.2)
- Sterile LB agar plates
- Disposable cuvettes (1.5 mL, d = 1 cm)
- DMSO
- Milli-Q water (autoclaved or filter sterilized)

Experimental methods for evaluating siderophore–antibiotic conjugates   35

- Sterile PCR strip tubes with lid
- Sterile transparent 96-well plates
- Eppendorf pipette tips (appropriate for pipettes used)
- Sterile 15 mL falcon tubes
- $10 \times$ Ent-Amp ($500\,\mu M$) working solution in 10% DMSO (aq) (prepared following steps 9–12 Section 2.3.3)
- $10 \times$ Amp (1 mM) working solution in 10% DMSO (aq)

Alternatives:

- This assay can also be performed in other growth media such as LB or 50% MHB supplemented with $200\,\mu M$ DP to induce Fe limitation (see Section 2.3.2 for more information).

2.4.3 Procedure

Days 1 and 2:

1. Prepare a starter culture as described in steps 1–5 of Section 2.3.3.

Day 3:

Prepare bacteria culture(s):

2. Using a 5 mL pipette, transfer 5 mL of modified M9 medium to a 15 mL culture tube.

3. Pipette $50\,\mu L$ of the starter culture into 5 mL of fresh medium (1:100 dilution).

4. Incubate the diluted culture at $37\,°C$ for 2–3 h until it reaches the beginning of log-phase of bacterial growth, OD_{600} ~0.3.

5. Using a 5 mL pipette, transfer a 3 mL aliquot of the culture to a falcon tube.

6. Centrifuge the culture at 3500 rpm, $4\,°C$, 10 min.

   Note: Ensure the centrifuge is appropriately balanced.

7. Remove the supernatant.

8. Resuspend the cell pellet in ~3 mL of fresh modified M9 medium.

9. Centrifuge at 3500 rpm, $4\,°C$, 10 min.

10. Remove the supernatant.

11. Repeat the steps 8–10 to appropriately wash the cell pellet from the spent medium.

   Note: Due to the higher starting cell density for this assay, it is necessary to wash the cells to avoid carryover from the spent medium. The wash step is particularly important when different growth media are used for the starter culture and the assay.

12. Resuspend the cell pellet in 3 mL fresh growth medium to $OD_{600} = 0.3$.

   Note: This relatively high inoculum (~3×10^8 CFU/mL for *E. coli* CFT073) enables monitoring of the cell killing by OD_{600} on a plate reader.

Set up the plate:

13. Using a permanent marker, label the lid of the 96-well plate with the compound(s), experiment number, bacterial strain, date, and time. Include wells for untreated control (1% DMSO) and be sure to dedicate at least two wells to each treatment condition (technical replicates).
14. Transfer 100 µL of growth medium to all peripheral wells as blanks (see Section 2.3.3 step 20)
15. In each well, pipette 90 µL of the resuspended bacterial culture.
16. Add 10 µL of the 10 × stock solution of Ent-Amp, Amp, or 10% DMSO to each appropriate well.

    Note: You may choose any concentration greater than MIC that results in a time-dependent cell killing. In this assay, there is no need for preparing a serial dilution of SAC/antibiotic.

    Note: if you choose to work with an Fe(III)-loaded SAC, see notes in Section 2.3.3. step 13.
17. Incubate the plate at 37 °C while shaking at 150 rpm.
18. Measure cell viability at appropriate time points, for example, every hour for 5 h.

    Note: Cell viability can be monitored either by OD_{600} values using a plate reader, or by CFU enumeration (Van Alst et al., 2023). The relevant time points may vary depending on the bacterial strain, medium conditions, and the conjugate used.
19. For each technical replicate subtract the value of the blank (average OD_{600} of cells with growth medium only). Calculate the average OD_{600} of wells with the same concentrations (see Section 2.3.3 step 25). Alternatively, calculate the CFU/mL of the culture at each time point from the number of colonies you have counted.
20. Plot the mean OD_{600} values or mean $Log_{10}(CFU/mL)$ values versus time and report the standard deviation as error bars. Refer to Fig. 3 as an example.

2.5 Competition with unmodified Ent

2.5.1 Introduction

The competition between SACs and native siderophores for cellular uptake can provide further evidence of receptor recognition and transport of SACs. The premise draws on principles from the siderophore literature, where administration of an unlabeled siderophore would decrease cellular accumulation of the labeled counterpart upon competition for uptake (Annamalai, Jin, Cao, Newton, & Klebba, 2004; Dertz, Xu, Stintzi, & Raymond, 2006). For SAC applications, bacterial growth can instead serve as the readout for the extent to which SACs are transported in comparison to native siderophores. Sufficient levels of the unmodified siderophore can compete with and prevent the uptake of the SAC and thus rescue the growth of the bacterial cells. We have modified the AMA assay (Section 2.3) to examine competitions between Ent-based conjugates and Ent, for uptake by OM receptors (e.g., FepA). Cultures are treated with a consistent, above MIC concentration of SAC in a medium supplemented with varying concentrations of exogenous Ent. The growth at different concentrations of Ent is compared to evaluate the extent of competition. This assay can provide key information about the receptor utilization of SACs because growth recovery with unmodified siderophore supplementation provides evidence of shared receptor recognition and uptake route(s) between the SAC and siderophore. For example, when *E. coli* CFT073 was treated with 0.1 μM of Ent-Amp (Fig. 2A), administration of 2 and 10 μM exogenous Ent recovered cell growth (Fig. 4) (Chairatana et al., 2015). We note that we have typically carried out this assay in an Fe-replete medium (50% MHB, supplemented with DP to limit Fe availability) to focus on the competition between siderophore moieties for receptor binding and transport, and minimize the Fe

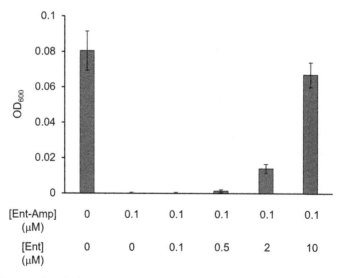

Fig. 4 Competition between Ent-Amp and exogenous Ent against *E. coli* CFT073. Performed in 50% MHB + 200 μM DP, 30 °C, 20 h, mean ± sd (n = 3). *Data taken and replotted from Chairatana, P., Zheng, T., & Nolan, E. M. (2015). Targeting virulence: Salmochelin modification tunes the antibacterial activity spectrum of β-lactams for pathogen-selective killing of Escherichia coli. Chemical Science, 6, 4458–4471.*

(III)-binding affinity as a factor effecting the uptake. Employing Fe-deficient medium may require preloading the siderophore moieties with Fe (Neumann et al., 2018). Extensions of this assay include comparing different siderophores for their receptor affinities relative to the native structure (Motz et al., 2024), or the administration of native siderophores with known receptor specificity in order to probe the receptors involved in uptake of the SAC (Chairatana et al., 2015). Variations of this competition assay have been used by others to provide evidence for the uptake pathways of siderophore conjugates (Caradec et al., 2023; Wencewicz et al., 2013). We note that it is also possible in some systems (e.g. *Pseudomonas aeruginosa*) for administration of a siderophore/xenosiderophore to sensitize bacterial cells to SACs utilizing the same transporter (Luscher et al., 2022). This finding is attributed to the upregulation of OM transporters based on two-component sensing systems of *P. aeruginosa* responsive to the presence of siderophores. The competition assay described here is the most appropriate for systems in which receptor expression is not inducible by substrate exposure.

Here we describe the competition assay for Ent-Amp and Ent in *E. coli* CFT073. The result of this experiment is summarized in Fig. 4.

2.5.2 Materials

Equipment:
- Innova 4000 Shaker (able to reach 37 °C and 250 rpm)
- Beckman Coulter DU800 spectrophotometer
- BioTek Synergy HT plate reader (capable of measuring at 600 nm for 96-well plates)
- P10, P20, P200, P1000 Eppendorf pipettes
- Pipette bulb; or electronic pipette controller
  Consumables:
- Freezer stock of *E. coli* CFT073 (ATCC 700928) in LB and 25% glycerol
- Sterile 5 mL pipettes
- Sterile 15 mL polystyrene or polypropylene culture tubes
- Sterile inoculating plastic loop; or metal loop sterilized by flame
- Sterile growth medium: LB (Becton Dickinson) and 50% MHB (Sigma-Aldrich)
- Sterile LB agar plates
- Disposable cuvettes (1.5 mL, d = 1 cm)
- DMSO
- Milli-Q water (autoclaved or filter sterilized)
- Sterile PCR strip tubes with lid

Experimental methods for evaluating siderophore–antibiotic conjugates 39

- Sterile transparent 96-well plates
- Sterile plastic 25 mL reservoir
- Eppendorf pipette tips (appropriate for pipettes used)
- Plastic wrap
- 200 mM DP stock solution in DMSO
- 20 × Ent-Amp working solution (2 µM) in 10% DMSO (aq) (prepared as described in steps 9–12, Section 2.3.3)
- 200 µM Ent working solution in 10% DMSO (aq) (prepared as described in steps 9–12, Section 2.3.3)

2.5.3 Procedure

Day 1:

1. Streak a LB agar plate from the freezer stock of *E. coli* CFT073 as described in steps 1–2 of Section 2.3.3.

Day 2:

2. Using a 5 mL pipette, transfer 5 mL of LB medium to a 15 mL culture tube.
   Note: LB medium without DP is used for the starter culture to ensure sufficient overnight growth. *E. coli* CFT073 growth is attenuated in 50% MHB supplemented with DP, which is used for the competition assay.

3. To prepare the starter culture, pick one of the well-isolated colonies using an inoculation loop and inoculate the 5-mL aliquot of LB medium. This culture will correspond to one biological replicate.

4. Place the tube, tilted, on a rack inside the shaker. Allow the culture to grow to the stationary phase at 37 °C for 14–16 h while shaking rapidly.

Day 3:

5. Using a 5 mL pipette, transfer 5 mL of LB medium to a 15 mL culture tube.

6. Add 5 µL of 200 mM DP solution in DMSO to the medium to achieve a final concentration of 200 µM.
   Note: LB and 50% MHB typically have a total Fe concentration of less than 10 µM. The DP is used in excess to ensure Fe limited conditions. Different bacterial strains may require different concentrations of DP as described in Section 2.3.2.

7. Pipette 50 µL of the starter culture into the 5-mL aliquot of LB supplemented with 200 µM DP (1:100 dilution).

8. Incubate the diluted culture at 37 °C for 3–4 h with shaking until it reaches the mid-log phase of bacterial growth, OD_{600} ~0.6 (see Section 2.3.3 step 8).

Prepare the Ent/Ent-Amp working solutions:

9. Serially dilute the stock solution of Ent to obtain $20 \times$ solutions with varying concentrations between 200 and $0\,\mu M$. Use 10% DMSO (aq) as solvent.

    Note: You may choose to use a different concentration range for the siderophore in your assay.

10. In a strip of PCR tubes, add $12.5\,\mu L$ of each $20 \times$ Ent solution to the same volume of $20 \times$ Ent-Amp to obtain $10 \times$ working solutions containing varying concentration of siderophore, and a constant concentration of SAC.

Prepare bacteria culture(s):

11. Transfer $700\text{--}800\,\mu L$ of the cell culture to a disposable cuvette and measure the OD_{600}. Ensure the culture is in the mid-log phase of the growth (see Section 2.3.3 step 8).

12. Dilute the mid-log phase culture into a sterile plastic reservoir with 50% MHB supplemented with $200\,\mu M$ DP to achieve a final OD_{600} of 0.001 ($\sim 10^6$ CFU/mL).

Prepare the plate:

13. Using a permanent marker, label the lid of the plate with the compound(s), experiment number, bacterial strain, date, and time. Include wells for the untreated contol (10% DMSO) and SAC only ($0\,\mu M$ Ent). Be sure to dedicate at least two wells to each treatment condition (technical replicates).

14. Transfer $100\,\mu L$ of growth medium to all peripheral wells as blanks (see Section 2.3.3 step 20).

15. Add $90\,\mu L$ of the diluted bacterial culture to each well.

16. Add $10\,\mu L$ of each $10 \times$ SAC+Ent working solution in appropriate wells of the 96-well plate.

17. Wrap the plate in a wet paper towel followed by plastic wrap and incubate at 30 °C with shaking (150 rpm) for 20 h.

Day 4:

18. Measure OD_{600} values for the plate via a plate reader.

19. Work up the data according to Section 2.3.3 step 25. Plot the average OD_{600} values versus Ent concentration as a bar graph and report the standard deviation using error bars. See Fig. 4 for an example.

2.6 ^{57}Fe uptake assay

2.6.1 Introduction

A variety of methods have been developed to probe siderophore uptake. Radioactive elements such as $^{55/59}$Fe (Stintzi, Barnes, Xu, & Raymond,

2000), $^{67/68}$Ga (Petrik, Pfister, Misslinger, Decristoforo, & Haas, 2020), and ^{89}Zr (Petrik et al., 2016) can be coordinated by siderophores and thus have been used to monitor siderophore uptake into bacterial cells by quantifying the uptake of radiolabels. Lanthanide ions such as Eu^{3+} or Tb^{3+} (Ferreira et al., 2017) enable luminescent imaging, and conjugation to fluorescent dyes (Hannauer et al., 2010; Lee et al., 2016; Weizman et al., 1996; Zheng & Nolan, 2012; Zheng et al., 2012) permits the use of fluorescence spectroscopy or cellular imaging to study siderophore uptake.

Here we describe an assay based on inductively-coupled plasma mass spectrometry (ICP-MS), where we utilize the non-radioactive isotope ^{57}Fe to monitor and quantify the uptake of ferric siderophores or SACs. ^{57}Fe is an ideal candidate for the uptake assay due to its low natural abundance (2.12% vs. 91.7% for ^{56}Fe) and the commercial availability of isotopically pure sources. Importantly, this assay does not require further chemical modification to the siderophore moiety (or SAC) or the use of radioactive elements and can be applied to any siderophore or SAC. Following the treatment of bacterial culture with siderophore or SAC preloaded with ^{57}Fe(III), ICP-MS is used to measure the concentration of each elemental isotope in a liquified cell sample. Consequently, ^{57}Fe is readily detected against a background of ^{56}Fe that represents the cell-associated Fe prior to Fe(III)-siderophore uptake, and the quantity of cell-associated ^{57}Fe serves as a measure of the amount of siderophore transported into the cells. Alternatively, the uptake of conjugates containing metal-based drugs such as Pt-based anticancer agents can be directly measured, without labeled Fe, relying on the negligible physiological abundance of the cargo (Guo & Nolan, 2022).

We recently employed the ^{57}Fe assay to quantify the amount of ^{57}Fe and hence siderophore taken up by *Salmonella enterica* serovar Typhimurium (*S*Tm) (Motz et al., 2024). The results from this assay were informative for comparing the uptake of TRENCAM (TC), a non-native synthetic analog of Ent, to that of Ent (Fig. 5). This experiment revealed excellent uptake of the two siderophore moieties and motivated further studies with TC-based SACs (Motz et al., 2024). We also extended this study to *S*Tm mutant strains lacking Ent OM receptors to gain information about the role of each receptor in the uptake of TC.

Here we describe the ^{57}Fe uptake assay for Ent and TC using *S*Tm. The result of this experiment is summarized in Fig. 5.

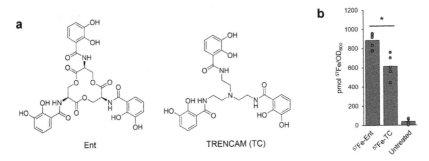

Fig. 5 (A) Structures of Ent and TRENCAM (TC), a synthetic mimic of Ent. (B) ^{57}Fe-Ent and ^{57}Fe-TC uptake by STm treated with 10 μM ^{57}Fe-siderophore. Performed in modified M9 medium, 30 °C, 1 h; *p < 0.005. Data taken and replotted from Motz, R. N., Guo, C., Sargun, A., Walker, G. T., Sassone-Corsi, M., Raffatellu, M., & Nolan, E. M. (2024). Conjugation to native and nonnative triscatecholate siderophores enhances delivery and antibacterial activity of a β-lactam to Gram-negative bacterial pathogens. *Journal of the American Chemical Society*, 146, 7708–7722.

2.6.2 Materials
Equipment:
- Innova 4000 Shaker (able to reach 37 °C and 250 rpm)
- Beckman Coulter DU800 spectrophotometer
- BioTek Synergy HT plate reader (capable of measuring at 600 nm for 96-well plates)
- P10, P20, P200, P1000 Eppendorf pipettes
- Pipette bulb or electronic pipette controller
- Beckman Coulter Allegra X-15R Centrifuge
- Beckman Coulter Microfuge 20R Centrifuge
- Agilent 7900 ICP-MS system in helium mode (or submit samples to a service facility)
- Milestone UltraWAVE Microwave digester

Consumables:
- Freezer stock of *Salmonella enterica* serovar Typhimurium IR715 Δ*entC* (MSC67) in LB and 25% glycerol (Motz et al., 2024)

    Note: the Δ*entC* mutant is deficient in the biosynthesis of Ent. This mutation is necessary to prevent Fe exchange between the native and synthetic siderophores.
- Sterile 5 mL pipettes
- Sterile 15 mL polystyrene or polypropylene culture tubes
- Sterile inoculating plastic loop; or metal loop sterilized by flame
- Sterile modified M9 (see Section 2.3.2)

Experimental methods for evaluating siderophore–antibiotic conjugates

- Sterile LB agar plates
- Disposable cuvettes (1.5 mL, d = 1 cm)
- DMSO
- Milli-Q water (autoclaved or filter sterilized)
- Sterile PCR strip tubes with lid
- Sterile transparent 96-well plates
- Eppendorf pipette tips (appropriate for pipettes used)
- Sterile 1.5 mL microcentrifuge tubes
- Sterile 5 mL falcon tubes
- Concentrated Ultra-high purity HNO_3
- Solid $^{57}Fe_2O_3$ (Cambridge Isotope Laboratories)
- 50 ppb solution of Tb in 2% HNO_3 prepared by diluting Agilent Technologies Terbium internal standard, (10 µg/mL Tb in 2% HNO_3; Part No. 5190-8590)
- Agilent Technologies ICP-MS standard solution containing 1000 µg/mL Fe and other metals (Part No. 5183-4688)
- 100 µM Ent working solution in 10% DMSO (see steps 9–12 Section 2.3.3)
- 100 µM TC working solution in 10% DMSO (see steps 9–12 Section 2.3.3)
  Note: the ε_{316} for TC in MeOH was found to be 10,100 $M^{-1}cm^{-1}$ (Motz et al., 2024).

2.6.3 Procedure

Days 1 and 2:

1. Prepare a starter culture from the freezer stock of *S*Tm Δ*entC* following the same steps described for *E. coli* in steps 1–5 Section 2.3.3.

Day 3:

2. Using a 5 mL pipette, transfer 5 mL of sterile modified M9 medium to a 15 mL culture tube.
3. Pipette 50 µL of overnight culture to the tube containing the fresh medium (1:100 dilution).
4. Incubate the culture for 2–4 h while shaking at 37 °C until the culture reaches the mid-log phase or OD_{600} of ~0.6 (see Section 2.3.3 step 8).

Meanwhile, prepare the $^{57}Fe(III)$-loaded siderophore solutions:

5. Dissolve ~4 mg of $^{57}Fe_2O_3$ in 125 µL 12 M HCl in a warm water bath. Add Milli-Q water to 5 mL. This is the ^{57}Fe stock solution and has a concentration of ~10 mM ^{57}Fe.

   Note: It is strongly recommended to use ICP-MS or another method to verify the concentration of the ^{57}Fe stock solution prior to moving forward with the assay.

6. Prepare a working solution of 1 mM ^{57}Fe by diluting the ^{57}Fe stock solution in water.

7. In a PCR tube, pipette the appropriate amount of Ent/TC working solution and add 9 μL of ^{57}Fe working solution per 100 μL of siderophore (0.9 eq Fe).

   Note: loading the siderophore with substoichiometric Fe ensures minimal unbound ^{57}Fe in the assay and thus avoids non-specific ^{57}Fe uptake by STm.

   Note: Highly acidic solutions can hydrolyze the trilactone ring of Ent.

Set up the plate:

8. Using a marker, label the lid of the plate with the compound(s), experiment number, bacterial strain, date and time. Include wells for untreated control (administered 10% DMSO), Ent, TC, and other controls as needed. Be sure to dedicate at least two wells to each treatment condition.

9. Transfer 100 μL of sterile modified M9 medium to all peripheral wells (see Section 3.2.2 step 20).

   Prepare and treat the cell cultures:

10. Transfer 700–800 μL of culture to a disposable cuvette. Measure the OD_{600} via UV–vis spectrophotometry to ensure the culture has reached mid-log phase.

11. Using a 5 mL pipette, transfer 4 mL of the culture to a 15 mL falcon tube.

12. Centrifuge at 4000 rpm, 10 min at 4 °C.

13. Remove the supernatant.

14. Resuspend the pellets into the appropriate volume of modified M9 medium to obtain an OD_{600} equivalent to ~2.4 (4-fold concentration of $OD_{600} = 0.6$ culture).

15. Pipette 90 μL of this cell culture to the appropriate wells on the plate.

16. Pipette 10 μL of ^{57}Fe-Ent/TC solution to the appropriate wells on the plate.

17. Secure the plate on a shaker, set to 30 °C and 150 rpm for 1 h.

18. After 1 h, remove the plate from the shaker and measure the OD_{600} of the cells on the plate reader.

   Note: the high cell density and short treatment time minimize the growth promoting or inhibiting effects of the siderophore/SAC treatment.

19. Using a pipette, transfer and combine the contents of wells with identical treatment conditions in sterile 1.5 mL microcentrifuge tubes. Label the tubes with the treatment condition (Ent, TC, untreated).

20. Centrifuge the tubes (4000 rpm, 10 min, 4 °C).

Experimental methods for evaluating siderophore–antibiotic conjugates 45

21. Remove the supernatant using a pipette.

22. Wash the cell pellets twice by suspending them into sterile modified M9 medium, centrifuging (4000 rpm, 10 min, 4 °C), and removing the supernatant.

23. Wash the cells once more with modified M9 medium with 2% EDTA (aq) added to remove any residual extracellular Fe.

24. Resuspend the cell pellets into 2 mL of 3–5% HNO_3 (aq) in 5 mL falcon tubes.

25. Add 40 μL of 50 ppb Tb solution to each sample.

Note: Tb is the internal standard used by the ICP-MS instrument.

26. Digest the samples by microwave digestion.

Note: alternatively, the cell pellets can be digested by adding 100 μL of concentrated HNO_3, and incubation at room temperature for 30 min. The addition of 1.9 mL Milli-Q water will result in a total acid content of 3–5%.

Prepare the ICP-MS standards and start the run:

27. Using 2–5% HNO_3 (aq) as solvent, prepare at least six 10-fold serial dilutions of the ICP-MS environmental standard solution with total volume of 2 mL. Include a blank with only the solvent.

Note: You may carry out the ICP-MS run and analysis on a separate day. Keep in mind the standard solutions must be prepared immediately prior to the experiment.

28. Add 40 μL of 50 ppb Tb solution to each standard.

29. Transfer samples to ICP-MS vials and place them on the autosampler of ICP-MS.

Note: Order the standards from low to high concentration.

30. Provide the software with the concentrations of standards, and the names of the samples.

31. Begin the ICP-MS run with 2 blank samples (3% HNO_3) to wash the tubing of any residual metals. The instrument reports the number of events that an isotope with a certain mass is detected and will report that as "counts/seconds". The fixed concentration of Tb is used as a normalizing factor for this measurement. The software uses the known concentration of elements in the standards and automatically prepares a calibration curve for each analyte to convert the "counts/s" to concentration in "ppb". You may also use the calibration curve to obtain these data manually.

32. Refer to the optical density recorded from the 96-well plate. Subtract the value of blank (sterile modified M9 medium) from each data point and calculate the average OD_{600} of each treatment condition (see Section 2.3.3 step 25).

Note: The Mg content of the samples can also be used to monitor the cell density. The Mg concentration is expected to be similar for all samples (within 20% error) provided they contained a comparable number of bacterial cells. This analysis can help identify whether the bacterial cell count of a sample changed during handling and the workup. Any replicates that have significantly different cell density, as indicated by analysis of Mg concentration, can be eliminated.

33. From the concentration of ^{57}Fe in ppb, determine the pmol of ^{57}Fe in each sample and divide by the average OD_{600} for the treatment condition. To compare ^{57}Fe uptake between different treatment conditions, plot the pmol ^{57}Fe/OD_{600} values as a bar graph. Refer to Fig. 5 as an example.

Note: Carry out Student's t test or alternative statistical tests to identify significant differences between the data sets.

3. Summary

This chapter summarizes four key microbiological assays used in our lab to evaluate the AMA and cellular uptake of SACs. We intend for our descriptions and protocols to be valuable resources for the community that motivate= further assay development, which we hope will expand and facilitate their application to a variety of SACs with different siderophores and cargos. Along these lines, we are excited by the recent fundamental and translational developments in the SAC field and look forward to its ongoing and future endeavors.

References

Almeida, M. C., Da Costa, P. M., Sousa, E., & Resende, D. I. S. P. (2023). Emerging target-directed approaches for the treatment and diagnosis of microbial infections. *Journal of Medicinal Chemistry, 66*, 32–70.

Annamalai, R., Jin, B., Cao, Z., Newton, S. M. C., & Klebba, P. E. (2004). Recognition of ferric catecholates by FepA. *Journal of Bacteriology, 186*, 3578–3789.

Aoki, T., Yoshizawa, H., Yamawaki, K., Yokoo, K., Sato, J., Hisakawa, S., ... Yamano, Y. (2018). Cefiderocol (S-649266), a new siderophore cephalosporin exhibiting potent activities against *Pseudomonas aeruginosa* and other Gram-negative pathogens including multi-drug resistant bacteria: Structure activity relationship. *European Journal of Medicinal Chemistry, 155*, 847–868.

Caradec, T., Anoz-Carbonell, E., Petrov, R., Billamboz, M., Antraygues, K., Cantrelle, F.-X., ... Hartkoorn, R. C. (2023). A novel natural siderophore antibiotic conjugate reveals a chemical approach to macromolecule coupling. *ACS Central Science, 9*, 2138–2149.

Chairatana, P., Zheng, T., & Nolan, E. M. (2015). Targeting virulence: Salmochelin modification tunes the antibacterial activity spectrum of β-lactams for pathogen-selective killing of *Escherichia coli. Chemical Science, 6*, 4458–4471.

CLSI. (2023). *Performance standards for antimicrobial susceptibility testing* (33th ed.). Clinical and Laboratory Standards Institute.

Darby, E. M., Trampari, E., Siasat, P., Solsona Gaya, M., Alav, I., Webber, M. A., & Blair, J. M. A. (2023). Molecular mechanisms of antibiotic resistance revisited. *Nature Reviews. Microbiology, 21*, 280–295.

Dertz, E. A., Xu, J., Stintzi, A., & Raymond, K. N. (2006). Bacillibactin-mediated iron transport in *Bacillus subtilis. Journal of the American Chemical Society, 128*, 22–23.

Diarra, M. S., Lavoie, M. C., Jacques, M., Darwish, I., Dolence, E. K., Dolence, J. A., ... Malouin, F. (1996). Species selectivity of new siderophore–drug conjugates that use specific iron uptake for entry into bacteria. *Antimicrobial Agents and Chemotherapy, 40*, 2610–2617.

EUCAST. (1998). Methods for the determination of susceptibility of bacteria to antimicrobial agents. Terminology. *Clinical Microbiology and Infection, 4*, 291–296.

EUCAST. (2020). *Breakpoint tables for interpretation of MICs and zone diameters* (10.0th ed.). http://www.eucast.org.

Fardeau, S., Dassonville-Klimpt, A., Audic, N., Sasaki, A., Pillon, M., Baudrin, E., ... Sonnet, P. (2014). Synthesis and antibacterial activity of catecholate–ciprofloxacin conjugates. *Bioorganic & Medicinal Chemistry, 22*, 4049–4060.

Ferreira, K., Hu, H.-Y., Fetz, V., Prochnow, H., Rais, B., Müller, P. P., & Brönstrup, M. (2017). Multivalent siderophore–DOTAM conjugates as theranostics for imaging and treatment of bacterial infections. *Angewandte Chemie International Edition, 56*, 8272–8276.

Gause, G. F. (1955). Recent studies on albomycin, a new antibiotic. *The British Medical Journal, 2*, 1177–1179.

Guo, C., & Nolan, E. M. (2022). Heavy-metal Trojan horse: enterobactin-directed delivery of platinum(IV) prodrugs to *Escherichia coli. Journal of the American Chemical Society, 144*, 12756–12768.

Guo, C., & Nolan, E. M. (2024). Exploring the antibacterial activity and cellular fates of enterobactin–drug conjugates that target Gram-negative bacterial pathogens. *Accounts of Chemical Research, 57*, 1046–1056.

Hagan, E. C., & Mobley, H. L. T. (2007). Uropathogenic *Escherichia coli* outer membrane antigens expressed during urinary tract infection. *Infection and Immunity, 75*, 3941–3949.

Hannauer, M., Barda, Y., Mislin, G. L. A., Shanzer, A., & Schalk, I. J. (2010). The ferrichrome uptake pathway in *Pseudomonas aeruginosa* involves an iron release mechanism with acylation of the siderophore and recycling of the modified desferrichrome. *Journal of Bacteriology, 192*, 1212–1220.

Hantke, K., Nicholson, G., Rabsch, W., & Winkelmann, G. (2003). Salmochelins, siderophores of *Salmonella enterica* and uropathogenic *Escherichia coli* strains, are recognized by the outer membrane receptor IroN. *Proceedings of the National Academy of Sciences of the United States of America, 100*, 3677–3682.

Ito, A., Nishikawa, T., Matsumoto, S., Yoshizawa, H., Sato, T., Nakamura, R., ... Yamano, Y. (2016). Siderophore cephalosporin cefiderocol utilizes ferric iron transporter systems for antibacterial activity against *Pseudomonas aeruginosa. Antimicrobial Agents and Chemotherapy, 60*, 7396–7401.

Ito, A., Sato, T., Ota, M., Takemura, M., Nishikawa, T., Toba, S., ... Yamano, Y. (2018). *In vitro* antibacterial properties of cefiderocol, a novel siderophore cephalosporin, against Gram-negative bacteria. *Antimicrobial Agents and Chemotherapy, 62*, e01454–17.

Ji, C., Miller, P. A., & Miller, M. J. (2012). Iron transport-mediated drug delivery: practical syntheses and in vitro antibacterial studies of tris-catecholate siderophore–aminopenicillin conjugates reveals selectively potent antipseudomonal activity. *Journal of the American Chemical Society, 134,* 9898–9901.

Lai, Y.-H., Franke, R., Pinkert, L., Overwin, H., & Brönstrup, M. (2023). Molecular signatures of the Eagle effect induced by the artificial siderophore conjugate LP-600 in *E. coli. ACS Infectious Diseases, 9,* 567–581.

Lee, A. A., Chen, Y.-C. S., Ekalestari, E., Ho, S.-Y., Hsu, N.-S., Kuo, T.-F., & Wang, T.-S. A. (2016). Facile and versatile chemoenzymatic synthesis of enterobactin analogues and applications in bacterial detection. *Angewandte Chemie International Edition, 55,* 12338–12342.

Lin, H., Fischbach, M. A., Liu, D. R., & Walsh, C. T. (2005). In vitro characterization of salmochelin and enterobactin trilactone hydrolases IroD, IroE, and Fes. *Journal of the American Chemical Society, 127,* 11075–11084.

Luscher, A., Gasser, V., Bumann, D., Mislin, G. L. A., Schalk, I. J., & Köhler, T. (2022). Plant-derived catechols are substrates of TonB-dependent transporters and sensitize *Pseudomonas aeruginosa* to siderophore-drug conjugates. *mBio, 13,* e01498–22.

McPherson, C. J., Aschenbrenner, L. M., Lacey, B. M., Fahnoe, K. C., Lemmon, M. M., Finegan, S. M., ... Tomaras, A. P. (2012). Clinically relevant Gram-negative resistance mechanisms have no effect on the efficacy of MC-1, a novel siderophore-conjugated monocarbam. *Antimicrobial Agents and Chemotherapy, 56,* 6334–6342.

Miethke, M., Hou, J., & Marahiel, M. A. (2011). The siderophore-interacting protein YqjH acts as a ferric reductase in different iron assimilation pathways of *Escherichia coli. Biochemistry, 50,* 10951–10964.

Miethke, M., & Marahiel, M. A. (2007). Siderophore-based iron acquisition and pathogen control. *Microbiology and Molecular Biology Reviews, 71,* 413–451.

Möllmann, U., Heinisch, L., Bauernfeind, A., Köhler, T., & Ankel-Fuchs, D. (2009). Siderophores as drug delivery agents: Application of the "Trojan Horse" strategy. *Biometals: An International Journal on the Role of Metal Ions in Biology, Biochemistry, and Medicine, 22,* 615–624.

Motz, R. N., Guo, C., Sargun, A., Walker, G. T., Sassone-Corsi, M., Raffatellu, M., & Nolan, E. M. (2024). Conjugation to native and nonnative triscatecholate siderophores enhances delivery and antibacterial activity of a β-lactam to Gram-negative bacterial pathogens. *Journal of the American Chemical Society, 146,* 7708–7722.

Neumann, W., Sassone-Corsi, M., Raffatellu, M., & Nolan, E. M. (2018). Esterase-catalyzed siderophore hydrolysis activates an enterobactin–ciprofloxacin conjugate and confers targeted antibacterial activity. *Journal of the American Chemical Society, 140,* 5193–5201.

Nikaido, H., & Rosenberg, E. Y. (1990). Cir and Fiu proteins in the outer membrane of *Escherichia coli* catalyze transport of monomeric catechols: Study with b-lactam antibiotics containing catechol and analogous groups. *Journal of Bacteriology, 172,* 1361–1367.

Pandey, A., Savino, C., Ahn, S. H., Yang, Z., Van Lanen, S. G., & Boros, E. (2019). Theranostic gallium siderophore ciprofloxacin conjugate with broad spectrum antibiotic potency. *Journal of Medicinal Chemistry, 62,* 9947–9960.

Pankey, G. A., & Sabath, L. D. (2004). Clinical relevance of bacteriostatic versus bactericidal mechanisms of action in the treatment of Gram-positive bacterial infections. *Clinical Infectious Diseases, 38,* 864–870.

Petrik, M., Pfister, J., Misslinger, M., Decristoforo, C., & Haas, H. (2020). Siderophore-based molecular imaging of fungal and bacterial infections—Current status and future perspectives. *Journal of Fungi, 6,* 73.

Petrik, M., Zhai, C., Novy, Z., Urbanek, L., Haas, H., & Decristoforo, C. (2016). *In vitro* and in vivo comparison of selected Ga-68 and Zr-89 labelled siderophores. *Molecular Imaging and Biology, 18,* 344–352.

Pinkert, L., Lai, Y.-H., Peukert, C., Hotop, S.-K., Karge, B., Schulze, L. M., ... Brönstrup, M. (2021). Antibiotic conjugates with an artificial MECAM-based siderophore are potent agents against Gram-positive and Gram-negative bacterial pathogens. *Journal of Medicinal Chemistry, 64*, 15440–15460.

Prinzi, A., & Rohde, R. (2020). Why differential & selective media remain invaluable tools. *American Society for Microbiology*. Retrieved from https://asm.org/articles/2020/september/why-differential-selective-media-are-invaluable-to.

Rayner, B., Verderosa, A. D., Ferro, V., & Blaskovich, M. A. T. (2023). Siderophore conjugates to combat antibiotic-resistant bacteria. *RSC Medicinal Chemistry, 14*, 800–822.

Sargun, A., Johnstone, T. C., Zhi, H., Raffatellu, M., & Nolan, E. M. (2021). Enterobactin- and salmochelin-β-lactam conjugates induce cell morphologies consistent with inhibition of penicillin-binding proteins in uropathogenic *Escherichia coli* CFT073. *Chemical Science, 12*, 4041–4056.

Sassone-Corsi, M., Nuccio, S.-P., Liu, H., Hernandez, D., Vu, C. T., Takahashi, A. A., ... Raffatellu, M. (2016). Microcins mediate competition among Enterobacteriaceae in the inflamed gut. *Nature, 540*, 280–283.

Scarrow, R. C., Ecker, D. J., Ng, C., Liu, S., & Raymond, K. N. (1991). Iron(III) coordination chemistry of linear dihydroxyserine compounds derived from enterobactin. *Inorganic Chemistry, 30*, 900–906.

Southwell, J. W., Black, C. M., & Duhme-Klair, A.-K. (2021). Experimental methods for evaluating the bacterial uptake of Trojan horse antibacterials. *ChemMedChem, 16*, 1063–1076.

Stintzi, A., Barnes, C., Xu, J., & Raymond, K. N. (2000). Microbial iron transport via a siderophore shuttle: A membrane ion transport paradigm. *Proceedings of the National Academy of Sciences of the United States of America, 97*, 10691–10696.

Thomas, X., Destoumieux-Garzón, D., Peduzzi, J., Afonso, C., Blond, A., Birlirakis, N., ... Rebuffat, S. (2004). Siderophore peptide, a new type of post-translationally modified antibacterial peptide with potent activity. *Journal of Biological Chemistry, 279*, 28233–28242.

Vassiliadis, G., Destoumieux-Garzón, D., Lombard, C., Rebuffat, S., & Peduzzi, J. (2010). Isolation and characterization of two members of the siderophore-microcin family, microcins M and H47. *Antimicrobial Agents and Chemotherapy, 54*, 288–297.

Van Alst, A. J., LeVeque, R. M., Martin, N., & DiRita, V. J. (2023). Growth curves: Generating growth curves using colony forming units and optical density measurements. *Journal of Visualized Experiments*. Cambridge, MA: JoVE Science Education Database.

Weizman, H., Ardon, O., Mester, B., Libman, J., Dwir, O., Hadar, Y., ... Shanzer, A. (1996). Fluorescently-labeled ferrichrome analogs as probes for receptor-mediated, microbial iron uptake. *Journal of the American Chemical Society, 118*, 12368–12375.

Wencewicz, T. A., Long, T. E., Möllmann, U., & Miller, M. J. (2013). Trihydroxamate siderophore–fluoroquinolone conjugates are selective sideromycin antibiotics that target *Staphylococcus aureus*. *Bioconjugate Chemistry, 24*, 473–486.

Zgurskaya, H. I., López, C. A., & Gnanakaran, S. (2015). Permeability barrier of Gram-negative cell envelopes and approaches to bypass it. *ACS Infectious Diseases, 1*, 512–522.

Zhao, S., Wang, Z.-P., Lin, Z., Wei, G., Wen, X., Li, S., ... He, Y. (2022). Drug repurposing by siderophore conjugation: Synthesis and biological evaluation of siderophore-methotrexate conjugates as antibiotics. *Angewandte Chemie International Edition, 61*, e202204139.

Zheng, T., Bullock, J. L., & Nolan, E. M. (2012). Siderophore-mediated cargo delivery to the cytoplasm of *Escherichia coli* and *Pseudomonas aeruginosa*: Syntheses of mono-functionalized enterobactin scaffolds and evaluation of enterobactin–cargo conjugate uptake. *Journal of the American Chemical Society, 134*, 18388–18400.

Zheng, T., & Nolan, E. M. (2012). Siderophore-based detection of Fe(III) and microbial pathogens. *Metallomics, 4*, 866–880.

Zheng, T., & Nolan, E. M. (2014). Enterobactin-mediated delivery of β-lactam antibiotics enhances antibacterial activity against pathogenic *Escherichia coli*. *Journal of the American Chemical Society, 136*, 9677–9691.

Zhu, M., Valdebenito, M., Winkelmann, G., & Hantke, K. (2005). Functions of the siderophore esterases IroD and IroE in iron-salmochelin utilization. *Microbiology (Reading, England), 151*, 2363–2372.

Zscherp, R., Coetzee, J., Vornweg, J., Grunenberg, J., Herrmann, J., Müller, R., & Klahn, P. (2021). Biomimetic enterobactin analogue mediates iron-uptake and cargo transport into *E. coli* and *P. aeruginosa*. *Chemical Science, 12*, 10179–10190.

CHAPTER THREE

A continuous fluorescence assay to measure nicotianamine synthase activity

Thiago M. Pasin, Kathleen M. Meneely, Deegan M. Ruiz, and Audrey L. Lamb*

Department of Chemistry, University of Texas at San Antonio, San Antonio, TX, United States
*Corresponding author. e-mail address: audrey.lamb@utsa.edu

Contents

1. Introduction	52
2. Assay design	54
3. Before you begin	55
4. Key resources table	55
5. Methods and equipment	56
5.1 Equipment	56
5.2 Reagents	56
6. Preparation of reagents	57
6.1 Purification of methylthioadenosine nucleosidase (MTAN)	57
6.2 Purification of adenine deaminase (AD)	60
6.3 Check for activity of MTAN and AD	64
7. Step-by-step method details	66
7.1 Prepare standard curve	66
7.2 Run the assay	68
8. Quantification and statistical analysis	69
9. Summary	72
References	73

Abstract

S-adenosylmethionine (SAM) is most widely known as the biological methylating agent of methyltransferases and for generation of radicals by the iron-sulfur dependent Radical SAM enzymes. SAM also serves as a substrate in biosynthetic reactions that harvest the aminobutyrate moiety of the methionine, producing methylthioadenosine as a co-product. These reactions are found in the production of polyamines such as spermine, siderophores derived from nicotianamine, and opine metallophores staphylopine and pseudopaline, among others. This procedure defines a highly sensitive, continuous fluorescence assay for the determination of steady state kinetic parameters for enzymes that generate the co-product methylthioadenosine.

Methods in Enzymology, Volume 702
ISSN 0076-6879, https://doi.org/10.1016/bs.mie.2024.06.013
Copyright © 2024 Elsevier Inc. All rights reserved, including those for text and data mining, AI training, and similar technologies.

1. Introduction

All organisms require iron to serve as a co-factor in redox active enzymes for critical reactions, including cellular respiration and nucleic acid synthesis. In plants, the enzymes of photosynthesis and nitrogen fixation are also iron-dependent, performing vital redox chemistry (Welch & Shuman, 1995). While iron is an essential nutrient, cellular and tissue concentrations must be strictly controlled, as high concentrations are toxic, due to Fenton chemistry (the generation of oxygen radicals that damage biomolecules) (Briat et al., 1995). Plants generate the molecule nicotianamine (NA), which serves as an inter- and intracellular chelator for transport and metal ion homeostasis. While iron is generally prevalent in the environment, ferrous iron (Fe^{2+}) readily oxidizes to the sparingly soluble ferric iron (Fe^{3+}). Therefore, graminaceous crop plants, such as wheat, rice, corn and barley, have developed a system whereby iron is scavenged from the rhizosphere. By biosynthesizing and secreting phytosiderophores (low molecular weight Fe^{3+}-chelating compounds) derived from nicotianamine, in particular the highly water soluble mugineic acids, a ferrous iron complex is taken up by dedicated importers (Marschner & Römheld, 1994).

Nicotianamine is generated by a single enzyme called nicotianamine synthase (NAS). This fascinating processive enzyme performs two different reactions in a single active site. Using three molecules of S-adenosylmethionine (SAM) as the substrate, the aminobutyrate moieties are processively linked together head-to-tail, and the initial aminobutyrate moiety is cyclized to form the azetidine ring (Scheme 1A). A co-product is

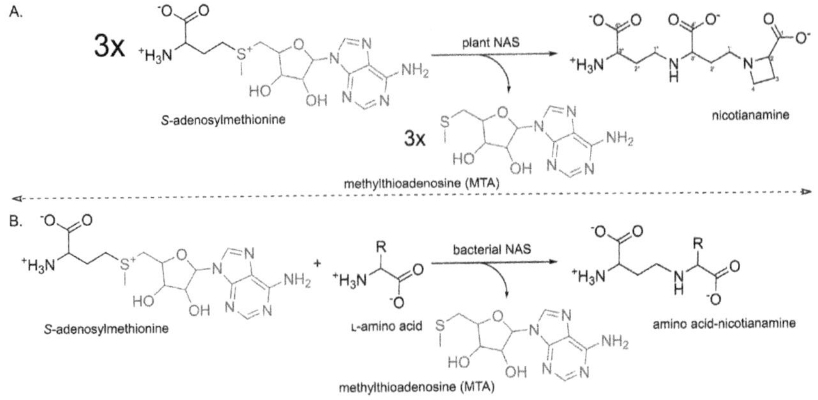

Scheme 1 Reactions catalyzed by plant (A) and bacterial (B) NAS enzymes.

three molecules of methylthioadenosine (MTA). Nicotianamine is further modified by an aminotransferase and a reductase to make deoxymugineic acid (DMA). The incorporation of the aminobutyrate moiety from SAM into mugineic acids derived from nicotianamine was determined by feeding studies. Selectively ^{13}C- and ^{14}C-labeled methionine SAM were administered to barley, which was subsequently grown in iron-poor conditions. These studies showed that each aminobutyrate moiety is derived from one methionine moiety of SAM (Shojima et al., 1990). An early crystal structure showed that the first aminobutyrate formed an azetidine ring (Sugiura & Tanaka, 1981).

Nicotianamine is made by plants and some fungi (Trampczynska, Bottcher, & Clemens, 2006). Archaea and some bacteria have an NAS-like enzyme in operons that generate opine metallophores for the import of metal ions from the environment. These NAS-like enzymes link an amino acid to one or two aminobutyrate moieties of SAM (Scheme 1B) (Ghssein et al., 2016; Gi et al., 2015; McFarlane & Lamb, 2017, 2020). In *Methanothermobacter thermautotrophicus*, two SAM aminobutyrates are linked to the primary amine of glutamate. Crystallographic analysis of the *M. thermautotrophicus* NAS suggests a processive mechanism in which a glutamic acid binds first followed by one SAM, resulting in aminoalkyl linkage and methylthioadenosine (MTA) release. The intermediate shifts deeper into the active site accommodating the binding of a second SAM followed by a second aminoalkyl linkage and then MTA and product release, suggesting an ordered sequential mechanism (Dreyfus et al., 2011; Dreyfus, Lemaire, Mari, Pignol, & Arnoux, 2009) In 2021, the structure of the nicotianamine synthase-like enzyme from the bacteria *Staphylococcus aureus* was determined (Luo et al., 2021). This enzyme links D-histidine to a single aminobuyrate from SAM, as the first step in a 2-step process to make the metallophore staphylopine (Ghssein et al., 2016; McFarlane & Lamb, 2020).

Only very recently has there been the addition of initial kinetic and mechanistic studies of the NAS enzymes to the literature. In April of 2023, using the NAS enzymes from the model organism *Arabidopsis thalania* (a non-graminaceous plant), the laboratory of Ute Krämer generated a coupled spectrophotometric assay, which showed that byproduct MTA was an inhibitor of the reaction, explaining the low turnover seen in previous work (Seebach et al., 2023). In addition, the assay allowed for continuous spectrophotometric monitoring of the reaction over time, a major advance compared to the discontinuous assay reliant on thin layer chromatography for detection (Herbik et al., 1999).

2. Assay design

We describe here a more sensitive continuous-detection steady state kinetic assay for nicotianamine synthase enzymes. The breakthrough by Krämer and colleagues was a big step forward. These authors developed an assay in which the byproduct, methylthioadenosine (MTA) was cleaved to generate adenine and methylthioribose, using a S-adenosylhomocysteine (SAH) nucleosidase and subsequently converted adenine to hypoxanthine by adenine deaminase. A decrease in absorbance at 265 nm due to the adenine to hypoxanthine conversion was monitored (Seebach et al., 2023). A significant increase in catalytic activity was observed compared to previous reports: the authors discovered that MTA is a competitive inhibitor and the buildup of the coproduct prevents product release and subsequent turnover. A major drawback to this assay is that the difference in absorbance is small, and can be masked by the absorbance of the proteins, potentially explaining why steady state kinetic parameters are not reported.

Building on the success of Krämer and co-workers, this new assay incorporates aspects of an assay developed by Vern Schramm in his work on purine biosynthesis and recycling (Brown et al., 2021). The assay relies on methylthioadenosine nucleosidase (MTAN), which converts methylthioadenosine to adenine and methylthioribose. Adenine deaminase (AD) converts adenine to hypoxanthine, and xanthine oxidase (XO) converts hypoxanthine to xanthine and xanthine to urate, with both reactions producing hydrogen peroxide. Hydrogen peroxide is a substrate for horseradish peroxidase (HRP), which converts the dye Amplex Red to resorufin, generating a strong fluorescent signal (excitation: 568 nm; emission: 590 nm) (Scheme 2). The Schramm lab uses this highly sensitive system to measure kinetic isotope effects of MTAN. While our measurements have not yet required that same level of precision, our adaption of this system will allow for highly effective measurements.

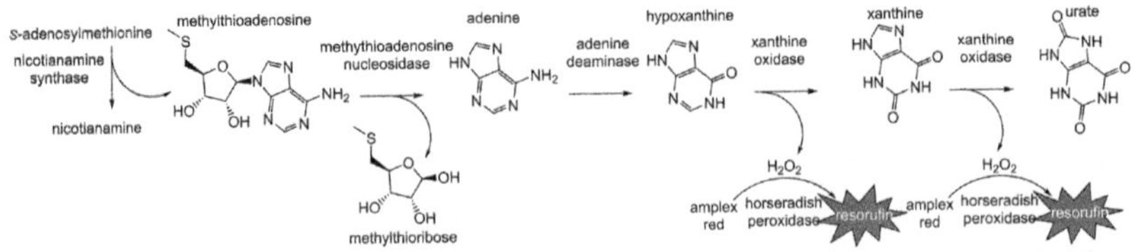

Scheme 2 Fluorescence-based continuous assay for measuring nicotianamine synthase activity (or potentially any enzymatic reaction that produces methylthioadenosine as a byproduct).

3. Before you begin

In order to carry out the assay, you will need to procure the 4 enzymes of the assay (MTAN, AD, XO and HRP), the substrate S-adenosyl methionine (SAM), the co-product methylthioadenosoine to generate a standard curve, and the reporter, Amplex Red. XO, HRP, SAM and Amplex Red are commercially available, as listed in the Key Resources Table. The overexpression vectors for MTAN and AD are available from AddGene and their straightforward preparations are described in the Step-By-Step Methods section.

4. Key resources table

Reagent or resource	Source	Identifier
Bacterial and Virus Strains		
Helicobacter pylori Methylthioadenosine nucleosidase plasmid in DH5alpha strain as agar stab	AddGene	69802
One Shot™ BL21(DE3) Chemically Competent *Escherichia coli*	Invitrogen™	C600003
Adenine deaminase overexpression plasmid in DH5alpha strain as agar stab	AddGene	218482
Enzymes		
Xanthine oxidase, Buttermilk	MilliporeSigma	682151-50U
Horseradish peroxidase	ThermoFisher Scientific	31491
Lysozyme, from chicken egg white	MilliporeSigma	12650-88-3
Critical Commercial Assays		
PureLink™ Quick Plasmid Miniprep Kit	ThermoFisher Scientific	K210010
Software and Algorithms		
Kaleidagraph	Synergy Software	5.0.3

The Adenine Deaminase overexpression plasmid was generated for this assay and is first reported here. The *adeC* gene from *Escherichia coli* K12 strain (NCBI reference sequence WP_001065718.1) was codon optimized by Genscript and inserted into the pET21b vector using *Nde*I and *Xho*I. The resulting plasmid encodes for AdeC with a C-terminal 6His-tag with ampicillin resistance.

5. Methods and equipment

5.1 Equipment

For this list we provide the generic name followed by the specific equipment we used in parentheses.
- Autoclave
- Floor Centrifuge (Beckman Coulter Avanti JXN-26 Floor Centrifuge).
- Microcentrifuge (Centrifuge 5425R, Eppendorf).
- Shaker-Incubator (Innova™ S44i - Stackable Incubator Shaker from Eppendorf)
- French Press (Glen Mills)
- SDS-PAGE electrophoresis (Invitrogen, A25977)
- UV-Vis Spectrophotometer (Shimadzu UV-2600)
- UV-Vis Quartz Cuvette, 10 mm (Starna Cells)
- Chromatography System (ÄKTA pure™ chromatography system, Cytiva)
- Ni-affinity column (Chelating Sepharose Fast Flow from Cytiva)
- Preparative size exclusion column for protein molecular weights up to 200 kDa (HiLoad Superdex 200 16/600 gel-filtration column from Cytiva)
- Amicon™ Ultra-15 Centrifugal Filter Units (Millipore-Sigma, UFC9010)
- Fluorimeter (Shimadzu RF-6000)
- 10 mm fluorimeter Micro Quartz cell (Starna Cells)

5.2 Reagents
- Ampicillin (GoldBio A-301-100) at 200 mg/mL (stock concentration)
- LB broth (MILLER) (ThermoFisher Scientific, BP1426-2)
- LB agar (ThermoFisher Scientific, BP1425-2)
- Sterile dH$_2$O
- PureLink™ Quick Plasmid Miniprep Kit (Thermo Fisher Scientific, K210010)
- Bradford Protein Assay (Bio-Rad, 5000006)
- Isopropyl β-D-1-thiogalactopyranoside (IPTG) (GoldBio, I2481C100)

- Tris-HCl (GoldBio, T-400-5)
- Imidazole (Acro Organics, 396740025)
- NaCl (Thermo Fisher Scientific S271-3)
- NaH$_2$PO$_4$ (Sigma-Aldrich, S0751-1KG)
- Na$_2$HPO$_4$ (Sigma-Aldrich, S9763-500G)
- DL-Dithiothreitol (DTT) (GoldBio, DTT100)
- Dimethyl sulfoxide (DMSO) (Sigma, D8418-100 mL)
- HEPES (GoldBio, H-401-2.5)
- 5′-Deoxy-5′-(methylthio)adenosine (MTA) (Sigma-Aldrich, D5011-100 mg)
- IgG (for Bradford) (Bio-Rad, 5000005)
- Premade SDS-PAGE gels 4–20% (Invitrogen, XP04205BOX)
- S-(5′-adenosyl)-L-methionine chloride (Millipore Sigma, A7007-100 mg)
- Amplex™ Red (ThermoFisher Scientific, A12222)

6. Preparation of reagents

6.1 Purification of methylthioadenosine nucleosidase (MTAN)

Timing: ~5 days.

Note: All broth and culture media must be autoclaved prior to use, and bacteria used for heterologous expression of proteins must be manipulated using sterile technique.

a) Grow the agar stab containing MTAN with a N-terminal His6-tag (procured from AddGene) in 100 mL LB broth containing 200 μg/mL ampicillin at 37 °C with 250 rpm shaking overnight.

b) Use a Plasmid Miniprep Kit to isolate the overexpression plasmid for transformation into One Shot™ BL21(DE3) Chemically Competent *E. coli* (use the manufacturer instructions for the miniprep and the transformation).

c) Use one colony from the BL21 (DE3) transformants to inoculate 100 mL of LB broth, with 200 μg/mL ampicillin, and incubate with shaking (250 rpm) at 37 °C overnight.

d) Add 10 mL of the overnight culture to 1 L of LB broth containing 200 μg/mL ampicillin. Incubate at 37 °C with 250 rpm shaking until the optical density at 600 nm (OD$_{600}$) reaches ~0.8. Add isopropyl β-d-1-thiogalactopyranoside (IPTG) to a final concentration of 0.5 mM, and adjust the temperature to 22 °C with 250 rpm shaking overnight.

e) Harvest the cells by centrifugation at 6000 ×g for 10 min at 4 °C. Decant and discard the broth.
f) Resuspend the cell pellet in 30 mL of 50 mM Tris-HCl (pH 8.0), 500 mM NaCl, 50 mM imidazole per liter of culture broth.
g) Lyse the resuspended cells with three passages through a French pressure cell at 13,000 psi.
h) Remove cell debris from the lysed cells by centrifugation at 12,000 × g for 20 min at 4 °C. Decant and retain the supernatant, discarding the cell debris pellet.
i) Load the supernatant onto a 25 mL Chelating Sepharose Fast Flow (GE Healthcare) column charged with nickel chloride and pre-equilibrated with 50 mM Tris-HCl pH 8.0, 500 mM NaCl, 50 mM imidazole. After loading, wash the column with 8 column volumes of the same buffer.
j) Elute the his-tagged MTAN with a step gradient using 5 column volumes of 50 mM Tris-HCl pH 8.0, 500 mM NaCl, 300 mM imidazole. Wash the column with 3 column volumes of 50 mM Tris-HCl pH 8.0, 500 mM NaCl, 500 mM imidazole to ensure all of the MTAN has eluted (Fig. 1).
k) Confirm that a 25 kDa protein eluted in the peak using 15% SDS-PAGE. MTAN has a molecular weight of 25 kDa, and is evident in the elution fractions (Fig. 2).

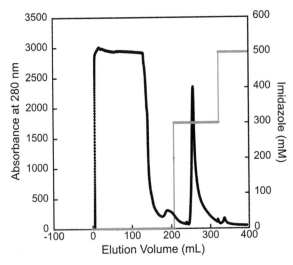

Fig. 1 Elution profile for histidine-tagged MTAN using nickel affinity chromatography. MTAN elutes using a step gradient of 300 mM imidazole.

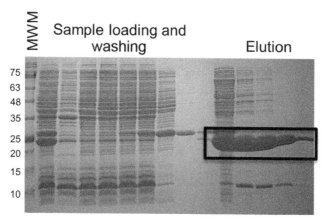

Fig. 2 Coomassie stained 15% SDS-PAGE of fractions eluted from nickel affinity column for histidine-tagged MTAN. The MTAN is highlighted with a box.

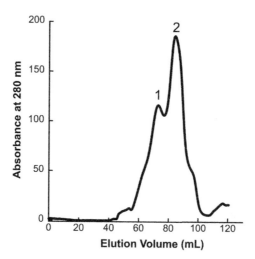

Fig. 3 Gel filtration elution profile for MTAN.

l) Pool the fractions containing the protein in the 300 mM imidazole step (~60 mL) and concentrate to ~30 mL using an Amicon centrifugal concentrations with a 10 kDa cutoff filter (using manufacturer instructions).

m) Inject 8 mL of the concentrated sample onto a Hi-Load Superdex 200 gel-filtration column (GE Healthcare), pre-equilibrated with 25 mM Tris-HCl pH 8.0 (Fig. 3). Elute the protein using one column volume of the same buffer as the pre-equilibration. Repeat this step as needed for all sample generated in step l.

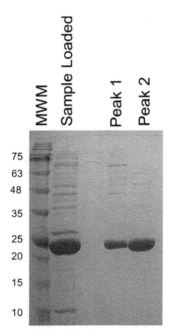

Fig. 4 Coomassie stained 15% SDS-PAGE of fractions eluted from gel filtration column for histidine-tagged MTAN.

n) MTAN elutes as a dimer, as estimated by using molecular weight standards on the gel filtration column. Pool peak 2 fractions and verify the presence of purified MTAN by SDS-PAGE (Fig. 4).

o) Concentrate the pooled fractions using an Amicon centrifugal concentrations with a 10 kDa cutoff filter to 10 mg/mL (400 μM). Determine protein concentration using Bradford reagent and a standard curve generated using IgG (using manufacturer's instructions).

p) Store the concentrated proteins in aliquots of 100 μL at −80 °C for later use. The purification should yield ~500 mg of MTAN per liter of culture.

6.2 Purification of adenine deaminase (AD)

Timing: ~5 days.

a) Grow the agar stab containing AD with a C-terminal His6-tag (procured from AddGene) in 100 mL LB broth containing 200 μg/mL ampicillin at 37 °C with 250 rpm shaking overnight.

b) Use a Plasmid Miniprep Kit to isolate the overexpression plasmid for transformation into One Shot™ BL21(DE3) Chemically Competent

E. coli (use the manufacturer instructions for the miniprep and the transformation).

c) Use one colony of the BL21 (DE3) transformants to inoculate 100 mL of LB broth, with 200 µg/mL ampicillin, and incubate with shaking (250 rpm) at 37 °C overnight.

d) Add 10 mL of the overnight culture to 1 L of LB broth containing 200 µg/mL ampicillin. Incubate at 37 °C with 250 rpm shaking until the OD_{600} reaches ~0.8. Add IPTG to a final concentration of 0.2 mM, and adjust the temperature to 25 °C with 250 rpm shaking overnight.

e) Harvest the cells by centrifugation at $6000 \times g$ for 10 min at 4 °C. Decant and discard the broth.

f) Resuspend the cell pellet in 30 mL of lysis buffer containing 50 mM NaH_2PO_4, 300 mM NaCl, 1 mg/mL lysozyme, pH 8.0, per liter of culture broth and place in the −80 °C freezer overnight.

g) Lyse the resuspended cells with three passages through a French pressure cell at 13,000 psi.

h) Remove cell debris from the lysed cells by centrifugation at $12,000 \times g$ for 20 min at 4 °C. Decant and retain the supernatant, discarding the cell debris pellet.

i) Load the supernatant onto a 25 mL Chelating Sepharose Fast Flow (GE Healthcare) column charged with nickel chloride and pre-equilibrated with 50 mM NaH_2PO_4 pH 8.0, 300 mM NaCl, 40 mM imidazole. After loading, wash the column with 6 column volumes of the same buffer.

j) Elute the his-tagged AD with a step gradient using 5 column volumes of 50 mM NaH_2PO_4 pH 8.0, 300 mM NaCl, 300 mM imidazole followed by 5 column volumes of 50 mM NaH_2PO_4 pH 8.0, 300 mM NaCl, 500 mM imidazole (Fig. 5). Alternatively, a single step of 500 mM imidazole could be used.

k) Confirm that a 63 kDa protein eluted in the 300 and 500 mM imidazole peaks using SDS-PAGE. AD has a molecular weight of 63 kDa and is evident in the elution fractions (Fig. 6).

l) Pool the fractions containing the protein in the 300 and 500 mM imidazole steps (~120 mL) and concentrate to ~20 mL using an Amicon centrifugal concentrations with a 10 kDa cutoff filter (using manufacturer instructions).

m) Inject 5 mL of the concentrated sample onto a 120 mL Superdex 200 gel-filtration column (GE Healthcare), pre-equilibrated with 50 mM

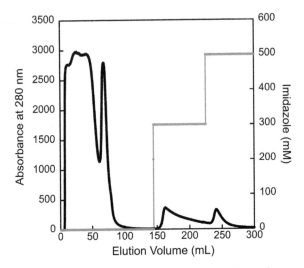

Fig. 5 Elution profile for histidine-tagged AD using nickel affinity chromatography. AD elutes using a step gradient of 300 and 500 mM imidazole.

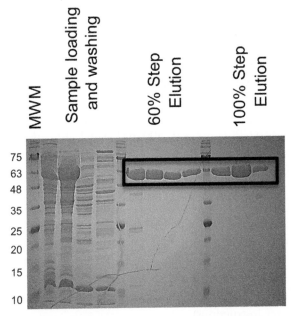

Fig. 6 Coomassie stained 15% SDS-PAGE of fractions eluted from nickel affinity column for histidine-tagged AD.

Tris-HCl, 1 mM DTT, 250 mM NaCl, pH 8.0. Elute the protein using one column volume of the same buffer as the pre-equilibration (Fig. 7). Repeat this step as needed for all sample generated in step 1.

n) AD elutes as a monomer, as estimated by using molecular weight standards on the gel filtration column. Pool fractions containing AD, as verified by SDS-PAGE (Fig. 8).

o) Concentrate the pooled fractions using an Amicon centrifugal concentrations with a 10 kDa cutoff filter to 20 mg/mL (317 µM).

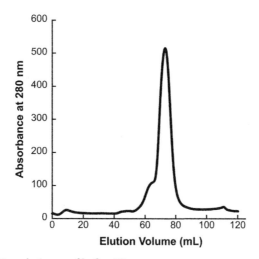

Fig. 7 Gel filtration elution profile for AD.

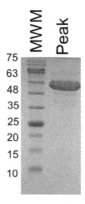

Fig. 8 Coomassie stained 15% SDS-PAGE of fractions eluted from gel filtration column for histidine-tagged AD.

Determine protein concentration using Bradford reagent and a standard curve generated using IgG (using manufacturer's instructions).

p) Store the concentrated proteins in aliquots of 100 µL at −80 °C for later use. The purification should yield ~200 mg of AD per liter of culture.

6.3 Check for activity of MTAN and AD

Before incorporating the newly purified MTAN and AD into the mastermix for the highly sensitive fluorescence assay, it is wise to check for activity. This can be done using methylthioadenosine as a substrate and measuring the decrease in absorbance at 265 nm, as proposed by Krämer and colleagues (Scheme 3) (Seebach et al., 2023).

Timing: ~30 min.
Stock solutions:
a) 50 mM HEPES pH 7.4
b) 50 mM MTA in 100% DMSO
c) 317 µM AD
d) 400 µM MTAN

Preparation of buffered 1 mM MTA:
a) Pipette 980 µL 50 mM HEPES pH 7.4 and 20 µL 50 mM MTA into a 1.5 mL Eppendorf tube to make a 1 mM solution, this solution should be enough for 5 reactions.

Negative Controls:
a) Add 180 µL of 50 mM HEPES pH 7.4 and 20 µL of buffered 1 mM MTA solution to a sub-micro quartz cell (minimum volume 40 µL) and baseline instrument at 265 nm.
b) No MTAN Control: As a separate sample, mix 149 µL of the 50 mM HEPES pH 7.4, with 20 µL buffered 1 mM MTA (from a) and 31 µL

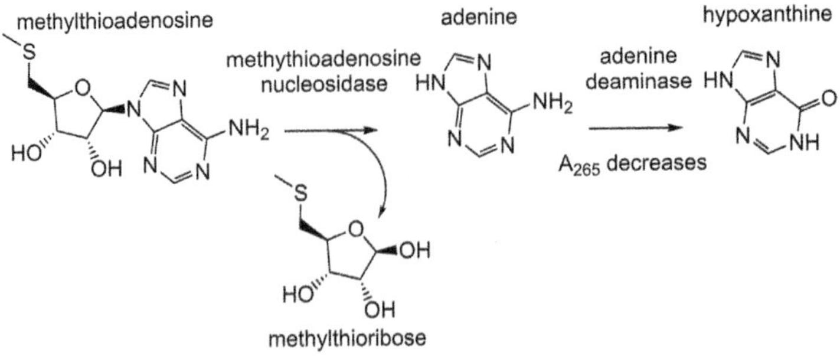

Scheme 3 Assay to test MTAN and AD activity.

317 μM AD and mix quickly with pipettor (do not vortex or use other methods that will introduce bubbles). Immediately place cuvette in spectrophotometer and measure the absorbance at 265 nm every second for 300 s

c) No AD Control: As a separate sample, mix 167.5 μL 50 mM HEPES pH 7.4, 20 μL buffered 1 mM MTA and 12.5 μL 400 μM MTAN and mix quickly with pipettor (do not vortex or use other methods that will introduce bubbles). Immediately place cuvette in spectrophotometer and measure the absorbance at 265 nm every second for 300 s

Full Reaction:

a) Pipette 136.5 μL 50 mM HEPES pH 7.4, 20 μL buffered 1 mM MTA and 31 μL 317 μM AD into a 1 mL quartz cuvette and baseline instrument at 265 nm.

b) Add 12.5 μL 400 μM MTAN to the same cuvette and mix quickly with pipettor (do not vortex or use other methods that will introduce bubbles). Immediately place cuvette in spectrophotometer and measure the absorbance at 265 nm every second for 300 s

Plot Data:

a) Plot data as "Change in Absorbance at 265 nm" versus time, normalizing the starting value to 0 (Fig. 9).

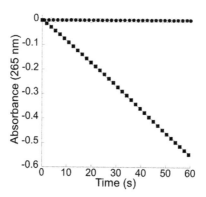

Fig. 9 Check for MTAN and AD activity observing decrease in absorbance at 265 nm (see Scheme 3). No MTAN control (closed circles) and No AD control (open circles) show no change in absorbance at 265 nm. The Full Reaction (with substrate MTA, and enzymes MTAN and AD, filled squares) shows a steady decrease in absorbance at 265 nm.

b) If the samples are working as expected, there should be no absorbance change for the two negative controls, whereas the full reaction should show a steady decrease in absorbance.

7. Step-by-step method details
7.1 Prepare standard curve
Timing: ~30 min.
Stock solutions:
a) 50 mM HEPES pH 7.4
b) 10 mM Amplex Red in 100% DMSO
c) 400 µM MTAN
d) 317 µM AD
e) 50 U XO Crystalline suspension in 60% ammonium sulfate (10 U/mL)
f) 120 U/mL HRP resuspended in 50 mM HEPES pH 7.4.
g) 50 mM MTA in 100% DMSO

Master Mix:
a) Mix together:
    a. 1376 µL 50 mM HEPES pH 7.4
    b. 6 µL 10 mM Amplex Red (30 µM final concentration)
    c. 250 µL 400 µM MTAN (50 µM final concentration)
    d. 315 µL 317 µM AD (50 µM final concentration)
    e. 12 µL 10 U/mL XO (60 mU final concentration)
    f. 1 µL 120 U/mL HRP (60 mU final concentration)
b) The total volume of master mix will be 1960 µL. This recipe is sufficient for 10 reactions if repetitions are needed.
c) Add the Amplex Red first to allow the DMSO to be diluted before adding the enzymes. Always keep the Master Mix on ice and in the dark. Make within 30–45 min of use, because the Amplex Red may photoinactivate.

MTA samples for Standard Curve:
a) Using the 50 mM MTA stock solution, make dilutions with final concentrations of 50, 100, 150, 200 and 250 µM MTA in 50 mM HEPES pH 7.4.

Take Measurements:
a) Mix 196 µL of the master mix with 4 µL of 50 mM HEPES pH 7.4 in a clean, dry sub-micro quartz fluorimeter cell (minimum volume 40 µL) and place in fluorimeter.

b) Wait for 1 min (with the shutter open) and excite the sample at 568 nm and read the emission 590 nm with the bandwidths set to 3 nm. This value serves as your 0 μM concentration.

c) To generate values for a standard curve, add 4 μL of the 50 μM MTA dilution to 196 μL master mix for a final concentration of 1 μM MTA. Repeat this process for the 100, 150, 200, and 250 μM dilutions of MTA to generate samples containing final concentrations of 2, 3, 4 and 5 μM standards, respectively.

d) Repeat steps a and b for each of the concentrations generated in step c. Between each of these readings, the cuvette should be washed with DI water and dried. Fresh master mix should be used for each new concentration.

Plot Data:

a) Subtract the 0 μM fluorescence value from all of the readings. This normalizes all values so that the data begins at the origin.

b) Plot relative fluorescence versus concentration (Fig. 10).

c) Determine the linear fit of the data to calculate concentrations in next section. Do not force the fit through the origin.

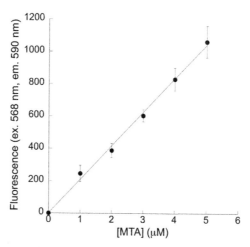

Fig. 10 Standard curve. The concentration of MTA generated by the NAS enzymes will be determined using this standard curve. The equation for the line in this plot is $y = 1.79 + 208x$, such that the fluorescence can be substituted for y to determine the concentration of MTA by solving for x.

7.2 Run the assay

Timing: 3 h.

Stock solutions:

a) 50 mM HEPES pH 7.4

b) 10 mM Amplex Red in 100% DMSO

c) 400 µM MTAN

d) 317 µM AD

e) 50 U XO Crystalline suspension in 60% ammonium sulfate (10 U/mL)

f) 120 U/mL HRP resuspended in 50 mM HEPES pH 7.4

g) 5 mM SAM

h) 1.5 µM of NAS to be assayed

Master Mix and Reaction Mix:

a) For the Master Mix, mix together:

    a. 1338 µL 50 mM HEPES pH 7.4

    b. 6 µL 10 mM Amplex Red (30 µM final concentration)

    c. 250 µL 400 µM MTAN (50 µM final concentration)

    d. 315 µL 317 µM AD (50 µM final concentration)

    e. 12 µL 10 U/mL XO (60 mU final concentration)

    f. 1 µL 120 U/mL HRP (60 mU final concentration)

b) The total volume of master mix will be 1922 µL. This recipe is sufficient for 10 reactions.

c) Add the Amplex Red first to allow the DMSO to be diluted before adding the enzymes. Always keep the Master Mix on ice and in the dark. Make within 30–45 min of use, because the Amplex Red can photoinactivate.

d) Make a serial dilution of equal parts 5 mM SAM and 50 mM HEPES pH 7.4 such that 8 dilutions are made ranging in final concentration from 5 to 0.019 mM.

e) For the Reaction Mix, mix together:

    a. 192 µL of Master Mix from part a

    b. 5 µL from a SAM serial dilution in part d (5–0.019 mM) or 5 µL of buffer (zero concentration)

f) Incubate in the dark and room temperature for 1 min

g) The incubation of the Reaction Mix with SAM is essential, since commercially available SAM frequently is contaminated with MTA as breakdown product, and SAM breaks down (slowly) in solution. This step scrubs contaminating MTA from the Reaction Mix before the addition of the protein to be assayed.

h) Do not add SAM to the master mix, make sure to prepare the reaction mix a minute before reading. It has been observed that SAM can breakdown when in the solution with the master mix for more than 30 min.

Take Measurements:

a) Add 3 µL of 1.5 µM NAS (for a final concentration of 20 nM) to the Reaction Mix previously prepared and mix quickly with pipettor (do not vortex or use other methods that will introduce bubbles).

b) Immediately place it in a cleaned and dried sub-micro quartz fluorimeter cell (minimum volume 40 µL) and in the fluorimeter. Excite the sample at 568 nm and read the emission 590 nm with the bandwidths set to 3 nm every second for 60 s. The reaction mix containing buffer instead of SAM solution will serve as your zero trace (no SAM added).

c) Rinse the cuvette with DI water, dry it and repeat steps a and b for each Reaction Mix with different SAM concentrations to generate sufficient points for purpose (for example, you likely want 10 traces for a Michaelis–Menten plot).

Plot Data:

a) Subtract the zero trace from each of the traces with different concentrations of SAM to remove background Amplex Red signal.

b) Convert relative fluorescence to MTA formed using the standard curve.

c) Plot MTA formed versus time (Fig. 11).

d) Determine the linear fit of the data at each concentration of SAM. Due to photo-oxidation of Amplex Red, to calculate the initial velocity (v_0), subtract the datapoints for the no SAM added sample from the remaining SAM concentrations to determine the actual initial velocities of each sample. The slope of the line represents v_0.

e) Divide each slope v_0 by the NAS enzyme concentration to get v_0/E.

f) Plot v_0/E versus [SAM] to generate Michaelis–Menten plot.

g) Fit data to Michaelis–Menten curve (Fig. 12).

h) Determine apparent k_{cat}, K_m and k_{cat}/K_m (Table 1).

8. Quantification and statistical analysis

The example data in Table 1 are for the nicotianamine synthase from barley (*Hordeum vulgare*) and the nicotianamine synthase-like enzyme from

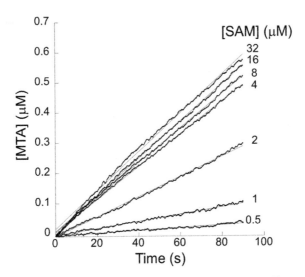

Fig. 11 Example primary data showing MTA formed versus time. The traces have all been normalized to start at zero (subtract initial timepoint of each trace from the remaining points of the trace) to make an easily interpretable graph.

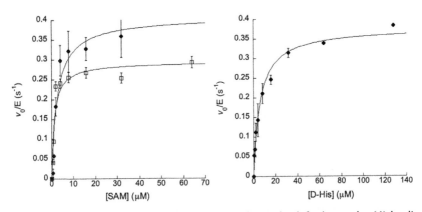

Fig. 12 Michaelis-Menten plot of NAS. The graph on the left shows the Michaelis-Menten as described above with S-adenosylmethionine as the varied substrate. The closed circles represent that data for *Hordeum vulgare* NAS (barley), which makes nicotianamine. The open squares represent the data for *S. aureus* NAS, which makes a D-histidine-aminobutyrate linkage that is NAS-like. The graph on the right is a modification of the above procedure with the SaNAS enzymes where D-histidine is the varied substrate.

Table 1 Apparent steady-state kinetic values for representative NAS (HvNAS) and NAS-like (SaNAS) enzymes.

	Varied substrate	K_m (mM)			k_{cat} (s^{-1})			k_{cat}/K_m (M^{-1}s^{-1})		
HvNAS	SAM	2.7	±	0.2	0.41	±	0.06	150,000	±	20,000
HvNAS[a]	SAM	~6.9			Not determined due to product inhibition					
SaNAS	SAM	1.3	±	0.1	0.29	±	0.04	230,000	±	40,000
SaNAS	D-His	7	±	3	0.38	±	0.02	60,000	±	30,000
SaNAS[b]	D-His	13.0	±	0.7	0.0298	±	0.0003	2300	±	100

[a]Herbik et al. (1999).
[b]McFarlane and Lamb (2017).

S. aureus involved in the production of the opine metallophore staphylopine. Because of the different biosynthetic outcomes (Scheme 1), we have to be careful in how we report kinetic values. HvNAS links together 3 aminobutyrates from 3 *S*-adenosylmethioniones, making 3 methylthioadenosines and 1 nicotianamine in a reaction that is hypothesized to be processive. In contrast, SaNAS links 1 aminobutyrate from 1 *S*-adenosylmethionione to D-histidine, making 1 methylthioadenosine and 1 nicotianamine-like product. We could report turnover as generation of nicotianamine and nicotianamine-like product; however, generation of nicotianamine requires 2 (potentially 3) reactions (2 linkages and 1 cyclization, which may or may not be concerted with a linkage step), whereas generation of the nicotianamine-like products generated by bacteria has only 1 chemical step (the linkage). In addition, there are nicotianimine synthase-like enzymes that have been identified by sequence and operonic similarity for which the specific opine biosynthetic product is not yet known. This means the number of chemical steps and the number of aminobutyrate moieties incorporated are unknown.

This assay directly measures the formation of the co-product methylthioadenosine, and thus indirectly the consumption of *S*-adenosylmethionione. Because of the inherent difficulty of comparing kinetic parameters for enzymes that do similar chemistry but with varying processivity per product made, we have instead chosen to report turnover as per aminobutyrate incorporated (from SAM generating methylthioadenosine). While this may be an average of chemical steps in the processive enzymes, it seems the most equivalent comparison.

9. Summary

To date, there are no kinetic parameters reported for nicotianamine synthase enzymes, except for the K_m estimate noted in Table 1 for HvNAS. Previous kinetic studies for the nicotianamine synthase-like enzymes were unable to provide a complete Michaelis-Menten analysis due to a weak absorbance signal (Seebach et al., 2023) or showed significant product inhibition (McFarlane & Lamb, 2017). Because the inhibitory molecule (methylthioribose) is consumed in the coupled assay, this inhibition is not present in the assay detailed here. Therefore, we provide a robust continuous steady state assay suitable for determining kinetic parameters and direct comparison of enzymes that generate nicotianamine or nicotianamine-like products, including the opine metallophores. This assay may also be beneficial for the study of polyamines, such as spermine and spermidine, or other processes that generate methylthioadenosine as a co-product.

References

Briat, J.-F., Fobis-Loisy, I., Grignon, N., Lobréaux, S., Pascal, N., Savino, G., ... Van Wuytswinkel, O. (1995). Cellular and molecular aspects of iron metabolism in plants. *Biology of the Cell, 84*(1), 69–81. https://doi.org/10.1016/0248-4900(96)81320-7.

Brown, M., Zoi, I., Antoniou, D., Namanja-Magliano, H. A., Schwartz, S. D., & Schramm, V. L. (2021). Inverse heavy enzyme isotope effects in methylthioadenosine nucleosidases. *Proceedings of the National Academy of Sciences USA, 118*(40), https://doi.org/10.1073/pnas.2109118118.

Dreyfus, C., Larrouy, M., Cavelier, F., Martinez, J., Pignol, D., & Arnoux, P. (2011). The crystallographic structure of thermonicotianamine synthase with a synthetic reaction intermediate highlights the sequential processing mechanism. *Chemical Communications (Camb), 47*(20), 5825–5827. https://doi.org/10.1039/c1cc10565e.

Dreyfus, C., Lemaire, D., Mari, S., Pignol, D., & Arnoux, P. (2009). Crystallographic snapshots of iterative substrate translocations during nicotianamine synthesis in Archaea. *Proceedings of the National Academy of Sciences USA, 106*(38), 16180–16184. https://doi.org/10.1073/pnas.0904439106.

Ghssein, G., Brutesco, C., Ouerdane, L., Fojcik, C., Izaute, A., Wang, S. L., ... Arnoux, P. (2016). Biosynthesis of a broad-spectrum nicotianamine-like metallophore in *Staphylococcus aureus*. *Science (New York, N. Y.), 352*(6289), 1105–1109. https://doi.org/10.1126/science.aaf1018.

Gi, M., Lee, K. M., Kim, S. C., Yoon, J. H., Yoon, S. S., & Choi, J. Y. (2015). A novel siderophore system is essential for the growth of *Pseudomonas aeruginosa* in airway mucus. *Scientific Reports, 5*, 14644. https://doi.org/10.1038/srep14644.

Herbik, A., Koch, G., Mock, H. P., Dushkov, D., Czihal, A., Thielmann, J., ... Baumlein, H. (1999). Isolation, characterization and cDNA cloning of nicotianamine synthase from barley. A key enzyme for iron homeostasis in plants. *European Journal of Biochemistry/FEBS, 265*(1), 231–239. https://doi.org/10.1046/j.1432-1327.1999.00717.x.

Luo, Z., Luo, S., Ju, Y., Ding, P., Xu, J., Gu, Q., & Zhou, H. (2021). Structural insights into the ligand recognition and catalysis of the key aminobutanoyltransferase CntL in staphylopine biosynthesis. *The FASEB Journal, 35*(5), e21575. https://doi.org/10.1096/fj.202002287RR.

Marschner, H., & Römheld, V. (1994). Strategies of plants for acquisition of iron. *Plant and Soil, 165*(2), 261–274. https://doi.org/10.1007/BF00008069.

McFarlane, J. S., & Lamb, A. L. (2017). Biosynthesis of an opine metallophore by *Pseudomonas aeruginosa*. *Biochemistry, 56*(45), 5967–5971. https://doi.org/10.1021/acs.biochem.7b00804.

McFarlane, J. S., & Lamb, A. L. (2020). Opine metallophore biosynthesis. 16 In H.-W. Liu, & T. Begley (Vol. Eds.), *Comprehensive natural products III: Chemistry and biology: 5*, (pp. 395–414). Elsevier 16.

Seebach, H., Radow, G., Brunek, M., Schulz, F., Piotrowski, M., & Kramer, U. (2023). Arabidopsis nicotianamine synthases comprise a common core-NAS domain fused to a variable autoinhibitory C terminus. *The Journal of Biological Chemistry, 299*(6), 104732. https://doi.org/10.1016/j.jbc.2023.104732.

Shojima, S., Nishizawa, N. K., Fushiya, S., Nozoe, S., Irifune, T., & Mori, S. (1990). Biosynthesis of phytosiderophores: In vitro biosynthesis of 2'-deoxymugineic acid from l-methionine and nicotianamine. *Plant Physiology, 93*(4), 1497–1503. https://doi.org/10.1104/pp.93.4.1497.

Sugiura, Y., & Tanaka, H. (1981). Structure, properties, and transport mechanism of iron (III) complex of mugineic acid, a possible phytosiderophóre. *Journal of the American Chemical Society, 103*, 4.

Trampczynska, A., Bottcher, C., & Clemens, S. (2006). The transition metal chelator nicotianamine is synthesized by filamentous fungi. *FEBS Letters, 580*(13), 3173–3178. https://doi.org/10.1016/j.febslet.2006.04.073.

Welch, R. M., & Shuman, L. (1995). Micronutrient nutrition of plants. *Critical Reviews in Plant Sciences, 14*(1), 49–82. https://doi.org/10.1080/07352689509701922.

CHAPTER FOUR

ITC-based kinetics assay for NIS synthetases

Katherine M. Hoffmann*, Jocelin D. Hernandez, Eliana G. Goncuian, and Nathan L. March

Department of Chemistry, California Lutheran University, Thousand Oaks, CA, United States
*Corresponding author. e-mail address: khoffmann@callutheran.edu

Contents

1. Introduction	76
2. General overview of the method	79
3. Before you begin	80
3.1 Protein preparation	80
4. Materials and equipment	81
5. Step-by-step method details	81
5.1 Optional recovery of siderophores	82
6. Expected outcomes	83
7. Quantification and statistical analysis	84
8. Optimization and troubleshooting	84
8.1 Problem: enzyme solubility	84
9. Summary	84
References	85

Abstract

NIS Synthetases are a widely distributed, novel superfamily of enzymes critical to stealth siderophore production—small molecules increasingly associated with virulence. Study of these enzymes for inhibition or utilization in biosynthesis of new antibiotics has been hindered by multiple kinetics assays utilizing different limiting reporters or relying on product dissociation as a precursor to signal. We present a label free, continuous readout assay optimized for NIS Synthetase systems utilizing an isothermal titration calorimetry instrument. This assay has been tested in an iterative system comparing multiple turnovers on a single substrate to a single bond formation event and is able to delineate these complex kinetics well. The ITC-based kinetic assay is the first label-free assay for the NIS field, which may allow for more detailed kinetic comparisons in the future, and may also have broader use for iterative enzymes in general.

Methods in Enzymology, Volume 702
ISSN 0076-6879, https://doi.org/10.1016/bs.mie.2024.06.017
Copyright © 2024 Elsevier Inc. All rights reserved, including those for text and data mining, AI training, and similar technologies.

1. Introduction

Siderophores are high-affinity, small molecule, iron chelators (Neilands, 1995) produced by aerobic and facilitative anaerobic bacteria to compete for and import critical iron ions into the cell (Neilands, 1993). Three functional groups commonly used for ferric iron sequestration are hydroxamates, hydroxycarboxylates, and catecholates (Barona-Gomez et al., 2006), which are incorporated into metabolic intermediates or derivatives and linked by amide bonds to form small hexa-dentate chelators (Patel, Song, & Challis, 2010). Several hundred variations of siderophores have been isolated, and at times several are attributed to production within the same bacterium. For instance, desferrioxamine E (dfoE) and G_1 (dfoG$_1$, Scheme 1) are both produced in *Streptomyces* strains (Patel et al., 2010) as well as *Hafnia alvei* and *Erwinia herbicola* within the family Enterobacteriaceae, members of which also synthesize the catecholate siderophore enterobactin (Patel et al., 2010).

Two main pathways exist for the creation of siderophores based on functional group. The catecholamate-based siderophores rely on large, multi-enzyme complexes, named for the remarkable creation of peptide bonds outside of a ribosome: the Non-Ribosomal Peptide Synthetases (NRPS). The NRPS pathways have been extensively studied for their role in the assembly of peptide-based pharmaceuticals, such as vancomycin and methicillin (Miethke & Marahiel, 2007) and remain a focus of carrier protein mechanisms in assembly line enzymology.

The second pathway, NRPS-Independent Siderophore (NIS) synthesis, utilizes hydroxamate or carboxylate functional groups instead, and the

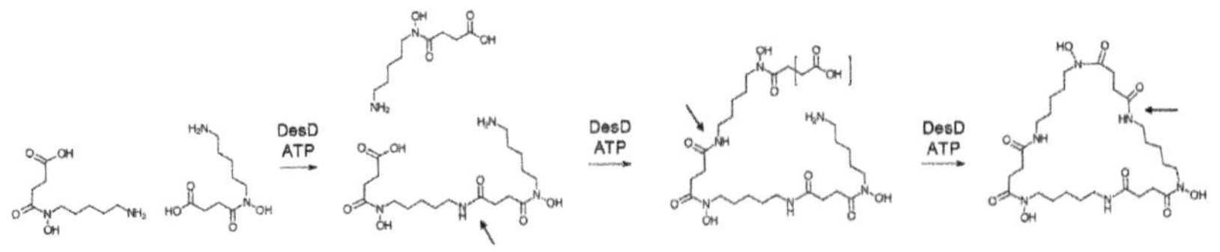

Scheme 1 Iterative Type C NIS synthetase DesD catalyzes three sequential peptide bonds (indicated with arrows) between *N*-hydroxy-*N*-succinyl cadaverine (HSC) substrates. In step one, two molecules of HSC are joined to form an intermediate dipeptide, which in step 2 is connected to a third molecule of HSC to form the tripeptide. The final step cyclizes the tripeptide to form the product siderophore desferrioxamine E (dfoE). Each bond formation event requires a stoichiometric equivalent of ATP to drive the reaction. A version of the tripeptide (dfoG) where the carboxyl group (in brackets) is missing is also commercially available and has been used since the 1970s to treat iron overload as deferoxamine B (Bergeron & Brittenham, 1994).

siderophores are often critical for survival (Challis, 2005; Sullivan, Jeffery, Shannon, & Ramakrishnan, 2006) and virulence (Cendrowski, MacArthur, & Hanna, 2004; Dale, Doherty-Kirby, Lajoie, & Heinrichs, 2004; Franza, Mahe, & Expert, 2005; Koppisch et al., 2005; Oves-Costales, Kadi, & Challis, 2009) in iron limiting conditions. Scientists have known about NIS siderophores for over forty years, and utilized the desferioxamine chelator (among others) in various functions, including treatment for iron toxicity (Bergeron & Brittenham, 1994; Bring, Partovi, Ford, & Yoshida, 2008; Pollack, Ames, & Neilands, 1970). Interest in NIS siderophore biosynthesis has only recently been jump-started by observations that NIS siderophores in particular are not recognized by the human immune system, conferring critical virulence to some of the most problematic pathogens (Clifton, Corrent, & Strong, 2009; Dellagi, Brisset, Paulin, & Expert, 1998; Hoette, Abergel, Xu, Strong, & Raymond, 2008).

NIS synthetases are a widely distributed, novel superfamily of enzymes critical to hydroxamate and carboxylate siderophore production. There are more than 80 proteins from at least 40 different bacteria which share up to 42% identity (Challis, 2005). A review by Challis in 2005, established the sub-groups of Types A, B and C based on carboxy source (citrate, a-ketoglutarate or derivatives, respectively). Studies since then have established that some of the enzymes and sub-types (Type C in particular) have unusual substrates (dual-functional substrates) and/or iterative behavior where substrates are linked sequentially and sometimes cyclized. DesD from *Streptomyces coelicolor*, exemplifies both qualities in its target substrates (Scheme 1).

Carboxylate substrates define the sub-types of the NIS synthetases (Challis, 2005), but they are joined via peptide bond to an amine substrate. In rare cases, the amine substrate is replaced by a hydroxy or second carboxy substrate and an ester bond is formed. These are often a second small molecule building block from metabolism or amino acid derivative, but some of the enzymes utilize multiple dual-functional substrates, and the macrocyclizing enzymes join the carboxylate and amine from the N and C terminus of their peptide-like large substrates.

NIS Synthetases are adenylate forming enzymes. The cofactor ATP binds in a unique nucleotide binding site at the bottom of the pocket, but all substrates and cofactor bind before chemistry occurs (Mydy, Bailey, Patel, Rice, & Gulick, 2020). ATP is converted to AMP and PPi over the course of the reaction, but neither are released until the end of the turnover event (Schmelz et al., 2009). Adenylating enzymes such as Acetyl-CoA

Synthase (Gulick, Starai, Horswill, Homick, & Escalante-Semerena, 2003) and 2-chlorobenzoate CoA ligase (Wu, Reger, Lu, Gulick, & Dunaway-Mariano, 2009) are known for catalyzing 2 successive reactions, adenylation and condensation, respectively, creating an acyl adenylate intermediate between substrate carboxylate and AMP moiety between steps. Informed by structural and biochemical characterization of the NIS Synthetase AcsD and variants, in 2009 Schmelz et al. proposed a molecular mechanism involving 2 steps: the first step (adenylation) is an S_N2-like nucleophilic attack by the carboxylate substrate on the alpha phosphate of ATP creating an acyl-adenylate intermediate and PP_i (Schmelz & Naismith, 2009). The second step (condensation) is a nucleophilic substitution where nucleophilic attack by the amine substrate on the acyl adenylate carbonyl forms a tetrahedral intermediate that collapses to an amide bond with release of free AMP.

Structural studies of NIS Synthetases IucA (Bailey, Drake, Grant, & Gulick, 2016), IucC (Bailey et al., 2018), AsbB (Nusca et al., 2012), AcsD (Schmelz et al., 2011), DfoC (Salomone-Stagni et al., 2018) and DesD (Hoffmann et al., 2020; Yang et al., 2022) have revealed a novel fold, and usually a homodimer oligomeric state, though not exclusively. Structures collectively have revealed the novel cofactor binding site, and one system (DesD) trapped the carboxylate substrate position with an acyl adenylate mimic. While there are some subtle movements and structural changes on binding, the NIS synthetases have so far not evidenced a large conformational change seen in other adenylating enzyme sub-families. However, no structures so far have described fully the amine binding site though compelling models have been proposed (Nusca et al., 2012) along with a hypothesized (Yang et al., 2022) flexible binding site that may not specifically trap a single position for the amine nucleophile. Given the inability to structurally characterize the substrate placement and interaction, kinetic and thermodynamic studies become increasingly important to describe specificity of recognition, turnover, and character.

Kinetic studies of NIS synthetases fall into 3 camps: a amine-analog (hydroxylamine) with chromophore reporter, (Kadi & Challis, 2009) two ATP turnover assays, and an ITC-based assay (Hoffmann et al., 2020; Hoffmann et al., 2022). The ATP coupled assays include an NADH coupled fluorescence assay to quantify AMP (product cofactor) production in the presence of various nucleophiles (Kadi, Oves-Costales, Barona-Gomez, & Challis, 2007), and a continuous coupled assay for PP_i production (Bailey et al., 2016, 2018; Nusca et al., 2012; Schmelz et al., 2009) to measure the rate of ATP turnover, both of which are dependent upon AMP or PP_i release and reporter speed. With these assays, the field has determined apparent catalytic rates ranging from 0.23 to $0.9 \, s^{-1}$ and apparent Michaelis constants (K_M) of 0.2–6 mM (Bailey et al., 2016; Kadi et al., 2007; Nusca et al., 2012). An exception is AcsD, whose carboxylate substrate kinetics revealed an apparent k_{cat} of $2.2 \times 10^3 \, s^{-1}$ and apparent K_M of 14.7 mM (Schmelz et al., 2009). In that case, the second order rate constants were nearly 50x what had been reported for the family on average, but also evidenced was significant apparent product-inhibition creating non-Michaelis Menten kinetics curves.

A common nucleophilic analog in these assays is hydroxylamine, which are used as a surrogate nucleophile to probe cofactor or carboxylate preference. However, dual function substrate binding enzymes cannot utilize the monofunctional hydroxylamine as a tolerable substitute, and the binding difference will influence the apparent K_M. The continuous coupled assay for PP_i production will report the rate of PP_i release as well as the rate of PP_i production, and these cannot easily be deconvoluted either. The limitations to each of these methods makes comparisons within the field extremely challenging given the additional variables at play, and underscores the need for a common assay as well.

The ITC based kinetics assay reported here relies on pre-established substrate identification, but utilizes the actual substrate with no analog, label, or reporter necessary. The high calorimetric signal associated with ATP dilution has been minimized by utilizing a reverse, single injection titration of the enzyme instead, and will not require dissociation of products. In our hands, this assay has been instrumental in determining critical catalytic residues (Hoffmann et al., 2020) and comparative iterative kinetics (Hoffmann et al., 2022). The iterative kinetics are notably complex, and the simplicity of direct calorimetric information has been key. The ITC-based kinetic assay is the first label-free assay for the NIS field, which may allow for more detailed kinetic comparisons in the future, and may also have broader use for iterative enzymes in general.

2. General overview of the method

Concentrated enzyme (100–300 μM) is loaded into the syringe as titrant and delivered in a singled injection to a final concentration of 1 μM in a reaction cell containing 0.5–10 mM substrates and cofactor (where the excess substrate is 5-10x the concentration of the limiting substrate). The calorimetry is monitored for up to 20 min (Fig. 1A) and continuously integrated to determine the activity over changing substrate concentrations (Fig. 1B). ATP is a challenging cofactor in ITC due to the highly anionic nature of the material and, as a result, enormous heats of dilution (and spontaneous hydrolysis) resulting from including it in the titrant. Therefore, this method places ATP in the cell, already at concentration, and adds enzyme catalyst to initiate the reaction.

3. Before you begin

3.1 Protein preparation

Timing: Day before/day of.

Protein Quality. Purify and dialyze all proteins for at least 8 h (overnight) in ITC buffer to ensure maximum buffer matching. Post-dialysis, establish the concentration of enzyme between 100 and 200 µM so that a few microliters may be injected into the 270 µL cell to initiate the reaction at 1 µM. Retain and utilize the dialysis ITC buffer to resuspend cofactor and substrates.

Instrument Preparation. Key to ITC experiments is the reference and reaction cell equilibrating to the same temperature. Most ITCs will default to 25 °C for experiments, so the day of, set the temperature to 37 °C,

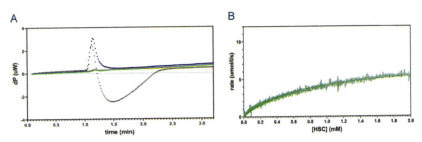

Fig. 1 Raw ITC data (A) and controls compared to the v vs. [S] of DesD catalytic turnover HSC to dfoE (B). (A) Continuous calorimetric data was ascertained from a single injection of wt DesD at 1 µM concentration into a cell containing 2 mM HSC/10 mM ATP. The exothermic reaction (negative dP required to maintain isothermal conditions) resulting from a single injection of wild type (black triangles), a loss of function variant (red triangles), heat-inactivated (green circles), or enzyme into buffer (blue squares). (B) Michaelis-Menten kinetics curves from continuous data (green). Each data point represents the average (small cross) and standard deviation (vertical lines) of three replicate trials, and for clarity, every other data point is shown. Data were analyzed in GraphPad Prism 8.2.

replace the reference water and fill the reaction cell with water or buffer leaving plenty of time for the instrument to establish temperatures. You may speed up this process by loading water and buffer that has been pre-heated to 37 °C.

Note: purified protein may be stored at 4 °C for up to six weeks without loss of activity (Yang et al., 2022).

4. Materials and equipment

- MicroCal PEAQ-Isothermal Titration Calorimeter with microcell (270 μL cell volume.)
- Purified enzyme (>95% pure) dialyzed overnight in ITC buffer and concentrated to 100–200 μM
- ITC buffer at 4 °C for stocks and 37 °C for dilutions: 50 mM HEPES buffer at pH 7.5, containing 150 mM NaCl, 5 mM TCEP, 15 mM $MgCl_2$ and 25% glycerol
- ATP in mg aliquots (Sigma)
- substrate in 1 mg aliquots
- Pierce Strong Cation Exchange Spin Columns, mini (Thermo Fisher Scientific)
- ProteoSIL 300 C8 analytical column 5.0 μm, 4.6 × 250 mm (GL Sciences, USA)

Alternatives: High percentages of glycerol in the buffer is necessary in our system, but it makes loading the cell without bubbles a test of the scientist's skill. The buffer glycerol content should be adjusted to the minimum necessary for solubility and activity over the 30 min assay. A simple test is to observe the product solution to ensure no visible precipitation of protein; alternatively, heat the enzyme mixture to 37 °C for 30 min in buffer, spin down, and ensure no loss of soluble protein.

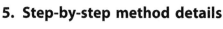

5. Step-by-step method details

A Single Injection Method for Determining Kinetic Constants by ITC.
Timing: 5 min before experiment.

1. Load 37 °C ITC buffer into the cell to pre-incubate.
2. Make a 10x stock solution of substrate. Store on ice.

3. Immediately before your run, prepare a fresh stock of 20 mM ATP in cold ITC buffer. Store on ice.
4. Prepare your cell solution (cofactor, substrate and buffer solution) at 1–2 mM substrate and either 0.5 mM ATP (cofactor limiting) or 5-10 mM ATP (substrate limiting). Dilute to volume with 37 °C ITC buffer. Spin this mixture at 11,000 rpm for 30 s
5. Load concentrated protein immediately into the syringe.
6. Load the 37 °C cell substrate mixture into the ITC microcell.
7. Calculate the amount of enzyme solution to inject to ensure a final protein concentration of 1 μM in the cell; for the Microcal PEAQ instrument, the cell volume is 270 μL. Inject the volume in a 1 injection experiment and monitor for 20 min, or until it returns to baseline.
8. After the experiment, optionally retain the product mixture for siderophore recovery (see below). Rinse the cell with 37 °C d_iH_2O three times and buffer once if preparing for another run.
9. Discard the ATP stock solution and any excess enzyme in the syringe depending on solubility at 37 °C. Fresh ATP stocks will need to be made before the next run, but substrate stocks may be used until depleted.
10. Before exporting for analysis, ensure the cell and syringe concentrations reflect that enzyme is in the syringe and adjust the enzyme concentration to reflect the final concentration after injection.

Critical: All ITC experiments are performed in ITC buffer, with all substrates and cofactors resuspended in samples of the actual dialysis buffer used for the enzyme.

5.1 Optional recovery of siderophores

Timing: post-run.
1. Precipitate and remove the enzyme from the kinetics solution post-experiment; this may require heat inactivation. HEPES buffer, glycerol, TCEP, siderophores, substrates, cofactors and adenylate intermediates will remain in solution; utilize a benchtop/centrifuge cation exchange column to remove anionic small molecules before attempting to create an iron chromophore for detection.
   a) Dilute assay conditions to <50 mM NaCl to enable binding.
   b) Bind in a low salt buffer, discarding flowthrough (contains glycerol, Hepes, and TCEP)
   c) Elute with up to 500 mM NaCl, at 10 mM siderophore concentrations, preferably.

d) Establish which fractions contain siderophores by mixing with an excess of freshly prepared FeCl$_3$. The chromophores will be deep red and may be monitored at 435 nm absorbance.

2. Purify siderophores over an isocratic reverse phase column in 8% Acetonitrile, 0.5% acetic acid running buffer. 20 min at 0.5 mL/min is sufficient to separate substrates, adenylates, and products on a 4 mm × 250 mM Prontosil column. Siderophores and adenylate intermediates may be monitored at 435 nm (Fig. 2) or appropriate λ_{max}, with adenylates additionally absorbing at 260 nm. Excess or partially chelated FeCl$_3$ can be monitored at 340 nm and will likely correlate with trailing shoulders of siderophore peaks.

Pause Point: Assay materials may be stored indefinitely at 4 °C and purified in batches.

6. Expected outcomes

V vs. [S] curves may be exported to analyze in the software of choice, but we find GraphPad Prism to be excellent for analyzing replicates and subtracting baseline data.

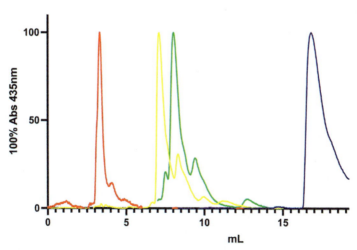

Fig. 2 HPLC purification of siderophores (from left to right: HSC, dfoB, dfoG, dfoE) in the presence of Fe^{3+} results in a chromophore with absorbance at 435 nm. The siderophores have good separation from each other, but consistent trailing shoulders that are likely underchelated Fe^{3+} in equilibria with the full chelate.

7. Quantification and statistical analysis

ITC proprietary software may need to be manipulated to accurately reflect the enzyme as titrant. Care should be taken to delineate data from baseline at especially the initial time of mixing—if the substrate concentration is higher than the maximum included than the initial data has been set too early.

8. Optimization and troubleshooting

8.1 Problem: enzyme solubility

If the enzyme will not stay soluble at high concentrations and precipitates before injection.

8.1.1 Exchange substrate and enzyme

To avoid solubility problems first ensure the buffer includes optimal glycerol and pH more than 1 pH unit from the pI. If precipitation or concentration is still problematic, the substrate and enzyme may be exchanged in the cell and syringe provided that a control run has acceptable heat of dilution for the substrate.

8.1.2 Problem: V vs. [S] curves are not optimal

If the signal to noise is insufficient, or if the curve fails to reach V_{max}.

8.1.3 Increase [substrate] and/or [enzyme]

Estimate the apparent k_{cat} and K_M values from initial runs and adjust substrate concentration to extend the data or increase enzyme concentrations to improve signal to noise.

9. Summary

The ITC based kinetics assay reported here is optimized for NIS Synthetases and may allow better comparison data within the family. It utilizes actual substrates with no analog, label, or reporter necessary, which will permit all sub-types of NIS Synthetases to use their cognate substrates. The high calorimetric background of ATP dilution has been minimized by utilizing a reverse, single injection titration of the enzyme instead, and will be broadly applicable across enzyme systems. Product siderophores may be recovered and identified using a short reverse phase separation method, also presented.

References

Bailey, D. C., Alexander, E., Rice, M. R., Drake, E. J., Mydy, L. S., Aldrich, C. C., & Gulick, A. M. (2018). Structural and functional delineation of aerobactin biosynthesis in hypervirulent *Klebsiella pneumoniae*. *The Journal of Biological Chemistry, 293*(20), 7841–7852. https://doi.org/10.1074/jbc.RA118.002798.

Bailey, D. C., Drake, E. J., Grant, T. D., & Gulick, A. M. (2016). Structural and functional characterization of aerobactin synthetase IucA from a hypervirulent pathotype of *Klebsiella pneumoniae*. *Biochemistry, 55*(25), 3559–3570. https://doi.org/10.1021/acs.biochem.6b00409.

Barona-Gomez, F., Lautru, S., Francou, F. X., Leblond, P., Pernodet, J. L., & Challis, G. L. (2006). Multiple biosynthetic and uptake systems mediate siderophore-dependent iron acquisition in *Streptomyces coelicolor* A3(2) and *Streptomyces ambofaciens* ATCC 23877. *Microbiology (Reading, England), 152*(Pt 11), 3355–3366. https://doi.org/10.1099/mic.0.29161-0.

Bergeron, R. J., & Brittenham, G. M. (1994). *The development of iron chelators for clinical use.* Boca Raton: CRC Press.

Bring, P., Partovi, N., Ford, J. A., & Yoshida, E. M. (2008). Iron overload disorders: Treatment options for patients refractory to or intolerant of phlebotomy. *Pharmacotherapy, 28*(3), 331–342. https://doi.org/10.1592/phco.28.3.331.

Cendrowski, S., MacArthur, W., & Hanna, P. (2004). Bacillus anthracis requires siderophore biosynthesis for growth in macrophages and mouse virulence. *Molecular Microbiology, 51*(2), 407–417. https://doi.org/10.1046/j.1365-2958.2003.03861.x.

Challis, G. L. (2005). A widely distributed bacterial pathway for siderophore biosynthesis independent of nonribosomal peptide synthetases. *Chembiochem: a European Journal of Chemical Biology, 6*(4), 601–611. https://doi.org/10.1002/cbic.200400283.

Clifton, M. C., Corrent, C., & Strong, R. K. (2009). Siderocalins: Siderophore-binding proteins of the innate immune system. *Biometals: An International Journal on the Role of Metal Ions in Biology, Biochemistry, and Medicine, 22*(4), 557–564. https://doi.org/10.1007/s10534-009-9207-6.

Dale, S. E., Doherty-Kirby, A., Lajoie, G., & Heinrichs, D. E. (2004). Role of siderophore biosynthesis in virulence of Staphylococcus aureus: Identification and characterization of genes involved in production of a siderophore. *Infection and Immunity, 72*(1), 29–37.

Dellagi, A., Brisset, M. N., Paulin, J. P., & Expert, D. (1998). Dual role of desferrioxamine in *Erwinia amylovora* pathogenicity. *Molecular Plant-Microbe Interactions: MPMI, 11*(8), 734–742. https://doi.org/10.1094/MPMI.1998.11.8.734.

Franza, T., Mahe, B., & Expert, D. (2005). Erwinia chrysanthemi requires a second iron transport route dependent of the siderophore achromobactin for extracellular growth and plant infection. *Molecular Microbiology, 55*(1), 261–275. https://doi.org/10.1111/j.1365-2958.2004.04383.x.

Gulick, A. M., Starai, V. J., Horswill, A. R., Homick, K. M., & Escalante-Semerena, J. C. (2003). The 1.75 A crystal structure of acetyl-CoA synthetase bound to adenosine-5'-propylphosphate and coenzyme A. *Biochemistry, 42*(10), 2866–2873. https://doi.org/10.1021/bi0271603.

Hoette, T. M., Abergel, R. J., Xu, J., Strong, R. K., & Raymond, K. N. (2008). The role of electrostatics in siderophore recognition by the immunoprotein Siderocalin. *Journal of the American Chemical Society, 130*(51), 17584–17592. https://doi.org/10.1021/ja8074665.

Hoffmann, K. M., Goncuian, E. S., Karimi, K. L., Amendola, C. R., Mojab, Y., Wood, K. M., & Orion, I. W. (2020). Cofactor complexes of DesD, a model enzyme in the virulence-related NIS synthetase family. *Biochemistry, 59*(37), 3427–3437. https://doi.org/10.1021/acs.biochem.9b00899.

Hoffmann, K. M., Kingsbury, J. S., March, N. L., Jang, Y., Nguyen, J. H., & Hutt, M. M. (2022). Chemoenzymatic synthesis of select intermediates and natural products of the desferrioxamine E siderophore pathway. *Molecules (Basel, Switzerland), 27*(19), https://doi.org/10.3390/molecules27196144.

Kadi, N., & Challis, G. L. (2009). Chapter 17. Siderophore biosynthesis a substrate specificity assay for nonribosomal peptide synthetase-independent siderophore synthetases involving trapping of acyl-adenylate intermediates with hydroxylamine. *Methods in Enzymology, 458*, 431–457. https://doi.org/10.1016/S0076-6879(09)04817-4.

Kadi, N., Oves-Costales, D., Barona-Gomez, F., & Challis, G. L. (2007). A new family of ATP-dependent oligomerization-macrocyclization biocatalysts. *Nature Chemical Biology, 3*(10), 652–656. https://doi.org/10.1038/nchembio.2007.23.

Koppisch, A. T., Browder, C. C., Moe, A. L., Shelley, J. T., Kinkel, B. A., Hersman, L. E., & Ruggiero, C. E. (2005). Petrobactin is the primary siderophore synthesized by *Bacillus anthracis* str. Sterne under conditions of iron starvation. *Biometals: An International Journal on the Role of Metal Ions in Biology, Biochemistry, and Medicine, 18*(6), 577–585. https://doi.org/10.1007/s10534-005-1782-6.

Miethke, M., & Marahiel, M. A. (2007). Siderophore-based iron acquisition and pathogen control. *Microbiology and Molecular Biology Reviews: MMBR, 71*(3), 413–451. https://doi.org/10.1128/MMBR.00012-07.

Mydy, L. S., Bailey, D. C., Patel, K. D., Rice, M. R., & Gulick, A. M. (2020). The siderophore synthetase IucA of the aerobactin biosynthetic pathway uses an ordered mechanism. *Biochemistry, 59*(23), 2143–2153. https://doi.org/10.1021/acs.biochem.0c00250.

Neilands, J. B. (1993). Siderophores. *Archives of Biochemistry and Biophysics, 302*(1), 1–3. https://doi.org/10.1006/abbi.1993.1172.

Neilands, J. B. (1995). Siderophores: Structure and function of microbial iron transport compounds. *The Journal of Biological Chemistry, 270*(45), 26723–26726.

Nusca, T. D., Kim, Y., Maltseva, N., Lee, J. Y., Eschenfeldt, W., Stols, L., & Sherman, D. H. (2012). Functional and structural analysis of the siderophore synthetase AsbB through reconstitution of the petrobactin biosynthetic pathway from *Bacillus anthracis*. *The Journal of Biological Chemistry, 287*(19), 16058–16072. https://doi.org/10.1074/jbc.M112.359349.

Oves-Costales, D., Kadi, N., & Challis, G. L. (2009). The long-overlooked enzymology of a nonribosomal peptide synthetase-independent pathway for virulence-conferring siderophore biosynthesis. *Chemical Communications (Camb), (*43), 6530–6541. https://doi.org/10.1039/b913092f.

Patel, P., Song, L., & Challis, G. L. (2010). Distinct extracytoplasmic siderophore binding proteins recognize ferrioxamines and ferricoelichelin in Streptomyces coelicolor A3(2). *Biochemistry, 49*(37), 8033–8042. https://doi.org/10.1021/bi100451k.

Pollack, J. R., Ames, B. N., & Neilands, J. B. (1970). Iron transport in Salmonella typhimurium: Mutants blocked in the biosynthesis of enterobactin. *Journal of Bacteriology, 104*(2), 635–639.

Salomone-Stagni, M., Bartho, J. D., Polsinelli, I., Bellini, D., Walsh, M. A., Demitri, N., & Benini, S. (2018). A complete structural characterization of the desferrioxamine E biosynthetic pathway from the fire blight pathogen *Erwinia amylovora*. *Journal of Structural Biology, 202*(3), 236–249. https://doi.org/10.1016/j.jsb.2018.02.002.

Schmelz, S., Botting, C. H., Song, L., Kadi, N. F., Challis, G. L., & Naismith, J. H. (2011). Structural basis for acyl acceptor specificity in the achromobactin biosynthetic enzyme AcsD. *Journal of Molecular Biology, 412*(3), 495–504. https://doi.org/10.1016/j.jmb.2011.07.059.

Schmelz, S., Kadi, N., McMahon, S. A., Song, L., Oves-Costales, D., Oke, M., & Naismith, J. H. (2009). AcsD catalyzes enantioselective citrate desymmetrization in siderophore biosynthesis. *Nature Chemical Biology, 5*(3), 174–182. https://doi.org/10.1038/nchembio.145 [pii].

Schmelz, S., & Naismith, J. H. (2009). Adenylate-forming enzymes. *Current Opinion in Structural Biology, 19*(6), 666–671. https://doi.org/10.1016/j.sbi.2009.09.004.

Sullivan, J. T., Jeffery, E. F., Shannon, J. D., & Ramakrishnan, G. (2006). Characterization of the siderophore of *Francisella tularensis* and role of fslA in siderophore production. *Journal of Bacteriology, 188*(11), 3785–3795. https://doi.org/10.1128/JB.00027-06.

Wu, R., Reger, A. S., Lu, X., Gulick, A. M., & Dunaway-Mariano, D. (2009). The mechanism of domain alternation in the acyl-adenylate forming ligase superfamily member 4-chlorobenzoate: Coenzyme A ligase. *Biochemistry, 48*(19), 4115–4125. https://doi.org/10.1021/bi9002327.

Yang, J., Banas, V. S., Patel, K. D., Rivera, G. S. M., Mydy, L. S., Gulick, A. M., & Wencewicz, T. A. (2022). An acyl-adenylate mimic reveals the structural basis for substrate recognition by the iterative siderophore synthetase DesD. *The Journal of Biological Chemistry, 298*(8), 102166. https://doi.org/10.1016/j.jbc.2022.102166.

CHAPTER FIVE

An in vitro assay to explore condensation domain specificity from non-ribosomal peptide synthesis

Minuri Ratnayake[a,b,c], Y.T. Candace Ho[d], Xinyun Jian[a,b,c], and Max J. Cryle[a,b,c,*]

[a]Department of Biochemistry and Molecular Biology, The Monash Biomedicine Discovery Institute, Monash University, Clayton, VIC, Australia
[b]EMBL Australia, Monash University, Clayton, VIC, Australia
[c]ARC Centre of Excellence for Innovations in Peptide and Protein Science
[d]Department of Chemistry, University of Warwick, Coventry, United Kingdom
*Corresponding author. e-mail address: max.cryle@monash.edu

Contents

1. Introduction	90
2. General method and statistical analysis	95
3. Molecular design of PCP$_2$-C$_3$ SpyCatcher and PCP$_3$ SpyTag constructs	96
3.1 Equipment	97
3.2 Reagents	97
3.3 Procedure	97
3.4 Notes	101
4. Protein expression and purification	102
4.1 Reagents	102
4.2 Procedure	102
4.3 Notes	105
5. Synthesis of chemical reagents	105
5.1 Equipment	106
5.2 Reagents	106
5.3 Synthesis	107
5.4 Notes	108
6. *In vitro* reconstitution of NRPS C-domain	108
6.1 Equipment	108
6.2 Reagents	108
6.3 Procedure	109
6.4 Notes	110
7. LC-HRMS/MS analysis of methylamine cleaved peptide products	111
7.1 Material and equipment	111
7.2 Buffer and reagents	111

Methods in Enzymology, Volume 702
ISSN 0076-6879, https://doi.org/10.1016/bs.mie.2024.06.010
Copyright © 2024 Elsevier Inc. All rights are reserved, including those for text and data mining, AI training, and similar technologies.

| 7.3 Procedures | 111 |
| 7.4 Notes | 114 |

8. Intact protein PPant ejection LC-ESI-Q-TOF-MS analysis of chemically stabilised peptide products — 114

8.1 Equipment	115
8.2 Buffers and reagents	115
8.3 Procedure	115
8.4 Notes	116

9. Conclusions	117
Acknowledgements	117
References	118

Abstract

Non-ribosomal peptide synthesis produces a wide range of bioactive peptide natural products and is reliant on a modular architecture based on repeating catalytic domains able to generate diverse peptide sequences. In this chapter we detail an in vitro biochemical assay to explore the substrate specificity of condensation domains, which are responsible for peptide elongation, from the biosynthetic machinery that produces from the siderophore fuscachelin. This assay removes the requirement to utilise the specificity of adjacent adenylation domains and allows the acceptance of a wide range of synthetic substrates to be explored.

1. Introduction

Non-ribosomal peptide synthetases (NRPSs) are important macro-molecular machines that produce a wide range of diverse and modified peptides that are often therapeutically important such as antibacterial, antifungal, antivirals and antitumour agents (Felnagle et al., 2008; Süssmuth & Mainz, 2017). The diversity of these molecules is due to the ability of NRPSs to incorporate a range of monomers along with extensive modification of the peptide that are performed during biosynthesis (Miller & Gulick, 2016; Süssmuth & Mainz, 2017). A typical NRPS consists of modules that are responsible for the addition of (typically) one specific amino acid to the growing peptide chain (Süssmuth & Mainz, 2017). An extension module contains three core domains that are essential to the elongation of the peptide (Fig. 1B): (1) Prior to chain elongation, the peptidyl carrier protein (PCP) is post-translationally modified via attachment of a phosphopantetheine (PPant) moiety derived from coenzyme A (CoA) by a 4′-phosphopantethienyl transferase (PPtase) (Miller & Gulick, 2016). (2) The adenylation domain selects and activates a specific amino

Condensation domain specificity 91

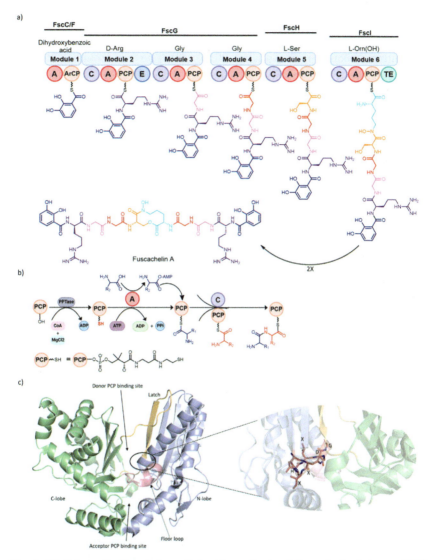

Fig. 1 (A) Schematic description of the biosynthesis of fuscachelin A by an NRPS assembly line. (B) The reaction catalysed by the three core domains (A, PCP and C-domains) of an NRPS assembly line forming a dipeptide. (C) Crystal structure of an archetypal C-domain (VibH from vibriobactin biosynthesis, PDB code: 1L5A) showing the V-shaped pseudo-dimer of chloramphenicol acetyl transferase (CAT) domains (green and blue). The active site motif HHxxDG is shown in red. Domain code: *A*, adenylation domain; *PCP*, peptidyl carrier protein; *C*, condensation domain; *E*, epimerization domain; *TE*, thioesterase domain.

acid building block together with consumption of ATP, before loading this amino acid onto the PPant moiety of the adjacent PCP domain (Ishikawa, Nakamura, Nakanishi, & Tanabe, 2024). (3) The condensation domain finally catalyses the formation of a peptide bond between the upstream PCP-bound aminoacyl/peptidyl (donor) substrate and the downstream PCP-bound aminoacyl (acceptor) substrate (Bloudoff & Schmeing, 2017). Typically within the final module of an NRPS assembly there is an additional thioesterase domain (TE) that is responsible for cleavage of the mature cyclic peptide from the assembly line, either in a cyclised or hydrolysed linear form (Horsman, Hari, & Boddy, 2016). In addition to these core domains, tailoring domains such as methylation (Mt) and epimerisation (E) domains are present that can introduce further diversification to the assembly line and hence peptide produced (Stachelhaus & Walsh, 2000; Süssmuth & Mainz, 2017).

The potential diversity of modular NRPS assembly lines as well as the comprehensive mechanistic and structural information obtained about these systems thus far makes them appealing from an engineering viewpoint, as engineered NRPSs could allow the biosynthesis of novel peptides and therapeutic agents such as new antibiotics (Camus, Truong, Mittl, Markert, & Hilvert, 2022; Süssmuth & Mainz, 2017). Individual domains and modules can be inserted, deleted or exchanged within an NRPS that can lead to the production of novel peptides (Baltz, 2014; Stachelhaus & Walsh, 2000). However, these techniques can also result in the disruption of interdomain contacts that in turn leads to reduced yields of the desired peptide product. Additionally, altering the natural substrate specificity of NRPS A-domains to a noncognate substrate is challenging. A-domains are considered to be the main source of specificity – and hence diversity – in non-ribosomal peptides, as they act as gatekeepers in selecting the substrate building blocks. As A-domains contain the binding site for substrates, modification of the NRPS machinery to produce new compounds has focused on reengineering A-domains (Miller & Gulick, 2016). In recent years, reengineering A-domains through mutagenesis has been shown to provide a potentially simple strategy that preserves the catalytic conformation and interactions between modules (Camus et al., 2022; Stanišić & Kries, 2019). Although there has been success in reengineering NRPS machineries, particularly in regards to A-domain specificity, the question of C-domain specificity for their substrates have not been widely explored. C-domain specificity could lead to complications in reengineering NRPS machineries as C-domains may fail to process noncognate substrates from modified A-domains. Therefore, gaining a deeper understanding of the C-domain function and specificity is key to the development of robust NRPS engineering strategies.

C-domains catalyse the nucleophilic attack of (typically) the NH_2 of the acceptor substrate on the thioester group of the PCP-bound donor substrate, resulting in the transfer of the donor substrate onto the acceptor substrate and thus elongating the peptide chain by one residue (Bloudoff & Schmeing, 2017). The first structurally characterised NRPS C-domain was VibH from vibriobactin biosynthesis, with the structure shown to comprise a pseudo-dimer of the chloramphenicol acetyl transferase (CAT) enzyme fold with two catalytic tunnels leading to the active site from both the donor and acceptor PCP binding sites (Fig. 1C) (Keating, Marshall, Walsh, & Keating, 2002). Additionally, C-domains have been shown to contain a HHxxxDG motif with the second histidine residue proposed to deprotonate the α-amino group of the acceptor aminoacyl-PCP during peptide bond formation. Insights obtained from mutation of these active site residues reinforced the importance of the second histidine residue for catalysis during the condensation reaction (Keating et al., 2002). Whilst not to the extent of A-domain specificity, C-domains have been shown to play a role in adding to the structural diversity of the assembly line. In addition to peptide bond formation, C-domains can play roles in ester bond formation, be involved in dehydration, controlling peptide stereochemistry in concert with E-domains, and β-lactam formation (Gaudelli, Long, & Townsend, 2015; Stachelhaus & Walsh, 2000; Wang et al., 2021). Early studies using chemically synthesised aminoacyl-CoA molecules and aminoacyl-N-acetylcysteamine thioesters (aminoacyl-SNACs) revealed the importance of C-domain specificity, showing that C-domains would only catalyse peptide bond formation between specific substrates (Belshaw, Walsh, & Stachelhaus, 1999; Ehmann, Trauger, Stachelhaus, & Walsh, 2000). However, the use of soluble substrates in this case limits the general applicability of these findings. C-domains have also been found to influence the activity of adjacent domains (Bloudoff & Schmeing, 2017), including A-domain specificity and activation rate via a process that is as yet unresolved (Kaniusaite et al., 2019; Li, Oliver, & Townsend, 2017; Meyer et al., 2016).

NRPSs are intricate complexes with domain-domain interactions requiring a detailed structural and biochemical understanding to fully elucidate selectivity within an NRPS. However, the role C-domains play in NRPS selectivity has largely remained elusive due in part to the challenges in structurally characterising C-domains as relevant complexes

(Bloudoff & Schmeing, 2017; Izoré et al., 2021). Using structural and biochemical characterisation, Izoré et al., 2021, investigated a PCP-bound analogue of the aminoacyl acceptor-PCP bound to the acceptor site of a C-domain. The NRPS system used in this study was taken from the biosynthesis machinery responsible for synthesis of the siderophore fuscachelin, which is produced by the thermophile *Thermobifida fusca* (Fig. 1A). The selectivity of this C-domain was investigated using a fused PCP-C-PCP construct that was generated by the use of the SpyCatcher/SpyTag system (Izoré et al., 2021; Zakeri et al., 2012). This system is based on an engineered CnaB2 domain of the FbaB protein found in the invasive strain of *Streptococcus pyogenes* (Zakeri et al., 2012). The C-terminal SpyTag, which is separated from the SpyCatcher, can rapidly form an intermolecular isopeptide bond between the two fragments, thus linking both proteins via a covalent bond (Fig. 2A). This allows for the reconstitution of the NRPS assembly line through separate loading of acceptor/donor substrates onto the PCP domain and the PCP-C domain constructs using chemically synthesised

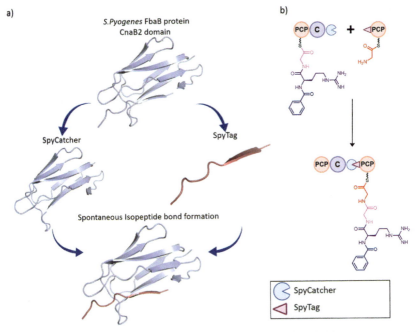

Fig. 2 (A) The SpyTag/SpyCatcher system used to reconstitute C-domain activity. (B) The condensation reaction using PCP$_2$C$_3$ SpyCatcher that is loaded by a dihydroxybenzoic acid (DHB)-D-Arg-Gly donor substrate and PCP$_3$ that is loaded with glycine. Domain code: *PCP*, peptidyl carrier protein; *C*, condensation domain.

Condensation domain specificity

Fig. 3 The general workflow described in this chapter, where the numbering corresponds to those found in the experimental procedures. Domain code: *PCP*, peptidyl carrier protein; *C*, condensation domain.

coenzyme A (CoA) substrates (Fig. 2B) (Izoré et al., 2021). Additionally, this assay bypasses substrate specificity of the downstream A-domain and allows exploration of C-domain acceptor site specificity (Bloudoff & Schmeing, 2017). The SpyCatcher/SpyTag system has revealed that – unlike A-domains – C-domains do not appear to possess a distinct pocket that encodes acceptor substrate specificity, although residues with large aromatic sidechains were not accepted. Furthermore, C-domains do not appear to generally possess highly stringent acceptor site specificity outside of specific examples where modifications of the acceptor bound PCP substrate are required prior to elongation (Kaniusaite et al., 2019).

In this chapter, we describe an experimental procedure for the biochemical exploration of C-domain specificity from the NRPS fuscachelin from the thermophile *T. fusca*. Specifically, a PCP_2-C_3 construct and downstream PCP_3 domain were used to probe the specificity of the acceptor site of the C_3-domain. The complete procedure is as follows: (i) molecular design and cloning of PCP_2-C_3 SpyCatcher and PCP_3 SpyTag, (ii) verifying the expression of the SpyCatcher/SpyTag complexed proteins, (iii) in vitro reconstitution of the NRPS using SpyCatcher/SpyTag, and (iv) mass spectrometry analysis of tetrapeptide formation performed by the C-domain (Fig. 3).

2. General method and statistical analysis

Security measures for a biosafety level-1 laboratory must be followed. Personal protective equipment (PPE) such as lab coats, safety glasses and chemically resistant gloves are required when performing the experiments. Fume hoods are required for the synthesis of aminoacyl-CoA and peptidyl-CoA. For the statistical analysis, the determination of activity and conversions level were performed either in duplicate or triplicate. Data was analysed using intact mass spectrometry and PPant ejection, with the results analysed using GraphPad Prism v6.05 for Windows (GraphPad Software, La Jolla, CA, United States).

3. Molecular design of PCP$_2$-C$_3$ SpyCatcher and PCP$_3$ SpyTag constructs

In this section we describe the use of In-Fusion cloning to generate PCP-C constructs linked to SpyCatcher and PCP linked to SpyTag. The SpyCatcher is linked C-terminally to the PCP-C protein of interest and the SpyTag is linked N-terminally to the PCP construct (Zakeri et al., 2012). No linker sequence was included between PCP$_2$-C$_3$ and the SpyCatcher, with a 6 amino acid linker (GSGESG) included between the SpyTag and PCP$_3$. In-Fusion cloning is highly versatile as it allows for accurate, seamless cloning that is ligase independent. This process is also highly efficient, possesses high fidelity and can be used in a wide array of applications such as high throughput cloning, site directed mutagenesis and single/multiple insert cloning. The homologous ends between insert(s) and vector are designed into the primers, and the insert(s) and vector are annealed during the In-Fusion cloning reaction. Because there is no need to digest the insert(s) prior to incubation with the vector, the In-Fusion cloning workflow is significantly shorter than that of traditional cloning using restriction enzymes (Fig. 4).

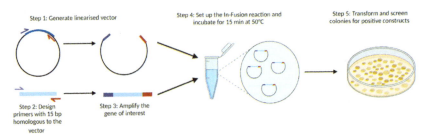

Fig. 4 Overview of the In-Fusion protocol.

3.1 Equipment

- PCR Thermocycler machine (T_{100} thermal Cycler, Bio-Rad).
- Thin-walled PCR tubes (DNAse free and/or sterile).
- Electrophoresis chamber (Bio-Rad).
- Water bath.
- UViDoc HD6 (Thermofisher Scientific).
- NanoDrop spectrophotometer (Thermofisher Scientific).

3.2 Reagents

- PCR oligonucleotides (see Section 3.3: steps 1 and 2).
- Phusion polymerase (as master mix, or with separate components: dNTPs
- mix, polymerase buffer) (NEB).
- *Dpn*I enzyme (NEB).
- Agarose gel (0.8%, w/v) from agarose powder.
- 1X TBE buffer (Tris/Borate/EDTA).
- Loading dye (X6) (NEB).
- Gel Green Nucleic acid stain (Biotium).
- Sterile water.
- PCR clean up kit (Thermofisher Scientific).
- Plasmid MiniPrep kit (Thermofisher Scientific).
- In-Fusion® HD Cloning Kit (Takara Bio).
- Agarose gel (1%, w/v) (from agarose powder).
- Sterile LB medium (for 1 L, 10 g tryptone, 5 g yeast extract and 5 g NaCl). For LB agar, add agar at 1% (w/v) final concentration (10 g for 1 L).
- *Escherichia coli* Stellar™ cells (TakaraBio).
- 50 mg/mL kanamycin stock solution (Sigma-Aldrich).
- 100 mg/mL ampicillin stock solution (Sigma-Aldrich).

3.3 Procedure

3.3.1 Generating PCP_2-C_3 and PCP_3 constructs

1. Design oligonucleotides for gene fragments encoding the desired reagents of FscG (UniProt ID Q47NR9) and the SpyCatcher and SpyTag. The 5′end of the primer contains 15 bases homologous to the construct to which it will be joined. The 3′end of the primers must be 18–25 bases in length and have a CG content of 40–60%. See examples in Table 1.
2. Solubilise the oligonucleotide stocks to 100 μM in milli-Q water. Then dilute to 10 μM for working stock. Store the 100 μM stock at −20 °C.
3. Set up PCR reaction using a PCR master mix, following the manufacturer's instructions, with an example shown in Table 2. The gene

Table 1 List of primer sequences used to generate SpyCatcher/ SpyTag constructs.

Construct	Primer	Sequence
PCP$_2$–C$_3$	1 (Fwd)	5'-GAACAGATCGGTGGTGTCACCGCCTACGAGGAGA-3'
	2 (Rev)	5'-GTCTAGAAAGCTCTATGCCCCGACACCACCT-3'
PCP$_3$	3 (Fwd)	5'-GGATCCCATCATCATCATCATCATTAAAAGCT-3'
	4 (Rev)	5'-ATGTATATCTCCTTCTTAAAGTTAAACAAAATTATTTCTAGAGGGA-3'
pOPINS	5 (Fwd)	5'-TAGAGCTTTCTAGACCATTTAAAACACCACCAC-3'
	6 (Rev)	5'-ACCACCGATCTGTTCGCG-3'
p17HIS	7 (Fwd)	5'-GGATCCCATCATCATCATCATCATTAAAAGCT-3'
	8 (Rev)	5'-ATGTATATCTCCTTCTTAAAGTTAAACAAAATTATTTCTAGAGGGA-3'
PCP$_2$–C$_3$ in pOPINS – SpyCatcher	9 (Fwd)	5'-TAGAGCTTTCTAGACCATTTAAACACCACCAC-3'
	10 (Rev)	5'-TGCCCCCGACACCACCTC-3'
SpyCatcher	11 (Fwd)	5'-GTGGTGTCGGGGGCAATGACAATTGAAGAAGATAGTGCTACCCA-3'
	12 (Rev)	5'-GTCTAGAAAGCTCTAAATATGAGCGTCACCTTTAGTTGCTTTGC-3'
PCP$_3$ in pHIS17 – SpyTag	13 (Fwd)	5'-GTCCGCGAACCCGCAACC-3'
	14 (Rev)	5'-CATATGTATATCTCCTTCTTAAAGTTAAACAAAATTATTTCTA GAGGGA-3'
SpyTag	15 (Fwd)	5'-GGAGATATACATATGGGAGCCCACATCGTG-3'
	16 (Rev)	5'-TGCGGGTTCGCGGACACCACTTTCACCACTACCCTT-3'

Table 2 PCR setup for a 50 μL reaction.

Components	Final concentration	Volume
5X Phusion HF or GC buffer	1X	10 μL
10 mM dNTPs	200 μM	1 μL
10 μM Forward primer	0.5 μM	2.5 μL
10 μM Reverse primer	0.5 μM	2.5 μL
Phusion DNA Polymerase	1.0 unit/50 μL PCR	0.5 μL
Template DNA	<250 ng	Variable
DMSO (optional)	3%	1.5 μL
Nuclease free water		To 50 μL

Table 3 Thermocycling conditions for PCR.

Steps	Time	Temperature (°C)
Initial denaturation	30 s	98
25–35 cycles	5–10 s	98
	10–30 s	45–72
	15–30/kb	72
Final extension	10 min	72
Hold	∞	12

fragments encoding PCP_2-C_3 were amplified using primers **1+2** and **3+4** for PCP_3. Target vectors pOPINS (for PCP_2-C_3) and PHIS17 (PCP_3) were linearised using primers **5+6** and **7+8** respectively.

4. Transfer the reaction to the thermocycler and start the PCR program (Table 3). Preheat the machine to 98 °C and ensure the lid is heated at 105 °C to prevent condensation.
5. Following completion of the PCR reaction, add 1 μL of Dpn1 to each reaction product to digest template DNA and incubate in the thermocycler for 2 h at 37 °C. After completion of reaction, store in ice or place tubes at −20 °C overnight.

6. Analyse the PCR products by loading them in a 0.8% agarose gel in TBE buffer (50 mL) with 0.7 µL of GelGreen nucleic acid stain. Mix 10 µL of PCR product and 2 µL of loading dye (1X) and load to the gel. Run samples at 120 V for 45 min and examine the gel in a UV gel doc.

7. Gel extract and purify the PCR products using the GeneJET Gel Extraction Kit (Thermo Fisher Scientific).

8. Measure the concentration of the DNA using the NanoDrop spectrophotometer.

9. Use PCR products (vector and insert) in the In-Fusion cloning reaction as per manufacturer's instructions.

10. Incubate reaction mixture for 15 min at 50 °C. Once this step is complete, the tubes can be stored on ice until required for transformation with competent *E. coli* Stellar™ cells (TakaraBio).

11. Thaw competent cells on ice for 20–30 min. Add 2.5 µL of the In-Fusion reaction mixture, gently mix the tube and place directly on ice for 30 min.

12. Heat shock the tube for 30 s at 42 °C. Place on ice for 5 min

13. Add 800 µL of LB media to the tube and incubate at 37 °C for 60 min while shaking.

14. Centrifuge the tube at 376 g for 2 min. Remove 600 µL of the supernatant and resuspend the pellet in the remaining in 200–300 µL and transfer to a LB agar plate with 100 mg/mL ampicillin (PHIS17) or 50 mg/mL kanamycin (pOPINS) respectively.

15. Incubate agar plates at 37 °C overnight.

16. Pick colonies from the respective plates and inoculate with 6 mL LB containing 100 100 µg/mL ampicillin (PCP$_3$ SpyTag) or 50 µg/mL kanamycin (PCP$_2$-C$_3$ SpyCatcher) and incubate overnight at 37 °C overnight with shaking.

17. Pellet the cultures via centrifugation at 1503 g for 10 min and decant the supernatant. Use the MiniPrep kit (Thermofisher Scientific) to purify the plasmid DNA from the pellet.

18. Send plasmid DNA for sequencing to ensure the correct construct is obtained.

3.3.2 Generating PCP$_2$-C$_3$ SpyCatcher and PCP$_3$ SpyTag constructs

1. Once constructs have been obtained, set up a PCR reaction to generate the PCP$_2$-C$_3$ SpyCatcher and PCP$_3$ SpyTag construct. The PCP$_2$-C$_3$ pOPINS plasmid was linearised using primers **9 + 10** and PCP$_3$ SpyTag

Condensation domain specificity

construct was linearised using primers **13 + 14**. Set up a PCR reaction as indicated above in steps 3–5.

2. The SpyCatcher insert was amplified from Addgene plasmid #35044 "pDEST14-SpyCatcher" using primers **11 + 12**. The SpyTag insert was amplified from the Addgene plasmid #35050 "pET28a-SpyTagMBP" using primers **15 + 16**. Set up a PCR reaction as indicated above in steps 3–5.

3. Analyse the PCR products by loading them in a 0.8% agarose gel in TBE buffer. Gel extract and purify the products using the GeneJET Gel Extraction Kit (Thermo Fisher Scientific). Measure the concentration of the fragments.

4. Perform the In-Fusion cloning reaction and transform into competent *E. coli* Stellar™ cells (TakaraBio) as indicated above in steps 9–15.

5. Pick colonies from the respective plates and inoculate with 6 mL LB containing 100 µg/mL ampicillin (PCP$_3$ SpyTag) or 50 µg/mL kanamycin (PCP$_2$-C$_3$ SpyCatcher) and incubate overnight at 37 °C overnight while shaking.

6. Obtain and verify the plasmid sequence as indicated above in steps 17–18.

3.4 Notes

1. If the PCR product does not contain a single distinct band, it is necessary to perform gel extraction as otherwise this would lead to your PCR product containing other DNA fragments and resulting in plasmid/s with incorrect insert/s.

2. If the PCR reaction results in a low yield, a gradient PCR can be performed to determine the optimal annealing temperature. GC buffer can be used for long GC rich DNA templates.

3. The In-Fusion reaction mixture can be stored in minus 20 °C until needed.

4. When using the In-Fusion kit for the first time, it is strongly recommended to perform the positive and negative control reactions in parallel with the In-Fusion cloning reaction to ensure the reaction was successful.

5. Ensure that no more than 5 µL of In-Fusion reaction mixture is added as more reaction mixture can inhibit the transformation.

6. For a cloning reaction with more than two inserts, it is recommended plating a larger volume (1/5-1/3 of the transformation reaction).

7. For a cloning reaction with more than two inserts, it is recommended plating a larger volume (1/5-1/3 of the transformation reaction).

4. Protein expression and purification

4.1 Reagents

- Plasmids of interest – PCP$_2$-C$_3$ SpyCatcher, PCP$_3$ SpyTag, *Cth* SUMO proteases (*Cth*), Sfp mutant R4-4 (Sfp) (Sunbul, Marshall, Zou, Zhang, & Yin, 2009).
- Competent *E. coli* BL21(DE3) cells.
- Competent *E. coli* BL21(DE3) (*entD-*) cells (from the laboratory of Prof. David Ackerley, Victoria University of Wellington, New Zealand).
- 50 mg/mL kanamycin stock solution (Sigma-Aldrich).
- 34 mg/mL chloramphenicol stock solution (Sigma-Aldrich).
- 100 mg/mL ampicillin stock solution (Sigma-Aldrich).
- Lysogeny broth (LB) media.
- Terrific broth (TB) media.
- Isopropyl β-d-1-thiogalactopyranoside (IPTG, Apollo Scientific).
- Protease inhibitor cocktail tablets (SIGMAFAST Protease Inhibitor Cocktail Tablets, EDTA-Free; Sigma-Aldrich).
- Benzonase® Nuclease (Sigma-Aldrich).
- Ni-NTA agarose (Macherey-Nagel).
- Dithiothreitol (DTT, Apollo Scientific).
- Sodium dodecyl sulfate (SDS) polyacrylamide gel.
- Precision Plus Protein™ Dual Colour Standards (Bio-Rad).
- InstantBlue (Thermofisher Scientific).
- 1.5 mL centrifuge tubes.
- 0.2 mL PCR tubes.

4.2 Procedure

Express and purify proteins PCP$_2$-C$_3$ SpyCatcher, PCP$_3$ SpyTag, *Cth* and Sfp for the enzyme assay.

4.2.1 Transformation

All plasmids are (co-)transformed into chemically competent *E. coli* BL21(DE3) (*entD-*) cell as follows:

1. (Co-)transform the desired expression plasmid(s) (1 μL) into chemically competent *E. coli* BL21(DE3) (*entD-*) cell. (See Note 4.3.1).
2. Incubate the mixture on ice for 30 min
3. Heat-shocks the bacterial cells at 42 °C for 30 s
4. Place the cells on ice for 2 min

Table 4 List of plasmids in vector and antibiotic resistant.

Plasmids	Vectors	Antibiotic resistance
PCP$_2$-C$_3$ SpyCatcher (see Note 4.3.2)	pET_SUMO	Kanamycin and chloramphenicol
PCP$_3$ SpyTag (see Note 4.3.2)	pET_His17	Ampicillin and chloramphenicol
SUMO protease (*cth*)	pET_21a	Ampicillin
Sfp R4-4 mutant	pET_28a	Ampicillin

5. Recover the bacterial cells by adding LB media (0.5 mL) and incubating at 37 °C for 45 min
6. Plate the cells on a LB agar plate supplemented with required antibiotics.
7. Incubate the plate at 37 °C overnight (Table 4).

4.2.2 Expression

All proteins are expressed as follows,
1. Pick a single colony from the agar plate and inoculate into a 200 mL LB pre-culture supplemented with required antibiotics.
2. Incubate the flask at 37 °C overnight with shaking at 180 rpm.
3. Use the pre-culture to inoculate 6×1 L of TB medium containing required antibiotics.
4. Incubate the culture at 37 °C with shaking at 180 rpm until the OD$_{600}$ nm reaches above 0.6.
5. Add IPTG to the TB culture to a final concentration of 0.1 mM.
6. Continue to shake at 180 rpm at 18 °C overnight.
7. Harvest the cells by centrifugation at 3064 g for 20 min at 4 °C.

4.2.3 Purification

All proteins are purified in the same manner unless stated otherwise.

4.2.3.1 Ni-NTA purification

1. Resuspend the cells in 200 mL of Ni–NTA buffer A supplemented with a protease inhibitor cocktail tablet and Benzonase® Nuclease (10 μL).
2. Lyse the cells by a cell disruptor operating at 14, 000–19,000 psi, and clarify the lysate by centrifugation at 22,680 g for 45 min at 4 °C.
3. Equilibrate 2 mL of bed volume Ni–NTA agarose with Ni–NTA buffer A.
4. Incubate the supernatant at 4 °C for 1 h with the Ni–NTA agarose with gentle stirring.

Table 5 List of Ni-NTA buffers used in each protein.

Protein	Ni-NTA buffer A	Ni-NTA buffer B
PCP_2-C_3 SpyCatcher	50 mM Tris-HCl pH 8.0, 300 mM NaCl, 10 mM imidazole	50 mM Tris-HCl pH 8.0, 300 mM NaCl, 1 M imidazole
PCP_3 SpyTag	50 mM Tris-HCl pH 8.0, 300 mM NaCl, 10 mM imidazole	50 mM Tris-HCl pH 8.0, 300 mM NaCl, 1 M imidazole
SUMO protease (*cth*)	50 mM Tris pH 7.4, 500 mM NaCl, 20 mM Imidazole, 0.5 mM EDTA, 1 mM DTT	50 mM Tris pH 7.4, 500 mM NaCl, 250 mM Imidazole, 0.5 mM EDTA, 0.5 mM DTT
Sfp R4-4 mutant	20 mM Tris-HCl pH 8.0, 500 mM NaCl, 5 mM imidazole	20 mM Tris-HCl pH 7.9, 500 mM NaCl, 1 M imidazole

5. After incubation, wash the beads with 20 bed volumes of Ni-NTA buffer A.
6. Elute bound protein with 5 bed volumes of Ni-NTA buffer B (Table 5).

4.2.4 PCP_2-C_3 SpyCatcher

1. Cleave the SUMO-tag by performing the following procedure after Ni-NTA purification and before size exclusion chromatography for the SpyCatcher protein.
2. Dialyse the protein against a buffer of 50 mM Tris-HCl pH 7.4, 300 mM NaCl and 1 mM DTT with SUMO protease overnight at 4 °C.
3. Equilibrate 2 mL of bed volume Ni-NTA agarose with Ni-NTA buffer A.
4. Incubate the protein at 4 °C for 20 min with the Ni-NTA agarose with gentle shaking.
5. Collect the flow-through of the Ni-NTA agarose and wash the beads with 2 bed volumes of Ni-NTA buffer A. (See Note 4.3.3).

4.2.5 Sfp mutant R4-4

1. Perform the following procedure after Ni-NTA purification.
2. Dialyse the elution protein twice against a buffer of 10 mM Tris-HCl pH 7.5, 1 mM EDTA and 10% glycerol.
3. Concentrate and aliquot the protein solution into 0.2 mL PCR tubes, flash-freeze the tubes in liquid nitrogen and store the tubes at −80 °C.

Table 6 List of gel filtration buffer for proteins.

Protein	Gel Filtration buffer
PCP$_2$-C$_3$ SpyCatcher	50 mM Tris-HCl, pH 7.4; 300 mM NaCl, 1 mM DTT
PCP$_3$ SpyTag	50 mM Tris-HCl, pH 7.4; 300 mM NaCl, 1 mM DTT
SUMO protease (*cth*)	50 mM Tris pH 7.4, 200 mM NaCl, 1 mM EDTA, 2 mM DTT, 5% (v/v) glycerol

4.2.5.1 Size exclusion chromatography

1. Further purify the protein by gel-filtration chromatography using a SRT 10 SEC 300 (105 mL) column connected to an ÄKTA PURE system.
2. Equilibrate the column with 1.2 column volumes of gel-filtration buffer.
3. Concentrate the protein to a desired volume and inject onto the column.
4. Fractionate the eluate into 1.5 mL fractions.
5. Analyse the elution fractions containing monomeric protein by sodium dodecyl-sulfate polyacrylamide gel electrophoresis (SDS-PAGE).
6. Combine and concentrate appropriate fractions using appropriate size of centrifugal filter units.
7. Determine the protein concentration by measuring protein absorbance at 280 nm using a NanoDrop One microvolume UV–vis spectrophotometer.
8. Concentrate and aliquot the protein into chilled 0.2 mL PCR tubes, flash freeze in liquid nitrogen, and store at −80 °C (Table 6).

4.3 Notes

1. To avoid producing *holo* PCP. The phosphopantetheine arm can be installed by the phosphopantetheinyl transferase EntD that adds a PPant group to certain PCP domains when they are expressed in *E. coli*.
2. pRARE plasmid is co-transformed due to the rare codons in the genomic DNA from which the constructs were amplified.
3. The cleaved SUMO tag is attached to the Ni-NTA agarose after incubation and the desired protein is found in the flow-through

5. Synthesis of chemical reagents

Studying the specificity of C-domains for their acceptor substrates can require the use of standard aminoacyl CoAs or stabilised substrates (using for example amide (Ho, Kaczmarski, et al., 2023) or thioether (Izoré et al., 2021) linked amino acids) depending on the desired experimental setup (e.g. stabilised constructs for crystallisation). These different substrates require different Mass Spectrometry experiments to analyse the results of elongation: the first involves Liquid Chromatography Mass Spectrometry (LCMS) to analyse cleaved peptides when using aminoacyl CoA substrates; while the second method utilises intact protein mass spectrometry coupled with PPant ejection to analyse peptides that are bound to the protein and cannot be cleaved due to the stabilised nature of their linker. Ester linked substrates possess intermediate stability and can be used for both structural and cleavage MS based analysis assays, although the synthesis of these substrates is often more challenging than the CoAs discussed here (Ho, Kaczmarski, et al., 2023). For these methods, different CoAs are synthesised, including peptidyl-CoA as a donor peptide and aminoacyl CoAs as receptor substrates. Here we describe the synthesis of tripeptidyl-CoAs as donor substrates, and Gly-CoA and stabilised Gly-CoA (Gly$_{stab}$-CoA, an analogue of the aminoacyl-CoA stabilised via a thioether bond) as acceptor substrates.

5.1 Equipment

- Rotary evaporator (Vaccubrand PC 3001 Vario pro).
- Freeze dryer (Christ Alpha 1-2 LD plus).
- Spark-proof floor centrifuge (Spintron GT-175).
- High performance liquid chromatography (HPLC, Shimadzu LC – 20AP) system equipped with analytical C_{18} columns (Agilent ZORBAX 300SB-C18, 7 μm, 212 × 250 mm).
- Liquid chromatography mass spectrometer (LCMS, Shimadzu LCMS-2020 LC/MS) system equipped with analytical C_{18} column (Agilent ZORBAX 300SB-C18, 5 μm, 4.6 × 250 mm).

5.2 Reagents

- Coenzyme A trilithium Salt (Chem-Impex).
- Ammonium bicarbonate (Sigma Aldrich).
- Ethylenediaminetetraacetic acid (EDTA, Sigma Aldrich).
- Tris (2-carboxyethyl)phosphine (TCEP, Sigma Aldrich).
- 2-Bromoethylamine hydrobromide (Sigma Aldrich).
- Acetonitrile (MeCN, Fisher Scientific).
- Boc-Gly-OH (Sigma Aldrich).
- Triethylamine (TEA, Sigma Aldrich).

- (1–Cyano-2-ethoxy-2-oxoethylidenaminooxy)dimethylamino-morpho-lino-carbenium. hexa-fluorophosphate (COMU, Oakwood Chemical).
- N, N-Dimethylformamide (DMF, Fisher Scientific).
- Diethyl ether (Fisher Scientific).
- Trifluoroacetic acid (TFA, Sigma Aldrich).
- Triisopropylsilane (TIS, Sigma Aldrich).

5.3 Synthesis

All synthesis must be conducted in a fume hood while wearing full PPE.

5.3.1 Peptidyl-CoA

Synthesise the desired peptidyl-CoA thioesters by generating peptidyl-hydrazides on solid phase, the subsequent hydrazide activation and displacement with Coenzyme A as previously stated (Ho, Zhao, Tailhades, & Cryle, 2023; Tailhades et al., 2018).

5.3.2 Aminoacyl-CoA synthesis

1. Dissolve Boc-Gly-OH (4.2 mg, 24 mmol, 2 equiv.), TEA (2 μL, 14.4 mmol, 1.2 equiv.) and COMU (6.12 mg, 14.4 mmol, 1.2 equiv.) in 5 mL of DMF.
2. Stir the mixture and cool it to 4 °C for 30 min
3. Add 5 mL of DMF containing CoA (10 mg, 12 mmol) dropwise to the solution.
4. Stir the mixture overnight at room temperature.
5. Concentrate the mixture using rotary evaporator.
6. Cleave the crude compound from the Boc-protecting group using a solution of TFA/TIS/H2O (95:2.5:2.5, v/v'/v", 5 mL).
7. Shake the reaction for 1 h at room temperature.
8. Concentrate the filtrate under a N_2 stream to ~1 mL.
9. Precipitate the compound with ice cold diethyl ether (9 mL).
10. Collect the compound by centrifugation in a flame-resistant centrifuge.
11. Wash the compound with ice cold diethyl ether (3 times).
12. Purify the aminoacyl-CoA using preparative RP-HPLC (gradient in MeCN; 0–40% over 30 min) (Fig. 5).

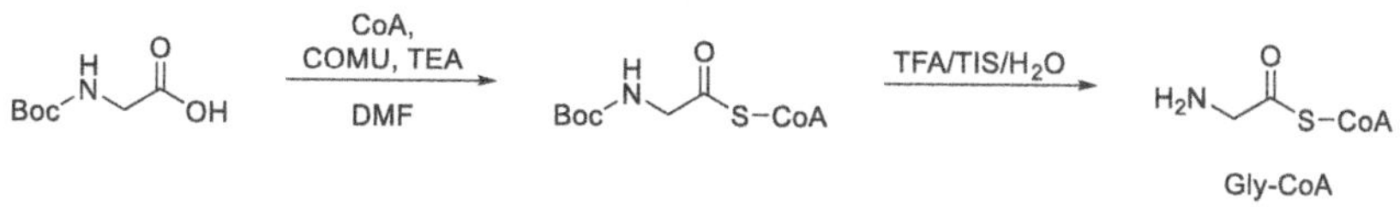

Fig. 5 Scheme of Gly-CoA synthesis.

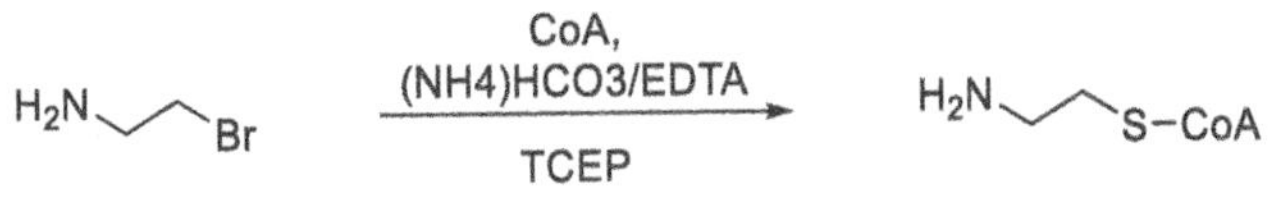

H₂N–CH₂–CH₂–Br →(CoA, (NH4)HCO3/EDTA, TCEP)→ H₂N–CH₂–CH₂–S–CoA

Fig. 6 Scheme of Gly_stab-CoA synthesis.

5.3.3 Stabilised aminoacyl-CoA

1. Dissolved CoA (10 mg, 12 mmol, 1 equiv.) in 10 mL of buffer (0.02 M (NH₄)HCO₃, 6.5 mM ethylenediaminetetraacetic acid (EDTA), pH 8) (See note 5.4.1).
2. Add TCEP (4.1 mg, 14.4 mmol, 1.2 equiv.) into the mixture and stir the reaction for 30 min
3. Add a 2 mL solution of 2-bromoethylamine hydrobromide in MeCN to the CoA solution. (See note 5.4.2).
4. Stir the reaction mixture at room temperature overnight.
5. Directly purify the crude stabilised aminoacyl-CoA using preparative RP-HPLC (gradient in MeCN; 0–40% over 30 min).
6. Aliquot the desired peptidyl-CoA into microtube tubes in an amount for an enzymatic reaction (100 μM), freeze dry and store at −20 °C. (See note 5.4.3) (Fig. 6).

5.4 Notes

1. Prepare fresh buffer each time.
2. 2-Bromoethylamine hydrobromide may cause skin, eye and respiratory irritation, refer to the chemical's safety data sheet for detailed information.
3. The stock of the substrate-CoA is prepared by weighting it out and dissolving it in water to a concentration of 100 μM. Each enzymatic assay is performed in a total working volume of 30 μL (Fig. 7).

6. In vitro reconstitution of NRPS C-domain

6.1 Equipment
- 0.5 mL centrifugal filters (Amicon® Ultra).
- Thermomixer (Eppendorf).

6.2 Reagents
- MilliQ.
- HEPES, pH 7.4.
- NaCl.
- MgCl₂.

Condensation domain specificity 109

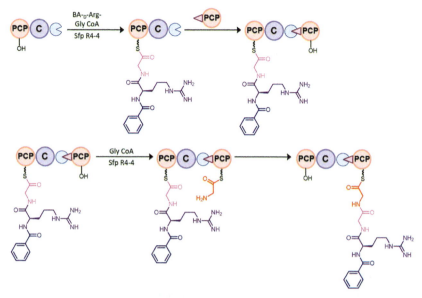

Fig. 7 Schematic overview of the condensation reaction using PCP$_2$C$_3$ SpyCatcher (which accepts the BA-D-Arg-Gly (peptide mimic) donor substrate) and PCP$_3$ (which accepts glycine).

- Dithiothreitol (DTT).
- Sfp R4-4.
- PCP-C catcher.
- Peptidyl-CoA.

6.3 Procedure
6.3.1 Loading of PCP-C with peptidyl-CoA

To perform this assay, first convert the *apo* PCP-proteins to their loaded *holo* forms using the phosphopantetheinyl transferase Sfp (R4-4 mutant) (Sunbul et al., 2009) and desired CoA/s. Each enzymatic assay is carried out in a 30 μL triplicate reaction as per the protocol below.

1. Set up the PCP-C loading reaction as follows in Table 7. The loading reaction utilises a 1:2:0.1 molar ratio of PCP-C to peptidyl-CoA to Sfp. BA-D-Arg-Gly peptidyl-CoA was used as the donor substrate.
2. Mix the reaction thoroughly.
3. Incubate the reaction at 30 °C for 1 h without shaking.
4. Add the reaction mixture to a centrifugal filter (30 kDa MWCO) with size exclusion. buffer (50 mM Tris-HCl, pH 7.4; 300 mM NaCl, 1 mM

Table 7 PCP-C loading reaction components.

Components	Stock concentration	Final concentration	30 μL reaction
HEPES (pH 7.4)	1000 mM	100 mM	3 μL
MgCl$_2$	1000 mM	10 mM	0.3 μL
NaCl	5000 mM	50 mM	0.3 μL
PCP-C catcher	Varies	100 μM	varies
peptidyl-CoA (donor substrate)	10,000 μL	200 μM	0.6 μL
Sfp R4-4	150 μL	10 μM	2.0 μL
Water	—	—	Top up to 30 μL

DTT) and spin at 13,523 g to reach the desired volume to remove unloaded peptidyl-CoA and Sfp mutant R4-4.

5. Repeat step 4 (×3).
6. Bring the volume of the reaction to 30 μL or 105 μL (for a 3.5× reaction).
7. Place on ice until use.

6.3.2 Reconstitution of NRPS peptide extension

1. For a reconstitution assay that uses Gly–CoA as the acceptor PCP$_3$ substrate, this requires offloading and analysis by HRMS. When Gly$_{stab}$–CoA was used as the acceptor PCP$_3$ substrate, PPant ejection analysis was required.
2. Add unloaded PCP$_3$ SpyTag (100 μM) to the reaction mix and incubate for 10 min at 30 °C for 10 min with no shaking.
3. Add MgCl$_2$ (10 mM), Gly–CoA (100 μM) and Sfp (10 μM) to the reaction mixture.
4. Mix well and incubate for 1 h 30 °C for 1 h with no shaking.
5. Following incubation, reactions were analysed by PPant ejection for thioether-tethered amino acids loaded PCP$_3$. Reactions that utilised thioester-tethered amino acids substrates were chemically cleaved by methylamine prior to analysis by HRMS.

6.4 Notes

1. Ensure CoA-peptide, CoA-amino acid and Sfp are added last to the reaction mixture.
2. Once the final reagent is added, ensure the tubes are spun down and thoroughly mixed prior to incubation.

7. LC-HRMS/MS analysis of methylamine cleaved peptide products

For peptide products with chemically cleavable linkage (e.g. thioester or ester) to the PPant arm of carrier proteins, the reconstitution reaction can be monitored via offloading peptide products using methylamine (see note 7.4), followed by purification of products from assays mixture and LC-HRMS/MS analysis.

7.1 Material and equipment
- 1.5 mL Eppendorf tubes.
- Solid phase extraction (SPE) column (Bond Elut Plexa 30 mg/mL), Agilent Technologies
- Freeze Dryer.
- LC-MS: Orbitrap Fusion mass spectrometer (Thermo Scientific) coupled online to a nano-LC (Ultimate 3000 RSLCnano; Thermo Scientific) via a nanospray source.
- Column: Reverse-phase C_{18} column (Acclaim PepMap RSLC, 75 μm × 50 cm, nanoViper, 2 μm, 100 Å; Thermo Scientific) connected to a trap column (Acclaim PepMap 100, 100 μm × 2 cm, nanoViper, C_{18}, 5 μm, 100 Å; Thermo Scientific).
- Data analysis software: QualBrowser (XCalibur 3.0.63, Thermo Scientific), MS-Product (ProteinProspector v5.22.1, UCSF).

7.2 Buffer and reagents
- 40% Methylamine solution (Sigma-Aldrich).
- H_2O (0.1% FA).
- MeOH (0.1% FA).
- 50% MeCN (0.1% FA).
- LC-MS grade MeCN (0.1% FA).
- LC-MS grade H_2O (0.1% FA).

7.3 Procedures

7.3.1 Methylamine cleavage of peptides from PCP domains
1. Upon completion of the reconstitution reaction (see 6.3.2), add 15 μL of 40% methylamine to the reaction mixture and incubated for 15 min at room temperature to liberate the methylamide peptide products.
2. Add 850 μL of H_2O (0.1% FA) to the above reaction mixture to quench the reaction.
3. Activate a SPE column by washing with 1 mL MeOH (0.1% FA) and equilibrating with 1 mL H_2O (0.1% FA).

4. Load the reaction mixture from step 2 to the SPE column by gravity.
5. Wash the column with 1 mL 0.1% FA in water and elute the column with 1 mL in 50% MeCN (0.1% FA), and freeze dry the eluent using freeze dryer at −50 °C.

7.3.2 LC-HRMS/MS analysis of cleaved peptide products

1. Redissolve the purified peptide products (from 7.3.1) in 50 μL of H_2O (0.1% FA) in 1.5 mL Eppendorf tubes and centrifuge at 15,000 g for 10 min to pellet any insoluble material.
2. Carefully transfer the supernatant into a LC autosampler compatible vial for LC-HRMS/MS analysis. See note for the LC-HRMS/MS analysis condition.
3. To analysis the raw LC-HRMS/MS data, *QualBrowser* is used to view spectra and generate extracted ion chromatograms (EICs) for the calculated mass of the singly charged species of substrate and product peptides at error range of 6 ppm, e.g. EIC @ *m/z*: 349.19620 – 349.2004 for substrate $[M+H]^+$ ion calculated for *m/z*: 349.1983 and EIC @ *m/z*: 406.21726 – 406.2214 for product $[M+H]^+$ ion calculated for *m/z*: 406.2197 (Fig. 8).
4. Compare the MS/MS ions are to the MS^2 fragments of the desired peptide products predicted using *MS-Product* (Fig. 9).

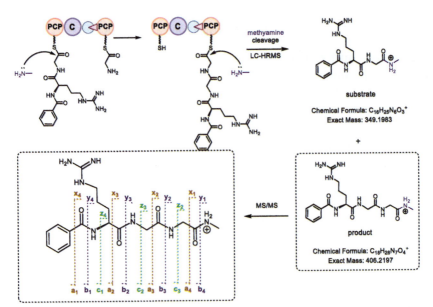

Fig. 8 Methylamine cleavage and HRMS/MS analysis of reconstitution assay of PCP$_2$-C$_3$::PCP$_3$ for BA-D-Arg-Gly donor peptide and Gly-CoA substrate. Domain code: *PCP*, peptidyl carrier protein; *C*, condensation domain.

Condensation domain specificity 113

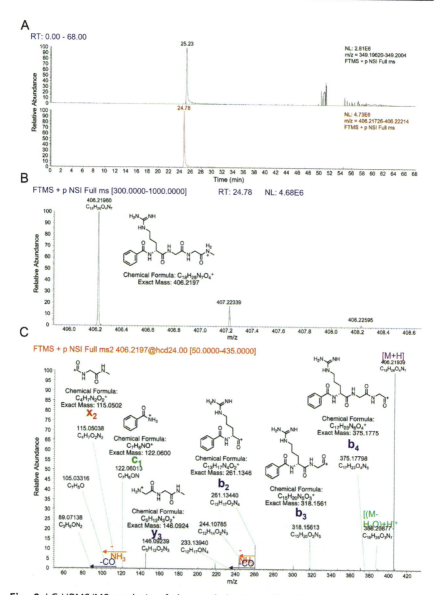

Fig. 9 LC-HRMS/MS analysis of the methylamine offloading products from reconstitution of PCP$_2$-C$_3$::PCP$_3$ for BA-D-Arg-Gly donor peptide and Gly-CoA substrate. (A) Extracted ion chromatograms of methylamine offloaded BA-D-Arg-Gly donor and BA-D-Arg-Gly-Gly product. (B) Mass spectrum of the methylamine offloaded BA-D-Arg-Gly-Gly product. (C) MS^2 spectrum of [M+H]$^+$ion calculated for BA-D-Arg-Gly-Gly product showing in agreement with selected MS^2 fragments.

5. The conversion of the extended peptide product can be calculated using the formula below:

$$\text{percentage conversion} = \frac{\text{peak area}\,(product)}{\text{peak area}\,(product) + \text{peak area}\,(donnor\ substrate)} \times 100$$

7.4 Notes

1. **Methylamine safety notes:** Associated risks include highly flammability and acute toxicity. General safe handling recommendations include keeping it away from heat, sparks, open flames, hot surfaces; using it in a fume hood and wearing protective gloves, clothing, eye protection, face protection. Refer to safety data sheet of methylamine for full information: https://www.sigmaaldrich.com/AU/en/sds/sial/426466?userType=undefined. Cysteamine can be used as an alternative if methylamine is inaccessible.
2. **LC elution condition:** Products were separated on the reverse-phase C_{18} column after binding to a trap column. Elution was performed on-line with a gradient from 6% MeCN to 30% MeCN in 0.1% FA over 30 min at 250 nL/min.
3. **Orbitrap Fusion mass spectrometer parameters:** Full scan MS was performed in the Orbitrap at 60,000 nominal resolutions, with targeted MS^2 scans of peptides of interest acquired at 15,000 nominal resolutions in the Orbitrap using HCD with stepped collision energy (24 ± 5% NCE).

8. Intact protein PPant ejection LC-ESI-Q-TOF-MS analysis of chemically stabilised peptide products

For peptide products with linkage to the PPant arm of carrier proteins that cannot be chemically cleaved (i.e. thioether and amide linkages), intact protein and tandem PPant ejection LC-HRMS/MS analysis provides an alternative approach to monitor the reconstitution assay. This approach allows the qualitative measurement of mass of intact protein with substrate/product attached after reaction, as well as the corresponding PPant fragment that is eliminated from carrier protein at gas phase during tandem mass spectrometry. Such approaches are valuable to confirm the catalytic activity of constructs used for structural studies (Fig. 10).

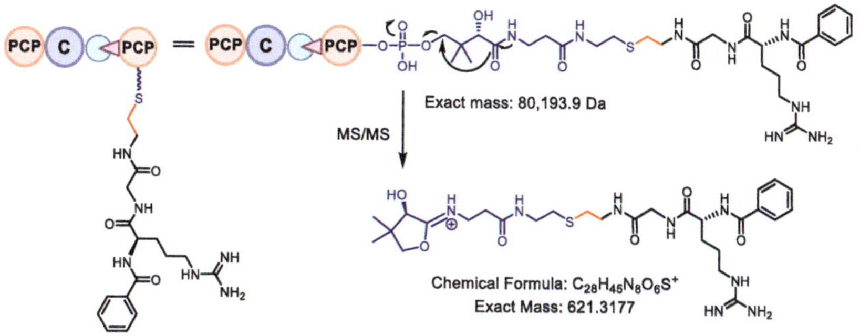

Condensation domain specificity

Fig. 10 Scheme showing PPant ejection ion activated during tandem mass spectrometry analysis of PCP_2-C_3::PCP_3 intact protein with thioether linkage (highlighted in red) to BA-D-Arg-Gly-Gly product.

8.1 Equipment

- LC–MS: Micro-TOFq mass spectrometer (Bruker Daltonics) coupled online to a 1200 series capillary/nano-LC (Agilent Technologies) via a Bruker nano ESI source.
- Columns: 150 mm reverse-phase column (ZORBAX 300SB–C18, 3.5 μm, 0.075 × 150 mm; Agilent Technologies) connected to a trap column (ZORBAX 300SB–C18, 5 μm, 0.30 × 5 mm cartridges; Agilent Technologies). MabPac SEC-1 5 μm 300 Å 50 × 4 mm (Thermo Scientific).
- Data analysis software: Data Explorer software version 3.4 build 192 (Bruker Daltonics, Bremen, Germany).

8.2 Buffers and reagents

- LC–MS grade MeCN (0.05% TFA and 0.05% FA).
- LC–MS grade H_2O (0.05% TFA and 0.05% FA).
- ESI-L Low Concentration Tuning Mix (Agilent Technologies, Santa Clara, CA, USA).

8.3 Procedure

1. Upon completion of the reconstitution assay (from 7.3.1), mix the reaction solution with one equivalent volume of H_2O (0.05% TFA and 0.05% FA) to dilute the protein to 50 μM, and centrifuge at 15,000 g, 4 °C for 10 min
2. Carefully transfer the supernatant to a LC auto sampler compatible vial for intact protein LC-ESI-QTOF-MS analysis.
3. The raw MS data is viewed and analysed using *Data Explorer*. Maximum Entropy algorithm is used for deconvolution of intact protein mass spectrum.
4. Intact protein mass is calculated for average mass using online tool *PeptideMass* (https://web.expasy.org/peptide_mass/) (Fig. 11).

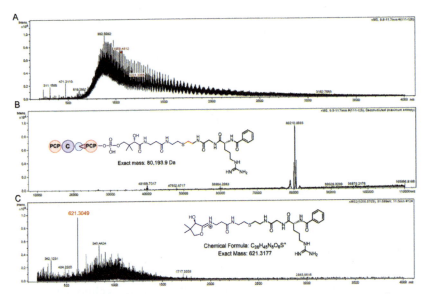

Fig. 11 Intact protein and PPant ejection LC-ESI-Q-TOF-MS/MS analysis of reconstitution of PCP$_2$-C$_3$::PCP$_3$ for BA-D-Arg-Gly donor peptide and Gly$_{stab}$-CoA substrate. (A) Raw MS spectrum of protein eluted between 9.9 and 11.7 min showing the charged states distribution with 76$^+$ charged ion with m/z: 1056.4812 highlighted. (B) Deconvoluted mass spectrum showing the observed [M]$^+$ mass of intact PCP$_2$-C$_3$::PCP$_3$ with tetrapeptide product bound, calculated for 80,193.3 Da. (C) MS^2 raw spectrum of the 76$^+$ charged ion with m/z: 1056.4812 showing the observed PPant ejection ion (highlighted in red), calculated for C$_{28}$H$_{45}$N$_8$O$_6$S$^+$ [M+H]$^+$: 621.3177.

8.4 Notes

1. **LC condition:** Proteins are separated on the reverse-phase column after binding to a trap column. Elution is performed on-line with a gradient from 4% MeCN to 60% MeCN in 0.1% FA over 30 min at 300 nL/min. Proteins > 20 kDa are separated on a MabPac SEC column with an isocratic gradient of 50% MeCN, 0.05% TFA and 0.05% FA at a flow rate of 50 μL/min. The protein is eluted over a 20-min run-time monitored by UV detection at 254 nm. After 20 min the flow path is switched to infuse Tune mix to calibrate the spectrum post acquisition. SEC column separation is optional when reaction sample can be cleaned up to remove excess of substrates and reaction buffer ingredients via buffer exchange to 0.05% TFA and 0.05% FA.
2. **Micro-TOF-Q mass spectrometer parameters:** The mass spectrometer was operated in positive ion mode with a scan range of

200–3000 m/z. Source conditions were: end plate offset at -500 V; capillary at -4500 V; nebuliser gas (N_2) at 0.6 bar; dry gas (N_2) at 6.0 L/min; dry temperature at 175 °C. Ion transfer conditions were: ion funnel 1 RF at 350 Vpp; ion funnel 2 RF at 400 Vpp; Hexipole RF at 400 Vpp; quadrupole low mass at 300 m/z; collision energy at 34.9 eV; collision RF at 300 Vpp; transfer time at 150.0 μs; pre-pulse storage time at 15.0 μs.

3. Intact protein and PPant ejection LC-HRMS analysis can also be applied to carrier proteins with PPant arm linked to chemically cleavable peptide products.

9. Conclusions

Non-ribosomal peptide synthesis is an important biosynthetic pathway able to generate a range of valuable peptide natural products. The modular nature of NRPSs more generally has inspired a range of engineering efforts to generate novel natural products, although the general success of these approaches requires a comprehensive understanding of the specificity enforced by the catalytic domains within these assembly lines. Here, we have detailed an in vitro biochemical assay that allows the specificity of the peptide forming C-domains to be assessed using synthetic donor/acceptor CoA substrates. Mass spectroscopic techniques can then be applied to determine the specificity of C-domains regarding the acceptance of varying donor/acceptor substrates, with the specific analysis method depending on the chemical linkage installed within the acceptor CoA substrate. This assay allows a wide range of donor and acceptor substrates to be tested, with the latter particularly important as this removes the requirement for an adjacent A-domain possessing the desired amino acid specificity to be present. This assay therefore allows the assessment of C-domain specificity within NRPS assembly lines, knowledge which is vital to ensure the success of future biosynthetic engineering strategies of these impressive biosynthetic machineries.

Acknowledgements

This work was supported by Monash Univer, EMBL Australia, the Australian Research Council (Discovery Project DP190101272 and DP210101752). This research was conducted by the Australian Research Council Centre of Excellence for Innovations in Peptide and Protein Science (CE200100012) and funded by the Australian Government.

References

Baltz, R. H. (2014). Combinatorial biosynthesis of cyclic lipopeptide antibiotics: A model for synthetic biology to accelerate the evolution of secondary metabolite biosynthetic pathways. *ACS Synthetic Biology, 3*(10), 748–758. https://doi.org/10.1021/sb3000673.

Belshaw, P. J., Walsh, C. T., & Stachelhaus, T. (1999). Aminoacyl-CoAs as probes of condensation domain selectivity in nonribosomal peptide synthesis. *Science (New York, N. Y.), 284*(5413), 486–489.

Bloudoff, K., & Schmeing, T. M. (2017). Structural and functional aspects of the non-ribosomal peptide synthetase condensation domain superfamily: Discovery, dissection and diversity. *Biochimica et Biophysica Acta (BBA)-Proteins and Proteomics, 1865*(11), 1587–1604.

Camus, A., Truong, G., Mittl, P. R. E., Markert, G., & Hilvert, D. (2022). Reprogramming nonribosomal peptide synthetases for site-specific insertion of α-hydroxy acids. *Journal of the American Chemical Society, 144*(38), 17567–17575. https://doi.org/10.1021/jacs.2c07013.

Ehmann, D. E., Trauger, J. W., Stachelhaus, T., & Walsh, C. T. (2000). Aminoacyl-SNACs as small-molecule substrates for the condensation domains of nonribosomal peptide synthetases. *Chemistry & Biology, 7*(10), 765–772.

Felnagle, E. A., Jackson, E. E., Chan, Y. A., Podevels, A. M., Berti, A. D., McMahon, M. D., & Thomas, M. G. (2008). Nonribosomal peptide synthetases involved in the production of medically relevant natural products. *Molecular Pharmaceutics, 5*(2), 191–211.

Gaudelli, N. M., Long, D. H., & Townsend, C. A. (2015). β-Lactam formation by a nonribosomal peptide synthetase during antibiotic biosynthesis. *Nature, 520*(7547), 383–387.

Ho, Y. C., Kaczmarski, J. A., Tailhades, J., Izoré, T., Steer, D. L., Schittenhelm, R. B., ... Cryle, M. J. (2023). Not always an innocent bystander: the impact of stabilised phosphopantetheine moieties when studying nonribosomal peptide biosynthesis. *Chemical Communications, 59*(53), 8234–8237.

Ho, Y. C., Zhao, Y., Tailhades, J., & Cryle, M. J. (2023). *A chemoenzymatic approach to investigate cytochrome P450 cross-linking in glycopeptide antibiotic biosynthesis. Non-Ribosomal Peptide Biosynthesis and Engineering: Methods and Protocols.* Springer, 187–206.

Horsman, M. E., Hari, T. P., & Boddy, C. N. (2016). Polyketide synthase and nonribosomal peptide synthetase thioesterase selectivity: logic gate or a victim of fate? *Natural Product Reports, 33*(2), 183–202.

Ishikawa, F., Nakamura, S., Nakanishi, I., & Tanabe, G. (2024). Recent progress in the reprogramming of nonribosomal peptide synthetases. *Journal of Peptide Science, 30*(3), e3545.

Izoré, T., Candace Ho, Y. T., Kaczmarski, J. A., Gavriilidou, A., Chow, K. H., Steer, D. L., ... Cryle, M. J. (2021). Structures of a non-ribosomal peptide synthetase condensation domain suggest the basis of substrate selectivity. *Nature Communications, 12*(1), 2511. https://doi.org/10.1038/s41467-021-22623-0.

Kaniusaite, M., Tailhades, J., Marschall, E. A., Goode, R. J., Schittenhelm, R. B., & Cryle, M. J. (2019). A proof-reading mechanism for non-proteinogenic amino acid incorporation into glycopeptide antibiotics. *Chemical Science, 10*(41), 9466–9482.

Keating, T. A., Marshall, C. G., Walsh, C. T., & Keating, A. E. (2002). The structure of VibH represents nonribosomal peptide synthetase condensation, cyclization and epimerization domains. *Nature Structural Biology, 9*(7), 522–526.

Li, R., Oliver, R. A., & Townsend, C. A. (2017). Identification and characterization of the sulfazecin monobactam biosynthetic gene cluster. *Cell Chemical Biology, 24*(1), 24–34.

Meyer, S., Kehr, J.-C., Mainz, A., Dehm, D., Petras, D., Süssmuth, R. D., & Dittmann, E. (2016). Biochemical dissection of the natural diversification of microcystin provides lessons for synthetic biology of NRPS. *Cell Chemical Biology, 23*(4), 462–471.

Miller, B. R., & Gulick, A. M. (2016). Structural biology of nonribosomal peptide synthetases. *Nonribosomal Peptide and Polyketide Biosynthesis: Methods and Protocols*, 3–29.

Stachelhaus, T., & Walsh, C. T. (2000). Mutational analysis of the epimerization domain in the initiation module PheATE of gramicidin S synthetase. *Biochemistry, 39*(19), 5775–5787.

Stanišić, A., & Kries, H. (2019). Adenylation domains in nonribosomal peptide engineering. *Chembiochem: A European Journal of Chemical Biology, 20*(11), 1347–1356.

Sunbul, M., Marshall, N. J., Zou, Y., Zhang, K., & Yin, J. (2009). Catalytic turnover-based phage selection for engineering the substrate specificity of Sfp phosphopantetheinyl transferase. *Journal of Molecular Biology, 387*(4), 883–898.

Süssmuth, R. D., & Mainz, A. (2017). Nonribosomal peptide synthesis—Principles and prospects. *Angewandte Chemie International Edition, 56*(14), 3770–3821.

Tailhades, J., Schoppet, M., Greule, A., Peschke, M., Brieke, C., & Cryle, M. J. (2018). A route to diastereomerically pure phenylglycine thioester peptides: Crucial intermediates for investigating glycopeptide antibiotic biosynthesis. *Chemical Communications, 54*(17), 2146–2149. https://doi.org/10.1039/C7CC09409D.

Wang, S., Fang, Q., Lu, Z., Gao, Y., Trembleau, L., Ebel, R., ... Deng, H. (2021). Discovery and biosynthetic investigation of a new antibacterial dehydrated non-ribosomal tripeptide. *Angewandte Chemie International Edition, 60*(6), 3229–3237.

Zakeri, B., Fierer, J. O., Celik, E., Chittock, E. C., Schwarz-Linek, U., Moy, V. T., & Howarth, M. (2012). Peptide tag forming a rapid covalent bond to a protein, through engineering a bacterial adhesin. *Proceedings of the National Academy of Sciences, 109*(12), E690–E697.

CHAPTER SIX

The production of siderophore analogues using precursor-directed biosynthesis

Tomas Richardson-Sanchez, Thomas J. Telfer, Cho Z. Soe, Kate P. Nolan, Michael P. Gotsbacher, and Rachel Codd*

School of Medical Sciences, The University of Sydney, Sydney, NSW, Australia
*Corresponding author. e-mail address: rachel.codd@sydney.edu.au

Contents

1. Introduction	122
1.1 The clinical natural product desferrioxamine B	122
1.2 Structural diversification of desferrioxamine B using precursor-directed biosynthesis	124
1.3 Generation of constitutional isomers of desferrioxamine B analogues	125
1.4 Mass spectrometry to identify desferrioxamine B analogues and constitutional isomers	126
1.5 Scope of precursor-directed biosynthesis in producing desferrioxamine B analogues	129
1.6 Theoretical maxima of desferrioxamine B analogues produced using precursor-directed biosynthesis	132
2. Materials and equipment	133
2.1 Bacteria and chemicals	133
2.2 Consumables	135
2.3 General equipment	135
2.4 LC-MS-Q instrumentation	135
2.5 LC-MS/MS-QQQ instrumentation	135
3. Protocol	136
3.1 Before you begin	136
3.2 Preparing solutions and use of high grade reagents	136
3.3 Preparing frozen stocks of *Streptomyces pilosus*	136
3.4 Precursor-directed biosynthesis using non-native diamine substrates	137
3.5 Measuring siderophore production	139
3.6 Siderophore purification	139
3.7 Analysing desferrioxamine B analogues and constitutional isomers	141
4. Summary	142
Acknowledgements	142
References	142

Methods in Enzymology, Volume 702
ISSN 0076-6879, https://doi.org/10.1016/bs.mie.2024.06.009
Copyright © 2024 Elsevier Inc. All rights are reserved, including those for text and data mining, AI training, and similar technologies.

Abstract

Siderophores are low-molecular-weight organic bacterial and fungal secondary metabolites that form high affinity complexes with Fe(III). These Fe(III)-siderophore complexes are part of the siderophore-mediated Fe(III) uptake mechanism, which is the most widespread strategy used by microbes to access sufficient iron for growth. Microbial competition for limited iron is met by biosynthetic gene clusters that encode for the biosynthesis of siderophores with variable molecular scaffolds and iron binding motifs. Some classes of siderophores have well understood biosynthetic pathways, which opens opportunities to further expand structural and property diversity using precursor-directed biosynthesis (PDB). PDB involves augmenting culture medium with non-native substrates to compete against native substrates during metabolite assembly. This chapter provides background information and technical details of conducting a PDB experiment towards producing a range of different analogues of the archetypal hydroxamic acid siderophore desferrioxamine B. This includes processes to semi-purify the culture supernatant and the use of liquid chromatography–tandem mass spectrometry for downstream analysis of analogues and groups of constitutional isomers.

1. Introduction
1.1 The clinical natural product desferrioxamine B

The trihydroxamic acid siderophore desferrioxamine B (DFOB) (Fig. 1, **1**) is a bacterial natural product with notoriety founded on its significant and sustained clinical impact. This secondary metabolite was identified in the late 1950s as a product of the soil bacterium *Streptomyces pilosus* and established as a high-affinity Fe(III) chelator. This useful property catapulted the non-toxic metabolite into the therapeutic arena, initially in its Fe(III)-bound form to supply iron to patients with anaemia, and subsequently and more successfully, as a free ligand to remove excess iron in patients with secondary iron overload disease resulting from transfusion-dependent hemoglobinopathies, including β-thalassemia (Codd, Richardson-Sanchez, Telfer, & Gotsbacher, 2018). Despite its challenging slow subcutaneous administration regimen, the application of DFOB (Desferal) from the 1960s transformed the clinical management and life expectancy of patients with β-thalassemia and remains in use today, alongside two other oral synthetic iron chelators.

The bacterium *S. pilosus* and other actinomycetes species produce DFOB as a siderophore metabolite for Fe(III) acquisition (Ejje, Soe, Gu, & Codd, 2013; Hider & Kong, 2010; Roberts, Schultz, Kersten, Dorrestein, & Moore, 2012). Sparingly soluble Fe(III) that exists as hydroxides/oxide species under aerobic and pH neutral environments is leached by DFOB to

The production of siderophore analogues using precursor-directed biosynthesis 123

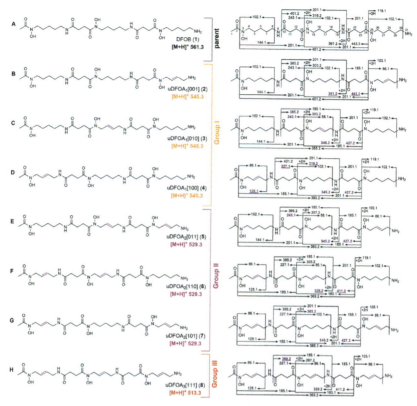

Fig. 1 DFOB (1 (A)) and analogues (2 (B), 3 (C), 4 (D), 5 (E), 6 (F), 7 (G), 8 (H)) produced from a PDB experiment using the exogenous diamine substrate *E*-DBE (left) with MS/MS fragmentation patterns (aligned at right) showing fragments unique to (purple, underlined) or useful (purple) (coupled with other observations) in assigning the constitutional isomers within Group I (2–4) or Group II (5–7). Adapted and reprinted with permission from Telfer, T. J., Gotsbacher, M. P., Soe, C. Z., & Codd, R. (2016). Mixing up the pieces of the desferrioxamine B jigsaw defines the biosynthetic sequence catalyzed by DesD. ACS Chemical Biology, 11(5), 1452–1462. https://doi.org/10.1021/acschembio.6b00056 (Copyright © 2016, American Chemical Society).

form a water soluble Fe(III)-DFOB complex which is returned to the source species and ultimately delivered *via* a set of protein-mediated uptake pathway to the cytosol for incorporation into survival-critical Fe-containing proteins. Siderophores are classified by the Fe(III) binding functional group and include hydroxamic acid, catechol, alpha-hydroxycarboxylic acid, thiazoli(*di*)ne, mixed groups, β-hydroxyaspartic acid (Hardy & Butler, 2018), and the most recently identified, *N*-nitroso-*N*-hydroxylamine group (Codd, 2023).

DFOB has three hard base bidentate O,O'-hydroxamic acid groups threaded along its backbone as evolved as an ideal ligand to bind the hard acid Fe(III) to form a 1:1 Fe(III)-DFOB complex with high thermodynamic stability (log $\beta_{110} = 30.5$) (Codd, 2008; Dhungana, White, & Crumbliss, 2001; Evers, Hancock, Martell, & Motekaitis, 1989). The discovery of high-impact natural products, such as DFOB, provides a platform and incentive to attempt to produce analogues with new structures and new properties.

1.2 Structural diversification of desferrioxamine B using precursor-directed biosynthesis

Precursor-directed biosynthesis (PDB) (also known as directed fermentation, feeding studies, or substrate feeding) is a method used to generate new analogues of a microbial secondary metabolite by culturing the producing species in medium augmented with a non-native precursor compound that competes against the native substrate during target biosynthesis (Kalkreuter, Carpenter, & Williams, 2018; Kirschning, Eichner, Hermane, & Knobloch, 2012; Kirschning, Taft, & Knobloch, 2007; Puja, Mislin, & Rigouin, 2023; Sattler, Grabley, & Thiericke, 1999; Thiericke & Rohr, 1993; Thiericke, Langer, & Zeeck, 1989). A PDB campaign might be expected to have a higher chance of success if knowledge of the biosynthetic mechanism of the natural product metabolite is available to support judicious selection of viable non-native substrates.

The clinical impact of DFOB prompted interest in understanding the biosynthesis of this and related hydroxamic acid siderophores (Barry & Challis, 2009; Challis, 2005; Codd, Soe, et al., 2018; Rütschlin & Böttcher, 2018, 2020; Rütschlin, Gunesch, & Böttcher, 2017, 2018; Schupp, Toupet, & Divers, 1988; Schupp, Waldmeier, & Divers, 1987; Soe & Codd, 2014; Soe, Pakchung, & Codd, 2012; Soe, Telfer, Levina, Lay, & Codd, 2016; Yang, Banas, Rivera, & Wencewicz, 2023). Early studies that used PDB to produce hydroxamic acid-based siderophores (Konetschny-Rapp, Jung, Raymond, Meiwes, & Zaehner, 1992; Meiwes, Fiedler, Zähner, Konetschny-Rapp, & Jung, 1990), together with foundational studies that deconvoluted the biosynthesis of DFOB and a macrocyclic analogue desferrioxamine E (DFOE) (Barona-Gómez, Wong, Giannakopulos, Derrick, & Challis, 2004; Kadi, Oves-Costales, Barona-Gómez, & Challis, 2007), inspired subsequent PDB studies to test the broader capacity of the *S. pilosus* DFOB biosynthetic enzyme cluster to produce non-native DFOB analogues as the topic of this chapter (Gotsbacher & Codd, 2020; Richardson-Sanchez & Codd, 2018;

Richardson-Sanchez, Nolan, & Codd, 2018; Richardson-Sanchez, Tieu, Gotsbacher, Telfer, & Codd, 2017; Telfer & Codd, 2018; Telfer, Gotsbacher, Soe, & Codd, 2016).

DFOB and other siderophores from this class are assembled from diamine-based precursors, including 1,5-diaminopentane (common name: cadaverine) and 1,4-diaminobutane (common name: putrescine), which are produced from the decarboxylation of L-lysine (L-lysine decarboxylase, LDC) (Schupp et al., 1987; Schupp et al., 1988) and L-ornithine (L-ornithine decarboxylase, ODC), respectively (Soe & Codd, 2014; Soe et al., 2012). Reactions downstream of the L-lysine decarboxylase (DesA) catalyse cadaverine mono-N-hydroxylation (DesB), acylation (N-acetyl or N-succinyl) of mono-N-hydroxy-cadaverine (DesC), and condensation reactions (DesD) between N-hydroxy-N-acetyl-cadaverine (HAC) and N-hydroxy-N-succinyl-cadaverine (HSC) units to give the HAC–HSC–HSC trimeric product DFOB (**1**) (Barona-Gómez et al., 2004; Kadi et al., 2007). This indicated that bacteriological medium inoculated with *S. pilosus* and supplemented with non-native diamines might be processed by the DesBCD enzyme cluster to offer a facile route to produce libraries of DFOB analogues with new properties. Setting the entry point of the native-non-native substrate competition at DesB (diamine mono-N-hydroxylase) opens scope given the range of commercially available diamine variants. Preliminary work showed it was challenging to out-compete reactions between DesA and abundant native L-lysine by augmenting culture medium with non-native L-lysine substrates, which themselves are few in number and costly.

1.3 Generation of constitutional isomers of desferrioxamine B analogues

Conducting a PDB experiment has inherent appeal, since it relies on the native biosynthetic machinery of the producing organism to provide access to structurally complex natural product analogues. In most instances, these analogues would be difficult to access using total chemical synthesis (Markham & Codd, 2024). Despite this appeal, PDB has some short-comings, including that analogues are produced in low yields and as mixtures, due to the inevitable competition between the native and non-native substrates during target assembly. Constructing mutant species of the producing organism to ablate the availability of the native substrate can reduce the complexity of the product profile (Kirschning et al., 2007; Kirschning et al., 2012), although this mutasynthesis approach has higher technical demands and can reduce organism fitness if the native substrate is required for multiple metabolic pathways.

This chapter retains its focus on the use of PDB as a facile approach towards generating DFOB analogues, and to recast the complex product profile as a knowledge benefit rather than an impediment. DFOB is an asymmetric trimer of three regions built from diamine precursors: the N-acetyl region (HAC-derived), the internal region (HSC-derived), and the primary amine region (HSC-derived). Introducing one non-native diamine substrate into the DFOB assembly process results in the production of DFOB itself as the parent metabolite (assembled entirely from native cadaverine: HAC–HSC–HSC) together with seven analogues. These seven analogues can be classified into three groups: Group I, II, III (Fig. 1, left column), as shown for the non-native substrate 1,4-diamino-2(E)-butene (E-DBE) (Telfer et al., 2016). Group I contains three members, each of which contain one diamine region built from the non-native precursor, and two diamine regions built from native cadaverine. Since DFOB is asymmetric, the diamine region built from the non-native precursor can be positioned either in the N-acetyl region, or the internal region, or the primary amine region, giving rise to three constitutional isomers. This constitutional isomerism prompted the design and use of a binary naming system, where the native diamine region was identified as '0', the non-native diamine region was identified as '1', and the order as written from left-to-right mapped to the N-acetyl region, the internal region, and the primary amine region (Telfer et al., 2016). Applying this system gives DFOB itself as [0 0 0], with the three constitutional isomers in Group I as [0 0 1], [0 1 0], and [1 0 0]. Group II also contains three constitutional isomers, with members in this case combining two diamine regions built from the non-native precursor and one diamine region built from native cadaverine, with the binary naming system identifying these as [0 1 1], [1 1 0], and [1 0 1]. Group III has one member in which all three of the diamine regions are built from the non-native precursor, named [1 1 1].

1.4 Mass spectrometry to identify desferrioxamine B analogues and constitutional isomers

Liquid chromatography-mass spectrometry traces from semi-purified culture supernatants from PDB experiments using either E-DBE (Fig. 2B) or oxybis (ethanamine) (OBEA) (Fig. 2D) as the exogenous diamine substrate show the partial resolution of the sets of constitutional isomers. Traces from the native control experiments for both systems (Fig. 2A and Fig. 2C) show a major signal at a retention time of 35 min attributable to DFOB (**1**). The lower intensity signal at retention time 34 min is attributable to $DFOA_1[0\,0\,1]$ in which the

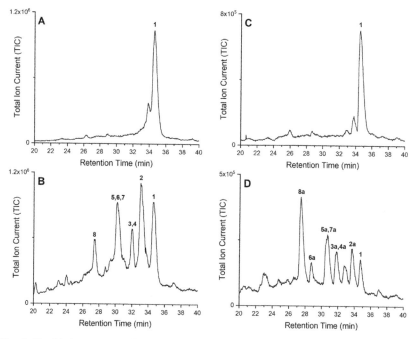

Fig. 2 Liquid-chromatography–mass spectrometry chromatogram by total ion current (TIC) detection from the semi-purified supernatant of *Streptomyces pilosus* cultured in the (**A**) absence, or (**B**) presence of exogenous 1,4-diamino-2(*E*)-butene (*E*-DBE); or the (**C**) absence, or (**D**) presence of exogenous oxybis(ethanamine) (OBEA). Native DFOB is **1**. Constitutional isomers containing *E*-DBE-derived inserts are **2–8**. Analogous constitutional isomers containing OBEA-derived inserts are **2a–8a**. From (A) Telfer, T. J., Gotsbacher, M. P., Soe, C. Z., & Codd, R. (2016). Mixing up the pieces of the desferrioxamine B jigsaw defines the biosynthetic sequence catalyzed by DesD. ACS Chemical Biology, 11(5), 1452–1462. https://doi.org/10.1021/acschembio.6b00056 (Copyright © 2016, American Chemical Society) and Figure adapted with permission from (B) Richardson-Sanchez, T., Tieu, W., Gotsbacher, M. P., Telfer, T. J., & Codd, R. (2017). Exploiting the biosynthetic machinery of Streptomyces pilosus to engineer a water-soluble zirconium(IV) chelator. Organic & Biomolecular Chemistry, 15(27), 5719–5730. https://doi.org/10.1039/C7OB01079F with permission from the Royal Society of Chemistry.

terminal amine region is occupied by *N*-hydroxy-*N*-succinyl-putrescine (HSP) assembled from putrescine derived from the decarboxylation of L-ornithine. In the *E*-DBE system, in addition to the signal at 35 min due to DFOB, there were four well-resolved signals containing the analogues **2–8** (Fig. 1, left column) either as unique species or as co-eluting mixtures. In the OBEA system, a signal for DFOB (**1**) was present at 35 min together with a set of well resolved signals representing analogues with native cadaverine-derived units replaced with OBEA-derived units (**2a–8a**). The distribution of constitutional isomers within a set of analogues can be influenced by the properties of the non-native substrate. The chromatographic profile of the *E*-DBE (Fig. 2B) and OBEA (Fig. 2D) analogue sets show differences in the resolution and relative concentrations of the constitutional isomers, which indicates subtle differences in the mechanistic processing of the respective non-native substrates by the DesBCD biosynthetic cluster.

MS/MS fragmentation can be used to inform the assignment of constitutional isomers (Fig. 1, right column, and Fig. 3). Selected ion monitoring (SIM) is first used to detect signals from the total ion current (TIC) by setting SIM values that correspond with the [M+H]$^+$ adduct of the proposed compound (Fig. 3, panel at left). MS/MS data acquisition across these signals gives the signal for the parent [M+H]$^+$ adduct (Fig. 3, panel at right) which matched **1** (Fig. 3A) or **8** (Fig. 3E) as unique species, or as in the case of **5–7**, as mixtures of constitutional isomers (Fig. 3D). The set of Group I constitutional isomers (**2–4**) was partially resolved into a major signal (Fig. 3B) and a minor signal (Fig. 3C). MS/MS fragmentation patterns can be used to assign the distribution of constitutional isomers present

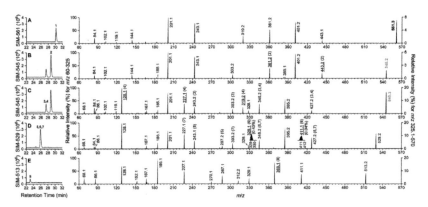

Fig. 3 MS/MS fragmentation spectra (column at right) from LC–MS signals (column at left) from the semi-purified supernatant of *Streptomyces pilosus* cultured in the presence of exogenous 1,4-diamino-2(*E*)-butene (*E*-DBE) measured at selected ion monitoring (SIM) values consistent with **1** (**A**), **2** (**B**), **3** and **4** (unresolved mixture of two species) (**C**), **5**, **6** and **7** (unresolved mixture of three species) (**D**), and **8** (**E**). The MS/MS panels show signals due to [M+H]$^+$ adduct(s) (**1**, black; **2–4**, orange, **5–7**, pink; **8**, red) and fragments unique to (purple, underlined) or useful (purple) (coupled with other observations) in assigning **2–8**. *Adapted and reprinted with permission from Telfer, T. J., Gotsbacher, M. P., Soe, C. Z., & Codd, R. (2016). Mixing up the pieces of the desferrioxamine B jigsaw defines the biosynthetic sequence catalyzed by DesD. ACS Chemical Biology, 11(5), 1452–1462. https://doi.org/10.1021/acschembio.6b00056 (Copyright © 2016, American Chemical Society).*

in each resolved signal. In the E-DBE system, the MS/MS fragmentation pattern of the major signal at 33 min ([M+H]$^+$, m/z 545.2) was characteristic of uDFOA$_1$[0 0 1] (**2**) (where u=unsaturated, A$_1$=one substrate exchange), with fragments at 361.2 and 443.2 unique to this Group I constitutional isomer. This isomer was resolved as a single species. The MS/MS fragmentation pattern of the signal at 23 min ([M+H]$^+$, m/z 513.2) gave a fragment at m/z 369.1 consistent with and unique to uDFOA$_3$[1 1 1] (**8**), which eluted as a single species. The remaining two constitutional isomers of the Group I set (uDFOA$_1$[0 1 0] (**3**) and uDFOA$_1$[1 0 0] (**4**)) co-eluted at 27.2 min. The three constitutional isomers of the Group II set (uDFOA$_2$[0 1 1] (**5**), uDFOA$_2$[1 1 0] (**6**) and uDFOA$_2$[1 0 1] (**7**)) co-eluted at 26 min.

1.5 Scope of precursor-directed biosynthesis in producing desferrioxamine B analogues

A set of studies has demonstrated the viability of the PDB approach, with several classes of DFOB analogues produced containing olefin bonds (Telfer et al., 2016), ether atoms (Richardson-Sanchez et al., 2017), a mixture of ether and thioether atoms (Richardson-Sanchez et al., 2018), fluorine atoms (Telfer & Codd, 2018), or an unusual disulphide bond (Richardson-Sanchez & Codd, 2018) (Fig. 4, Table 1). This demonstrates the broad substrate selectivity of the DesBCD biosynthetic cluster. PDB studies using exogenous 1,4-diaminobutane (DB) (putrescine) have been conducted with *S. pilosus*, which generated higher relative concentrations of DFOA-type analogues in which the native 1,5-diaminopentane (cadaverine) substrate was exchanged for the non-native DB-processed substrate (Gotsbacher & Codd, 2020; Telfer et al., 2016).

Since the trimeric structure and overall charge of a given analogue is the same within an analogue set (Fig. 2B: **1–8**; Fig. 2D: **1**, **2a–8a**) it is reasonable to use the area under the curve (AUC) of the LC-MS trace as a surrogate of concentration. The resolution of the analogues from parent DFOB by reverse-phase LC enabled an estimate of compound yield. For the E-DBE system, the isolated yield from a 50-mL culture of analogues (excluding DFOB, produced at 4.1 mg) was about 20 mg, with uDFOA$_1$[0 0 1] (**2**) at 5.5 mg, uDFOA$_1$[0 1 0] (**3**) and uDFOA$_1$[1 0 0] (**4**) at 3.7 mg (mixture of two unresolved analogues), uDFOA$_2$[0 1 1] (**5**), uDFOA$_2$[1 1 0] (**6**) and uDFOA$_2$[1 0 1] (**7**) at 6.4 mg (mixture of three unresolved analogues), and uDFOA$_3$[1 1 1] (**8**) at 4.4 mg (Telfer et al., 2016). These isolated yields were broadly similar with the predicted yields from AUC measurements. While

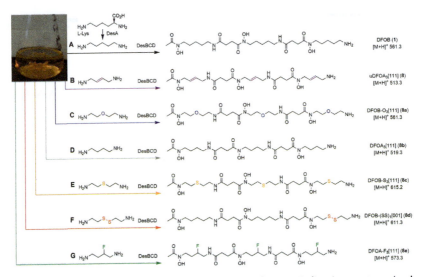

Fig. 4 The decarboxylation of L-lysine by DesA produces 1,5-diaminopentane (cadaverine, A) as the native diamine precursor processed by DesBCD to give DFOB (1). The addition of a native diamine (putrescine, D) or non-native diamines, containing an olefin region (B), or ether (C), thioether (E), disulphide (F) or fluorine (G) atoms, to culture medium inoculated with *Streptomyces pilosus*, produced a range of constitutional DFOB isomers.

the *E*-DBE system produced uDFOA₃[1 1 1] (8) in a relatively modest yield, the equivalent analogue showing the complete exchange of native for non-native substrate in a PDB experiment using OBEA yielded DFOB-O₃[1 1 1] (8a) as the major species (Fig. 2D). Analogue DFOB-O₃[1 1 1] (8a) was isolated from a 50-mL culture in a yield of 8.6 mg. Processing multiple 50-mL cultures enabled sufficient yields of DFOB-O₃[1 1 1] (8a) be obtained to conduct ^{1}H and ^{13}C nuclear magnetic resonance (NMR) spectroscopic characterisation and to measure water solubility. The project design selected OBEA as a substrate predicted to produce ether-containing analogues of DFOB with greater water solubility than DFOB itself. The rationale was that subsequent semi-synthetic amidation reactions using DFOB-O₃[1 1 1] (8a) would offset the significant reduction in water solubility observed for parallel reactions using DFOB (1). The water solubility of DFOB-O₃[1 1 1] (8a) was 45-times greater than DFOB (1), which met the project intent and supported the production of a water-soluble chain-extended chelator with potential for use with zirconium-89 which is a radionuclide in development for use in positron emission tomography imaging (Richardson-Sanchez et al., 2017).

Table 1 Overview of precursor-directed synthesis (PBD) in the production of analogues of desferrioxamine B (DFOB).

Non-native substrate	Abbreviation	Motif	No[a]	References
1,4-Diamino-2(*E*)-butene	*E*-DBE	Olefin bond	8	Telfer et al. (2016)
1,4-Diaminobutane	DB	Methylene reduction	8	Gotsbacher and Codd (2020) and Telfer et al. (2016)
Oxybis(ethanamine)	OBEA	Ether	8	Richardson-Sanchez et al. (2017)
Thiobis(ethanamine)	TBEA	Thioether	8	Richardson-Sanchez et al. (2018)
Oxybis(ethanamine) and Thiobis(ethanamine)	OBEA, TBEA	Ether, and/or thioether	27	Richardson-Sanchez et al. (2018)
Disulfanebis(ethanamine) (cystamine)	CS	Disulphide	2	Richardson-Sanchez and Codd (2018)
rac-1,4-Diamino-2-fluorobutane	*rac*-FDB	Fluorine	>8[b]	Telfer and Codd (2018)

[a]Number of analogues produced including DFOB itself.
[b]Additional sets of analogues produced as regio-isomers due to the asymmetric substitution pattern in this non-native substrate.

The use of disufanebis(ethanamine) (common name: cystamine (CS)) as a non-native substrate in a DFOB analogue production campaign was higher-risk, since the properties of CS (molecular volume, $149 \, \text{Å}^3$; dipole moment, 2.83 D) differed significantly from native 1,5-diaminopentane (molecular volume, $128 \, \text{Å}^3$; dipole moment, 1.57 D). The most significant variation was the bent structure of CS (C–S–S–C dihedral angle, 100.1°) compared to linear DP, which might seem to challenge the DesBCD biosynthetic cluster. In the CS experiment, DFOB was produced in addition to only one analogue, in which the CS-processed substrate was inserted at the amine-containing region. Analogue DFOB-$(SS)_1[0\,0\,1]$ was incubated in the presence of dithiothreitol, which cleaved the embedded disulphide bond to liberate 2-aminoethanethiol and a modified DFO unit that retained functional integrity in binding Fe(III) (Richardson-Sanchez & Codd, 2018). This approach indicated potential for using the reduced form of DFOB-$(SS)_1[0\,0\,1]$ in gold-based self-assembled monolayer immobilisation platforms for biosurveillance or biosensing. Due to limited yields of DFOB-$(SS)_1[0\,0\,1]$ accessible from a PDB approach, subsequent work used a synthetic approach to generate a thiol-derivatised DFOB analogue to enable resin immobilisation for a pulldown study (Ni et al., 2023). Since the CS substrate was installed at the amine region of DFOB, it might also be useful to append an antibiotic to this region, with the possibility of seeing a reduction-triggered release of antibiotic following uptake of the assembly by a given pathogen. This strategy harbours risk should the reduction occur premature to reaching its destination (Neumann & Nolan, 2018).

1.6 Theoretical maxima of desferrioxamine B analogues produced using precursor-directed biosynthesis

The wide success in generating analogues of DFOB using PDB prompted consideration of the mathematical relationship between the maximum number of analogues that could be produced as a function of the number of exogenous substrates employed and the number of unique regions in the product for substrate insertion. The following discussion relates to the assembly of DFOB analogues built from substrates with C_{2v} symmetry (Fig. 4A–F). In the case of a PDB experiment with DFOB and one exogenous non-native diamine substrate, the maximum number of target compounds that can be produced is 8 (2^3) (Fig. 5A). The base term describes the total number of available substrates (as a sum of native and non-native substrates), and the power term describes the number of structurally unique regions present in the target metabolite that can be occupied by processed substrates. Conducting the PDB experiment focused on producing DFOB

The production of siderophore analogues using precursor-directed biosynthesis

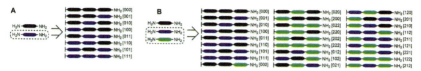

Fig. 5 Theoretical maxima and experimentally observed constitutional isomers of desferrioxamine B (DFOB) produced from PDB campaigns using (**A**) one (blue), or (**B**) two (blue, green) exogenous non-native diamine substrates in competition with the native substrate (black). Substrates (C_{2v} symmetry) are coded 0 (black), 1 (blue) or 2 (green), with the sequence defined in square brackets as ordered from the *N*-acetyl group to the terminal amine.

analogues using two exogenous non-native diamine substrates can furnish a theoretical maximum number of 27 (3^3) DFOB analogues (Fig. 5B). In both the above cases, one member is the parent metabolite itself. In the case where a 1:1 mixture of OBEA and TBEA was supplemented into bacteriological medium inoculated with *S. pilosus*, a total of 27 DFOB analogues were generated (one of which was DFOB itself), with the pattern of native (cadaverine=0) and non-native (OBEA=1, TBEA=2) substrate inserts as predicted by the mathematical relationship (Fig. 5B). This set of mixed ether and thioether DFOB analogues was produced in analytical yields, but nonetheless conveyed the notion of using PDB to dial up a particular mixed-substrate analogue, akin to a Rubik's cube (Richardson-Sanchez et al., 2018). The use of substrates with lower symmetry or with chiral centres (Fig. 4G) introduces a further level of regio-isomerism, in which the substituent (eg, fluorine atom) can appear either proximal or distal to a reference point within a given insert region. This generates a larger population of analogues that may not be resolvable by reverse phase liquid chromatography (Telfer & Codd, 2018).

Here, we detail the technical aspects of conducting a PDB experiment to generate non-native analogues of DFOB, with these methods likely to have general applicability to other classes of secondary metabolites assembled from diamine building blocks.

2. Materials and equipment

2.1 Bacteria and chemicals

- *S. pilosus* Ettlinger et al. (ATCC 19797)
- Difco yeast mold (YM) broth (BD Biosciences; product code: 271120)
- Hydrochloric acid (37%; Sigma-Aldrich; product code: 320331)

- Sodium hydroxide (≥99%; Sigma–Aldrich; product code: S5881)
- Dimethyl sulfoxide (DMSO) (≥99.9%; Sigma–Aldrich; product code: D8418)
- Chelex 100 sodium form (Sigma–Aldrich; product code: C7901)
- Potassium phosphate monobasic (≥99.0%; Sigma–Aldrich; product code: P5655)
- Sodium phosphate dibasic (≥99.0%; Sigma–Aldrich; product code: S9763)
- L-Threonine (≥99.5%; Sigma–Aldrich; product code: 89179)
- Magnesium sulphate heptahydrate (≥98%; Ajax Finechem; product code: AJA302)
- Calcium chloride (≥98%; Ajax Finechem; product code: AJA960)
- Zinc sulphate heptahydrate (≥98%; Sigma–Aldrich; product code: Z0251)
- Tris(hydroxymethyl)aminomethane (Trizma base) (≥99.9%; Sigma–Aldrich; product code: T6066)
- 1,4-Diamino-2(*E*)-butene dihydrochloride (*E*-DBE) (>96%; Small Molecules Inc; 18-2859)
- 1,4-Diaminobutane dihydrochloride (putrescine; DB) (≥98%; Sigma–Aldrich; product code: P7505)
- 2,2′-Oxybis(ethanamine) dihydrochloride (OBEA) (97%; Sigma–Aldrich; product code: 176095)
- 2,2′-Thiobis(ethanamine) (TBEA) (>98.0%; Tokyo Chemical Company; product code: T1538)
- 2,2′-Disulfanebis(ethanamine) dihydrochloride (cystamine; CS) (≥98%; Sigma–Aldrich; product code: C8707)
- *rac*-1,4-Diamino-2-fluorobutane dihydrochloride (*rac*-DFB) (Innovapharm; product code: RC280011541)
- Iron(III) chloride hexahydrate (≥98%; Ajax Finechem; AJA743)
- Chromeazurol S (CAS) (45%; Sigma–Aldrich; product code: 199532)
- Piperazine (99%; Sigma–Aldrich; product code: P45907)
- 5-Sulfosalicylic acid dihydrate (≥99%; Sigma–Aldrich; product code: S2130)
- Iron(III) perchlorate hydrate (≥99%; Sigma–Aldrich; product code: 309281)
- Perchloric acid (70%; Sigma–Aldrich; product code: 244252)
- Amberlite XAD−2 (Sigma Aldrich; product code: SU853005)
- Methanol (≥99%; Ajax Finechem; product code: AJA318)
- Ni(II) Sepharose 6 Fast Flow (GE Healthcare; product code: 17531802)
- 4-(2-Hydroxyethyl)−1-piperazineethanesulfonic acid (HEPES) (AMRESCO; product code: 0.511)
- Sodium chloride (≥99%; Sigma–Aldrich; product code: S5886)
- Methanol (≥99%; Sigma–Aldrich; product code: 439193)
- Acetonitrile (190 grade; Ajax Finechem; AJA2315)

2.2 Consumables

- Eppendorf tubes
- Plastic Erlenmeyer flask (250 mL; VWR International; product code: 89095-266)
- Plastic Erlenmeyer flasks (125 mL; VWR International; product code: 89095-260)
- Minisart syringe filters (0.22 µm; Sartorius; product code: 16532)
- Econo-Column (2.5 cm i.d.×30 cm; BioRad; product code: 7372532)

2.3 General equipment

- FreeZone 4.5 L, −105 °C Benchtop Freeze-dryer (Labconco, Kansas City, MO, USA)
- Hanna HI2210 pH Metre with an HI1083 microelectrode and three calibration standards (pH 4.0, 7.0, 10.0) (Hanna Instruments, Keysborough VIC, Australia)
- FLUOstar Omega microplate reader (BMG Labtech, Mornington, VIC, Australia)
- New Brunswick Innova 42 Shaking Incubator (Eppendorf, Hamburg, Germany)

2.4 LC-MS-Q instrumentation

- Agilent system comprising a 1260 series quaternary pump with an inbuilt degasser, a 1200 series autosampler, a temperature-controlled column compartment, a diode array detector, a fraction collector, and a 6120 series single quadrupole mass spectrometer.
- Agilent OpenLAB chromatography data system ChemStation Edition (B.04.02) software for data acquisition and processing.
- Analytical Agilent Eclipse XDB-C18 reverse-phase prepacked column (particle size: 5 µm, i.d. 4.6×150 mm). More recent work uses an Agilent Eclipse XDB-C18 reverse-phase prepacked column better suited for analytical work (particle size: 3 µm, i.d. 2.1×150 mm).
- Gradient of 0–30% acetonitrile:water (0.1% formic acid v/v) over 40 min and a 0.2 mL min^{-1} flow rate with an injection volume of 10 µL, and a capillary voltage of 4000 V.

2.5 LC-MS/MS-QQQ instrumentation

- Agilent system consisting of a 1290 series quaternary pump with inbuilt degasser, a 1200 series autosampler, a temperature-controlled column compartment, a diode array detector, and a 6460 series triple quadrupole mass spectrometer with jet stream technology, operated in positive

polarity mode. Collision energy voltages were optimised for individual precursor ions and ranged from 16 to 28 V. The fragmentor voltage, drying gas flow, temperature and nebuliser of the mass spectrometer were set to 150 V, 10 mL min^{-1}, 300 °C and 25 psi, respectively.
- The column and LC conditions were identical to those used for LC-MS-Q analysis.
- Agilent MassHunter Workstation LC/MS Data Acquisition and Qualitative Analysis (v. 10.1) software for data acquisition and processing.

3. Protocol

3.1 Before you begin

Conduct your work according to the biosafety classification requirement of the bacterial strain being used, and as guided by the Biosafety in Microbiological and Biomedical Laboratories (BMBL) published by the U.S. Department of Health and Human Services. *S. pilosus* Ettlinger et al. (ATCC 19797) is biosafety level 1 (BSL-1), which describes a low-risk microbe enabling laboratory work to be conducted on benches with no need for isolation or special containment facilities. Any spills should be immediately decontaminated. Bacteriological waste should be treated and disposed of according to the biological waste and safety protocols and regulations established at your organisation, generally involving the use of an autoclave.

3.2 Preparing solutions and use of high grade reagents

Prepare solutions using ultrapure water (deionized water purified to ultrapure water) (18.2 MΩ cm at 25 °C) and using analytical or higher-grade reagents containing the lowest possible metal content.

3.3 Preparing frozen stocks of *Streptomyces pilosus*

1. Use aseptic techniques for the following steps. Open the ampoule containing freeze dried *S. pilosus* Ettlinger et al. ATCC 19797 (American Type Culture Collection) by scoring with a sharp file, disinfecting with a gauze dampened with ethanol, and then snaping the ampoule at the scored area.
2. Sterilise YM broth (BD Biosciences, product number: 271120) (2.1 g of YM powder per 100 mL ultrapure water) by autoclaving (121 °C, 100 kPa, 15 min). Add sterile YM broth (0.5 mL, 2.1% w/v, pH 6.5) to rehydrate the lyophilised *S. pilosus* culture and mix well.

The production of siderophore analogues using precursor-directed biosynthesis   137

3. Transfer the rehydrated culture into sterile YM broth (100 mL, 2.1% w/v, pH 6.5) and incubate the culture on a rotary shaker (160 rpm, 28 °C) for 3 days.

4. Transfer aliquots (1 mL) of the *S. pilosus* culture into Eppendorf tubes. Add 10% v/v of DMSO as a cryoprotectant and store the bacterial stocks at −80 °C..

3.4 Precursor-directed biosynthesis using non-native diamine substrates

1. Add Chelex resin (Chelex 100 sodium form, Sigma-Aldrich, product C7901) to YM broth (2.1% w/v) at 1 g per 100 mL broth. Add a clean magnetic stirrer bar to the flask, cover the flask opening with aluminium foil, and stir the solution using a stirring plate for 2 h.

   Pause point: Siderophores are biosynthesized under conditions of limited iron. To promote the production of siderophores, media and other solutions as specified should be treated with Chelex resin to deplete the concentration of Fe(III).

2. Decant the YM broth from the Chelex resin and adjust the pH value of the broth (using solutions of HCl or NaOH in Chelex-treated water) to pH 6.5.

3. Pour the YM broth (100 mL) into a plastic Erlenmeyer flask (250 mL, VWR International, product number 89095-266) that has been pre-washed with water, seal the flask, and autoclave (121 °C for 15 min).

   Pause point: Siderophores can leach Fe(III) and Al(III) from borosilicate glass, which prescribes the use of plastic flasks as more suitable for culturing siderophore-producing species.

4. Defrost an aliquot (1 mL) of the frozen *S. pilosus* stock and inoculate into the flask containing the sterilised YM broth and incubate for 4 d (160 rpm, 28 °C).

5. Centrifuge the preculture ($3838g$, 20 min) to sediment the cells, and decant the supernatant.

6. Resuspend the cells in an enriched medium (preparation detailed below in steps **7–10**) and distribute evenly into plastic Erlenmeyer flasks (125 mL, VWR International, product number 89095-260) containing enriched medium (48 mL).

7. Prepare YM broth (4.2% w/v) and treat with Chelex resin (as above) and adjust the pH value of the broth to pH 6.5 prior to autoclaving (121 °C for 15 min).

8. Prepare a phosphate buffer solution containing (per 1 L of water): $KH_2PO_4 \cdot 3H_2O$ (85.8 g) and Na_2HPO_4 (2.63 g); and treat with Chelex resin (as above) and adjust the pH value to pH 6.5 before autoclaving.

9. Prepare a solution of enrichment components containing (per L of Chelex-treated water): L-Threonine (0.04 g), $MgSO_4 \cdot 7H_2O$ (0.958 g), $CaCl_2$ (2.42 g), $ZnSO_4 \cdot 7H_2O$ (0.0064 g) and tris(hydroxymethyl) aminomethane (trizma base) (0.993 g) and sterilise filter using a 0.22 μm filter unit (Sartorius Minisart Syringe Filters).

   Pause point: The enrichment components solution is prepared using Chelex-treated water but the final solution itself is not treated with Chelex-resin, since this would remove the required trace metal ions.

10. Prepare the enriched medium by combining the YM media (4.2% w/v), the phosphate buffer solution, and the solution of enrichment components in a 2:1:1 volumetric ratio in a biosafety cabinet.

    Pause point: A preliminary study should be conducted to determine the tolerance of the bacterium towards the chosen non-native diamine substrate. Cultures (50 mL) should be grown in medium covering a range of non-native diamine substrate concentrations (e.g., 2, 5, 10, and 20 mM) and the production of siderophores measured using a Chrome Azurol S (CAS) or Fe(III) addition assay (Section 3.5). This will guide the maximal non-native diamine substrate concentration used to promote the production of new analogues. In the case of *S. pilosus*, a concentration of 20 mM of the set of non-native diamine substrates evaluated to date (Fig. 4) can be tolerated (e.g., *E*-DBE, 20 mM; DB, 20 mM; OBEA, 20 mM; and TBEA, 20 mM). In some instances, the cost of the substrate could prompt a decision to operate using a lower concentration (e.g., OBEA, 10 mM; CS, 10 mM; and *rac*-FDB, 10 mM).

11. Prepare a stock solution of the exogenous non-native diamine substrate in 2 mL ultrapure water at a final concentration specific to the given substrate (*e.g.*, *E*-DBE·2HCl: 0.1591 g to give a 500 mM stock solution; or OBEA·2HCl: 0.0885 g to give a 250 mM stock solution), adjust the pH value to pH 6.5 and sterilise filter the solution using a 0.22 μm filter unit.

12. Add an aliquot (2 mL) of the non-native diamine substrate stock solution to the culture medium (48 mL) inoculated with *S. pilosus* to give the desired final concentration of the non-native diamine substrate (eg, *E*-DBE, 20 mM; OBEA, 10 mM) and incubate for 8–10 d (160 rpm, 28 °C).

The production of siderophore analogues using precursor-directed biosynthesis   139

Pause point: The optimal pH value for the growth of *S. pilosus* and the production of siderophores is pH 6.5. The addition of solutions of diamine precursors (often provided as hydrochloride salts) to bacteriological medium inoculated with *S. pilosus* requires careful attention to maintain this optimal pH value.

3.5 Measuring siderophore production

Siderophores can be detected using a CAS assay (Schwyn & Neilands, 1987; Singh, Hider, & Porter, 1990) or by an Fe(III) addition assay. Although the Fe(III) addition assay is simpler to conduct, it is not suitable for use with all sample matrices, due to interference effects. In these cases, the CAS assay is preferred. Literature protocols should be followed to prepare the solutions required for the CAS assay and to conduct the CAS assay, which will indicate the presence of siderophores in a sample solution upon a colour change from blue to pink, as measured by visual change or using a quantitative approach with a plate reader (wavelength 630 nm). The steps to conduct the Fe(III) addition assay are given below.

1. Add an aliquot (100 μL) of Fe(III) addition assay solution (10 mM ferric perchlorate in 0.2 M perchloric acid) to a sub-sample (200 μL) of culture supernatant in a 1.5-mL Eppendorf tube.
2. Centrifuge (13,800g for 15 min) and transfer the supernatant to a 96-well plate and after a 10 min incubation period at room temperature read the absorbance at 500 nm using a plate reader. An alternative is to replace the perchlorate system with 10 mM ferric chloride in 0.2 M hydrochloric acid.
3. Remove an aliquot (1 mL) of the culture supernatant each day from day 0 (time of inoculation) to day 8 and monitor the production of siderophores using the CAS assay or the Fe(III) addition assay as described in steps 1 and 2 above.

3.6 Siderophore purification

1. Harvest the cultures by centrifugation (3838*g*, 20 min) on day 8 or when the production of siderophores has reached a plateau.
2. Add a sample of Amberlite XAD-2 resin (100 mL) to methanol (200 mL) and stir the suspension for 15 min and collect the resin by suction filtration.
3. Wash the XAD-2 resin with water (2×200 mL, 15 min) and then pour the suspension into a Econo-Column (2.5 cm i.d.×30 cm, BioRad) and allow the resin bed to settle under gravity flow. Use a backwash

approach (inject a syringe filled with ultrapure water to the exit nozzle of the column to mobilise the resin beads and allow to settle) to give a homogeneous resin bed distribution.

4. Wash the XAD-2 resin with water (200 mL) with a flow rate (under gravity, adjusted manually by the exit nozzle) of 5 mL min^{-1}.

5. Adsorb the culture supernatant onto the XAD-2 resin bed (50 mL supernatant per 100 mL XAD-2 resin bed) and wash the resin bed with water (200 mL) to remove unbound components.

6. Wash the resin bed with aqueous methanol (50% v/v, 200 mL) collecting fractions of 40 mL containing the siderophores.

   Pause point: This procedure can be adapted from a column setting to a batch extraction mode for large-scale siderophore production.

7. Analyse the fractions for the presence of siderophores using the CAS assay and lyophilise the siderophore-positive fractions to dryness.

8. Prepare a column (Econo-Column: 2.5 cm i.d.×30 cm, BioRad) containing Ni(II) Sepharose resin (column volume=40 mL) and wash the resin bed with water (200 mL) and then with binding buffer (200 mL: 10 mM HEPES, 0.2 M NaCl, pH 9.0) to equilibrate the resin bed.

   Pause point: Hydroxamic acid-based siderophores can be purified using immobilised metal affinity chromatography (IMAC) operating by the metal binding property of siderophores (Braich & Codd, 2008; Gotsbacher & Codd, 2020; Richardson-Sanchez & Codd, 2018; Richardson-Sanchez et al., 2017; Richardson-Sanchez et al., 2018; Telfer & Codd, 2018; Telfer et al., 2016).

9. Dissolve the lyophilised powder (step 7) in 2 mL of binding buffer and adsorb the solution onto the Ni(II) Sepharose resin bed.

10. Wash the resin bed with binding buffer (200 mL) to remove unbound components and then wash the resin bed with elution buffer to displace bound siderophores (200 mL, 10 mM HEPES, 0.2 M NaCl, pH 5.5) and collect fractions of 40 mL.

11. Analyse the fractions for siderophores using the CAS assay. Pool the siderophore-positive fractions and remove the solvent *in vacuo* using a clean rotary evaporator (Buchi Rotovapor R-300).

    Pause Point: This procedure can be adapted from a column setting as described to the use of spin columns pre-packed with Ni(II)-iminodiacetic acid IMAC resin (G-Biosciences).

12. Add HPLC-grade methanol (10 mL) to the residue (containing siderophores and buffer components) from Step **11** and mix thoroughly by

The production of siderophore analogues using precursor-directed biosynthesis 141

hand to extract the siderophores into the methanol phase. Transfer the methanol phase into a clean round bottom flask and repeat the methanol extraction procedure three times (total of 4×10 mL) before removing the methanol *in vacuo* using a rotary evaporator (Buchi Rotovapor R-300).

13. Dissolve the residue in a minimal volume of ultrapure water (guide: 0.3–1 mL depending on siderophore content) in preparation for analysis using LC–MS.

3.7 Analysing desferrioxamine B analogues and constitutional isomers

1. Inject 10 µL samples of the desalted semi-purified siderophore samples prepared in ultrapure water into the LC-MS system set to operate using a 0–30% acetonitrile:water gradient over 40 min with a flow rate of 0.2 mL min^{-1}.

2. Analyse individual signals by MS/MS fragmentation by running a product ion method in positive polarity. The m/z value corresponding to the exact mass of a singly protonated target compound ($[M+H^+]^+$) is denoted as "precursor ion" (for MS1). For MS/MS parameters, the "product ion" scan range is from m/z 50 to roughly 50 Da above the precursor m/z, with a step size of 0.1 Da in "profile" mode and a scan time of 500 µsec. The ideal collision energy (CE) is specific for each precursor ion and needs to be determined individually through repeat injections or the integrated MassHunter Optimiser application. For desferrioxamine-related metabolites the CE ranges from 16 to 28 V under the employed conditions, to achieve near-complete fragmentation of the precursor ion. In the resulting MS/MS spectrum, the relative peak height of the precursor ion should sit at 5–10% of the most abundant (100%) fragment ion.

3. View the acquired data in MassHunter Qualitative Analysis software, export as .csv files and analyse/display further in the preferred graphing software.

Pause point: The Fe(III) binding property of the mixture of siderophores can be measured by adding Fe(III) to the mixture and re-injecting the sample into the LC–MS system. In most cases, the retention time of the Fe (III)-siderophore complex will be different from the free siderophore. It is not possible to obtain MS/MS fragmentation signatures of Fe(III)-siderophore complexes.

4. Summary

The production of analogues of bacterial metabolites using PDB has a rich history, with more recent use of this method declining likely due to a narrow range of viable competitor substrates, modest yields, and the need to resolve analogues from the parent metabolite. Despite these shortcomings, the intellectual benefits of compound discovery, new insight into the nuances of biosynthetic pathways, and developing rigorous bioanalytical measurements, analyses, and nomenclature systems to identify and classify analogues and regio-isomers, maintains the appeal inherent to the method, coupled with the advantage of requiring readily accessible technical requirements. This body of work and technical description has focused on hydroxamic acid siderophores, in particular analogues of DFOB. Other work has used PDB to produce analogues of catechol-based siderophores (Reitz & Butler, 2020), as a demonstration of scope. Ultimately, the PDB method harbours rich potential for both academic research and for engaging undergraduate students in the wonders of natural products and chemical biology research.

Acknowledgements

The Australian Research Council (DP180100785, DP220100101) and the University of Sydney (MBI Seed Grant 2018, Drug Discovery Institute-Sydney ID Seed Funding 2021) are acknowledged for financial support.

References

Barona-Gómez, F., Wong, U., Giannakopulos, A. E., Derrick, P. J., & Challis, G. L. (2004). Identification of a cluster of genes that directs desferrioxamine biosynthesis in *Streptomyces coelicolor* M145. *Journal of the American Chemical Society, 126*(50), 16282–16283. https://doi.org/10.1021/ja045774k.

Barry, S. M., & Challis, G. L. (2009). Recent advances in siderophore biosynthesis. *Current Opinion in Chemical Biology, 13,* 205–215. https://doi.org/10.1016/j.cbpa.2009.03.008.

Braich, N., & Codd, R. (2008). Immobilized metal affinity chromatography for the capture of hydroxamate-containing siderophores and other Fe(III)-binding metabolites from bacterial culture supernatants. *Analyst, 133*(7), 877–880. https://doi.org/10.1039/B802355G.

Challis, G. L. (2005). A widely distributed bacterial pathway for siderophore biosynthesis independent of nonribosomal peptide synthetases. *Chembiochem: A European Journal of Chemical Biology, 6*(4), 601–611. https://doi.org/10.1002/cbic.200400283.

Codd, R. (2008). Traversing the coordination chemistry and chemical biology of hydroxamic acids. *Coordination Chemistry Reviews, 252*(12–14), 1387–1408. https://doi.org/10.1016/j.ccr.2007.08.001.

Codd, R. (2023). Siderophores and iron transport. In V. L. Pecoraro, & Z. Guo (Vol. Eds.), *Comprehensive inorganic chemistry III: Vol. 2,* (pp. 3–29). (pp. 3)Oxford: Elsevier. https://doi.org/10.1016/B978-0-12-823144-9.00044-3.

Codd, R., Richardson-Sanchez, T., Telfer, T. J., & Gotsbacher, M. P. (2018). Advances in the chemical biology of desferrioxamine B. *ACS Chemical Biology, 13*(1), 11–25. https://doi.org/10.1021/acschembio.7b00851.

Codd, R., Soe, C. Z., Pakchung, A. A. H., Sresutharsan, A., Brown, C. J. M., & Tieu, W. (2018). The chemical biology and coordinaton chemistry of putrebactin, avaroferrin, bisucaberin, and alcaligin. *Journal of Biological Inorganic Chemistry, 23*(7), 969–982. https://doi.org/10.1007/s00775-018-1585-1.

Dhungana, S., White, P. S., & Crumbliss, A. L. (2001). Crystal structure of ferrioxamine B: A comparative analysis and implications for molecular recognition. *Journal of Biological Inorganic Chemistry, 6*(8), 810–818. https://doi.org/10.1007/s007750100259.

Ejje, N., Soe, C. Z., Gu, J., & Codd, R. (2013). The variable hydroxamic acid siderophore metabolome of the marine actinomycete *Salinispora tropica* CNB-440. *Metallomics, 5*(11), 1519–1528. https://doi.org/10.1039/C3MT00230F.

Evers, A., Hancock, R. D., Martell, A. E., & Motekaitis, R. J. (1989). Metal ion recognition in ligands with negatively charged oxygen donor groups. Complexation of Fe (III), Ga(III), In(III), Al(III), and other highly charged metal ions. *Inorganic Chemistry, 28*(11), 2189–2195. https://doi.org/10.1021/ic00310a035.

Gotsbacher, M. P., & Codd, R. (2020). Azido-desferrioxamine siderophores as functional click chemistry probes generated in culture upon adding a diazo-transfer reagent. *Chembiochem: A European Journal of Chemical Biology, 21*(10), 1433–1455. https://doi.org/10.1002/cbic.201900661.

Hardy, J., & Butler, A. (2018). β-Hydroxyaspartic acid in siderophores: Biosynthesis and reactivity. *Journal fo. Biological Inorganic Chemistry, 23*, 957–967. https://doi.org/10.1007/s00775-018-1584-2.

Hider, R. C., & Kong, X. (2010). Chemistry and biology of siderophores. *Natural Product Reports, 27*(5), 637–657. https://doi.org/10.1039/b906679a.

Kadi, N., Oves-Costales, D., Barona-Gómez, F., & Challis, G. L. (2007). A new family of ATP-dependent oligomerization-macrocyclization biocatalysts. *Nature Chemical Biology, 3*(10), 652–656. https://doi.org/10.1038/nchembio.2007.23.

Kalkreuter, E., Carpenter, S. M., & Williams, G. J. (2018). Precursor-directed biosynthesis and semi-synthesis of natural products. In N. J. Westwood, & A. Nelson (Eds.). *Chemical and biological synthesis: Enabling approaches for understanding biology* (pp. 275–312) Cambridge, UK: Royal Society of Chemistry. https://doi.org/10.1039/9781788012805-00275.

Kirschning, A., Eichner, S., Hermane, J., & Knobloch, T. (2012). Mutant manufacturers. In O. Genilloud, & F. Vicente (Vol. Eds.), *Drug discovery from natural products: Vol. 25*, (pp. 58–78). (pp. 58)Cambridge, UK: The Royal Society of Chemistry. https://doi.org/10.1039/9781849733618-00058.

Kirschning, A., Taft, F., & Knobloch, T. (2007). Total synthesis approaches to natural product derivatives based on the combination of chemical synthesis and metabolic engineering. *Organic & Biomolecular Chemistry, 5*, 3245–3259. https://doi.org/10.1039/B709549J.

Konetschny-Rapp, S., Jung, G., Raymond, K. N., Meiwes, J., & Zaehner, H. (1992). Solution thermodynamics of the ferric complexes of new desferrioxamine siderophores obtained by directed fermentation. *Journal of the American Chemical Society, 114*(6), 2224–2230. https://doi.org/10.1021/ja00032a043.

Markham, T. E., & Codd, R. (2024). A mild and modular approach to the total synthesis of desferrioxamine B. *Journal of Organic Chemistry.* https://doi.org/10.1021/acs.joc.3c02739.

Meiwes, J., Fiedler, H.-P., Zähner, H., Konetschny-Rapp, S., & Jung, G. (1990). Production of desferrioxamine E and new analogues by directed fermentation and feeding fermentation. *Applied Microbiology and Biotechnology, 32*(5), 505–510. https://doi.org/10.1007/BF00173718.

Neumann, W., & Nolan, E. M. (2018). Evaluation of a reducible disulfide linker for siderophore-mediated delivery of antibiotics. *Journal of Biological Inorganic Chemistry*, *23*(7), 1025–1036. https://doi.org/10.1007/s00775-018-1588-y.

Ni, J., Wood, J. L., White, M. Y., Lihi, N., Markham, T. E., Wang, J., et al. (2023). Reduction-cleavable desferrioxamine B pulldown system enriches Ni(II)-superoxide dismutase from a *Streptomyces* proteome. *RSC Chemical Biology*, *4*(12), 1064–1072. https://doi.org/10.1039/d3cb00097d.

Puja, H., Mislin, G. L. A., & Rigouin, C. (2023). Engineering siderophore biosynthesis and regulation pathways to increase diversity and availability. *Biomolecules*, *13*(6), 959. https://doi.org/10.3390/biom13060959.

Reitz, Z. L., & Butler, A. (2020). Precursor-directed biosynthesis of catechol compounds in *Acinetobacter bouvetii* DSM 14964. *Chemical Communications*, *56*(81), 12222–12225. https://doi.org/10.1039/D0CC04171H.

Richardson-Sanchez, T., & Codd, R. (2018). Engineering a cleavable disulfide bond into a natural product siderophore using precursor-directed biosynthesis. *Chemical Communications*, *54*(70), 9813–9816. https://doi.org/10.1039/C8CC04981E.

Richardson-Sanchez, T., Nolan, K. P., & Codd, R. (2018). Rubik's cube of siderophore assembly established from mixed-substrate precursor-directed biosynthesis. *ACS Omega*, *3*(12), 18160–18169. https://doi.org/10.1021/acsomega.8b02803.

Richardson-Sanchez, T., Tieu, W., Gotsbacher, M. P., Telfer, T. J., & Codd, R. (2017). Exploiting the biosynthetic machinery of *Streptomyces pilosus* to engineer a water-soluble zirconium(IV) chelator. *Organic & Biomolecular Chemistry*, *15*(27), 5719–5730. https://doi.org/10.1039/C7OB01079F.

Roberts, A. A., Schultz, A. W., Kersten, R. D., Dorrestein, P. C., & Moore, B. S. (2012). Iron acquisition in the marine actinomycete genus *Salinispora* is controlled by the desferrioxamine family of siderophores. *FEMS Microbiology Letters*, *335*(2), 95–103. https://doi.org/10.1111/j.1574-6968.2012.02641.x.

Rütschlin, S., & Böttcher, T. (2018). Dissecting the mechanism of oligomerization and macrocyclization reactions of NRPS-independent siderophore synthetases. *Chemistry - A European Journal*, *24*(60), 16044–16051. https://doi.org/10.1002/chem.201803494.

Rütschlin, S., & Böttcher, T. (2020). Engineering siderophores. In A. K. Shukla (Vol. Ed.), *Chemical and synthetic biology approaches to understand cellular functions—Part C: Vol. 633*, (pp. 29–47). (pp. 29). https://doi.org/10.1016/bs.mie.2019.10.030.

Rütschlin, S., Gunesch, S., & Böttcher, T. (2017). One enzyme, three metabolites: *Shewanella algae* controls siderophore production *via* the cellular substrate pool. *Cell Chemical Biology*, *24*(5), 598–604. https://doi.org/10.1016/j.chembiol.2017.03.017.

Rütschlin, S., Gunesch, S., & Böttcher, T. (2018). One enzyme to build them all: Ring-size engineered siderophores inhibit the swarming motility of *Vibrio*. *ACS Chemical Biology*, *13*(5), 1153–1158. https://doi.org/10.1021/acschembio.8b00084.

Sattler, I., Grabley, S., & Thiericke, R. (1999). Structure modification *via* biological derivitization methods. In S. Grabley, & R. Thiericke (Eds.). *Drug Discovery from Nature* (pp. 191–214)Berlin: Springer-Verlag. https://doi.org/10.1007/978-3-642-60250-4_11.

Schupp, T., Toupet, C., & Divers, M. (1988). Cloning and expression of two genes of *Streptomyces pilosus* involved in the biosynthesis of the siderophore desferrioxamine B. *Gene*, *64*(2), 179–188. https://doi.org/10.1016/0378-1119(88)90333-2.

Schupp, T., Waldmeier, U., & Divers, M. (1987). Biosynthesis of desferrioxamine B in *Streptomyces pilosus*: Evidence for the involvement of lysine decarboxylase. *FEMS Microbiology Letters*, *42*(2–3), 135–139. https://doi.org/10.1111/j.1574-6968.1987. tb02060.x.

Schwyn, B., & Neilands, J. B. (1987). Universal chemical assay for the detection and determination of siderophores. *Analytical Biochemistry*, *160*(1), 47–56. https://doi.org/10.1016/0003-2697(87)90612-9.

Singh, S., Hider, R. C., & Porter, J. B. (1990). Separation and identification of desferrioxamine and its iron chelating metabolites by high-performance liquid chromatography and fast atom bombardment mass spectrometry: Choice of complexing agent and application to biological fluids. *Analytical Biochemistry, 187*(2), 212–219. https://doi.org/10.1016/0003-2697(90)90446-g.

Soe, C. Z., & Codd, R. (2014). Unsaturated macrocyclic dihydroxamic acid siderophores produced by *Shewanella putrefaciens* using precursor-directed biosynthesis. *ACS Chemical Biology, 9*(4), 945–956. https://doi.org/10.1021/cb400901j.

Soe, C. Z., Pakchung, A. A. H., & Codd, R. (2012). Directing the biosynthesis of putrebactin or desferrioxamine B in *Shewanella putrefaciens* through the upstream inhibition of ornithine decarboxylase. *Chemistry & Biodiversity, 9*(9), 1880–1890. https://doi.org/10.1002/cbdv.201200014.

Soe, C. Z., Telfer, T. J., Levina, A., Lay, P. A., & Codd, R. (2016). Simultaneous biosynthesis of putrebactin, avaroferrin and bisucaberin by *Shewanella putrefaciens* and characterisation of complexes with iron(III), molybdenum(VI) or chromium(V). *Journal of Inorganic Biochemistry, 162*, 207–215. https://doi.org/10.1016/j.jinorgbio.2015.12.008.

Telfer, T. J., & Codd, R. (2018). Fluorinated analogues of desferrioxamine B from precursor-directed biosynthesis provide new insight into the capacity of DesBCD. *ACS Chemical Biology, 13*(9), 2456–2471. https://doi.org/10.1021/acschembio.8b00340.

Telfer, T. J., Gotsbacher, M. P., Soe, C. Z., & Codd, R. (2016). Mixing up the pieces of the desferrioxamine B jigsaw defines the biosynthetic sequence catalyzed by DesD. *ACS Chemical Biology, 11*(5), 1452–1462. https://doi.org/10.1021/acschembio.6b00056.

Thiericke, R., Langer, H.-J., & Zeeck, A. (1989). Studies of precursor-directed biosynthesis with Streptomyces. Part 2. New and unusual manumycin analogues produced by *Streptomyces parvulus. Journal of the Chemical Society, Perkin Transactions, 1*(5), 851–855. https://doi.org/10.1039/P19890000851.

Thiericke, R., & Rohr, J. (1993). Biological variation of microbial metabolites by precursor-directed biosynthesis. *Natural Product Reports, 10*(3), 265–289. https://doi.org/10.1039/np9931000265.

Yang, J., Banas, V. S., Rivera, G. S. M., & Wencewicz, T. A. (2023). Siderophore synthetase DesD catalyzes N-to-C condensation in desferrioxamine biosynthesis. *ACS Chemical Biology, 18*(6), 1266–1270. https://doi.org/10.1021/acschembio.3c00167.

CHAPTER SEVEN

Preparation of coenzyme F430 biosynthetic enzymes and intermediates

Prosenjit Ray, Chelsea R. Rand-Fleming, and Steven O. Mansoorabadi*

Department of Chemistry and Biochemistry, Auburn University, Auburn, AL, United States
*Corresponding author. e-mail address: som@auburn.edu

Contents

1. Introduction	148
2. Expression and purification of the coenzyme F430 biosynthesis enzymes	151
2.1 HemC	152
2.2 HemD	153
2.3 SirA	154
2.4 SirC	155
2.5 CfbA	157
2.6 CfbB	158
2.7 CfbCD	159
2.8 CfbE	160
2.9 McrD	161
3. Synthesis and purification of coenzyme F430 biosynthetic intermediates	162
3.1 Sirohydrochlorin	163
3.2 Ni-sirohydrochlorin	163
3.3 Ni-sirohydrochlorin a,c-diamide	164
3.4 $15,17^3$-seco-F430-17^3-acid	165
3.5 Coenzyme F430	166
4. Concluding remarks	167
Acknowledgments	167
References	168

Abstract

Methyl-coenzyme M reductase (MCR) is the key enzyme in pathways for the formation and anaerobic oxidation of methane. As methane is a potent greenhouse gas and biofuel, investigations of MCR catalysis and maturation are of interest for the development of both methanogenesis inhibitors and natural gas conversion strategies. The activity of MCR is dependent on a unique, nickel-containing coenzyme F430, the most highly reduced tetrapyrrole found in nature. Coenzyme F430 is biosynthesized from

---

Methods in Enzymology, Volume 702
ISSN 0076-6879, https://doi.org/10.1016/bs.mie.2024.06.008
Copyright © 2024 Elsevier Inc. All rights are reserved, including those for text and data mining, AI training, and similar technologies.

sirohydrochlorin in four steps catalyzed by the CfbABCDE enzymes. Here, methods for the expression and purification of the coenzyme F430 biosynthesis enzymes are described along with conditions for the synthesis and purification of biosynthetic intermediates on the milligram scale from commercially available porphobilinogen.

1. Introduction

Methanogenic archaea (methanogens) are an important group of microorganisms that inhabit a wide range of anoxic environments, from marine sediments to the gastrointestinal tracts of animals (Garcia, Gribaldo, & Borrel, 2022). Methanogens produce nearly a billion tons of methane per year as an obligate byproduct of their energy metabolism through the pathway of methanogenesis (Conrad, 2009). As methane is both the major component of natural gas, which accounts for more than a quarter of the nation's energy demands, and a potent greenhouse gas with as much as ~85 times the global warming potential of carbon dioxide, investigations of enzymes essential for methanogenesis are of great current interest (Moore, Zielinska, Pétron, & Jackson, 2014). The key, methane-forming enzyme of methanogenesis is methyl–coenzyme M reductase (MCR) (Ermler & Grabarse, 1997). MCR catalyzes the conversion of coenzyme B and methyl–coenzyme M to the mixed heterodisulfide CoB-S-S-CoM and methane (Fig. 1) using a unique, nickel-containing coenzyme F430 (Thauer, 2019).

MCR is an $\alpha_2\beta_2\gamma_2$ heterohexamer with two active sites that each contain coenzyme F430 and several unprecedented post-translational modifications (PTMs) of unknown function (Fig. 2) (Mansoorabadi, Zheng, & Ngo, 2017). Homologs of MCR have recently been reported to catalyze the anaerobic oxidation of methane and other short-chain alkanes in methanotrophic and alkanotrophic archaea, respectively (Lemaire & Wagner, 2022). The structure of a MCR homolog isolated from Black Sea mats enriched in anaerobic methanotrophic archaea (ANME) was solved and shown to contain a distinct set of PTMs and a modified coenzyme F430 containing a methylthio group at $C17^2$ of the carbocyclic E ring (Fig. 2)

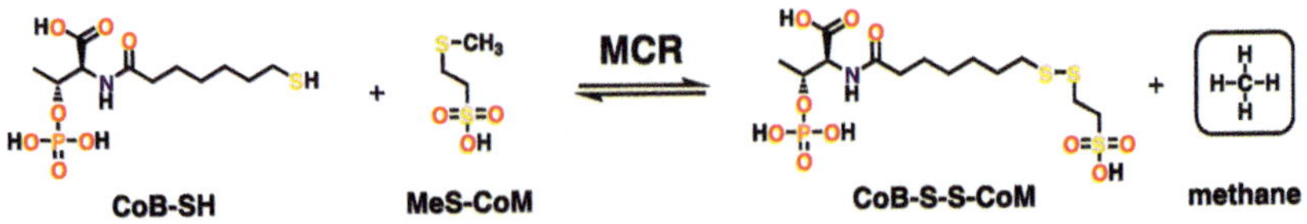

Fig. 1 MCR-catalyzed reaction leading to the formation of methane.

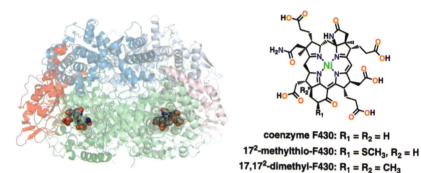

Fig. 2 Structures of MCR (PDB ID: 1mro), coenzyme F430, and variants found in anaerobic alkanotrophic archaea.

(Shima et al., 2011). Similarly, the structure of a MCR homolog catalyzing the anaerobic activation of ethane from *Candidatus Ethanoperedens thermophilum* contains two additional unique PTMs and a coenzyme F430 variant with methylations at C17 and C17^2 (Fig. 2) (Hahn et al., 2021). Additional modified F430 coenzymes were also identified in cell extracts of selected methanogens and ANME, although the biochemical roles of these variants and the pathways involved in their biosynthesis remain unknown (Allen, Wegener, & White, 2014).

Coenzyme F430 is a nickel-chelated tetrahydrocorphin and the most highly reduced tetrapyrrole found in nature (Pfaltz et al., 1982). Tetrapyrroles, the "pigments of life," are an important class of biomolecules that play essential roles in many fundamental biological processes, such as oxygen and electron transport (heme) and photosynthesis (chlorophyll) (Battersby, 2000). There are two principal pathways for the biosynthesis of all tetrapyrroles starting from the amino acids glycine or glutamate (Fig. 3) (Oh-hama, 1997). Both pathways converge at the intermediate 5-amino-levulinic acid, which is converted to uroporphyrinogen III, the last common precursor of all tetrapyrroles, by the enzymes HemBCD (Dailey et al., 2017). Uroporphyrinogen III can then be decarboxylated by HemE to coproporphyrinogen III, leading to the formation of heme and chlorophyll, or methylated and oxidized to sirohydrochlorin by a S-adenosyl-L-methionine (SAM)-dependent methyltransferase (SirA) and a NAD(P)$^+$-dependent dehydrogenase (SirC), respectively, which is the precursor for the biosynthesis of C2 and C7 methylated tetrapyrroles such as siroheme and cobalamin (Bryant, Hunter, & Warren, 2020). As coenzyme F430 has methyl groups at C2 and C7, sirohydrochlorin was predicted to be an intermediate in its biosynthesis (Gilles & Thauer, 1983).

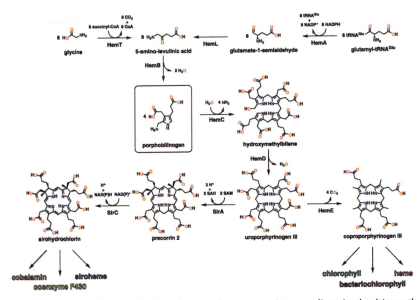

Fig. 3 Tetrapyrrole biosynthetic pathways. A common intermediate in the biosynthesis of all tetrapyrroles, porphobilinogen, is commercially available.

The coenzyme F430 biosynthetic pathway (Fig. 4) was elucidated by Mansoorabadi and coworkers (Zheng, Ngo, Owens, Yang, & Mansoorabadi, 2016) and independently confirmed by the Warren and Layer groups (Moore et al., 2017). These studies demonstrated that sirohydrochlorin was indeed converted to coenzyme F430, a process involving 4 enzymatic steps catalyzed by the products of the coenzyme F430 biosynthesis genes *cfbABCDE*. The pathway begins with nickelochelation by CfbA, followed by CfbB-catalyzed amidation to form Ni-sirohydrochlorin *a,c*-diamide. The CfbCD complex, a primitive homolog of nitrogenase, then mediates a net six-electron reduction and γ-lactamization reaction to form 15,17^3-seco-F430-17^3-acid. Finally, a Mur ligase homolog CfbE catalyzes carbocyclic ring formation to complete the synthesis of coenzyme F430. The yield of coenzyme F430 produced in the CfbE reaction was also found to be enhanced by the inclusion of McrD, a protein of unknown function from the *mcr* operon (Cram et al., 1987), suggesting it may play a role as a coenzyme F430 chaperone (Zheng et al., 2016). Consistent with this hypothesis, a structure of *apo* and semi-*apo* MCR was recently solved wherein McrD was asymmetrically bound, improving active site accessibility for the delivery of coenzyme F430 (Chadwick, Joiner, Ramesh, Mitchell, & Nayak, 2023).

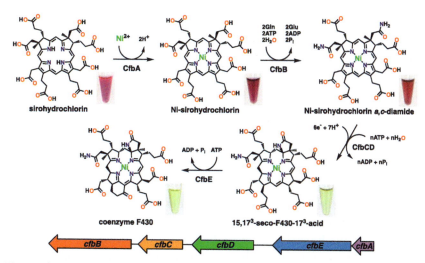

Fig. 4 The coenzyme F430 biosynthetic pathway and the *cfb* gene cluster from *Methanosarcina acetivorans*.

Given the importance of MCR, and by extension the enzymes involved in its maturation, to the global carbon cycle, protocols to prepare the coenzyme F430 biosynthesis enzymes and intermediates are of interest. Here, methods are described for the preparation of the tetrapyrrole biosynthetic enzymes required for the synthesis of coenzyme F430 from commercially available porphobilinogen (HemCD, SirAC, CfbABCDE, and McrD). Additionally, conditions for the enzymatic synthesis and purification of sirohydrochlorin, Ni-sirohydrochlorin, Ni-sirohydrochlorin *a,c*-diamide, 15,17^3-seco-F430-17^3-acid, and coenzyme F430 on the milligram scale are reported.

2. Expression and purification of the coenzyme F430 biosynthesis enzymes

In this section, methods for the expression and purification of the enzymes required for the biosynthesis of coenzyme F430 from porphobilinogen are described. These include (in order of action): HemC, HemD, SirA, SirC, CfbA, CfbB, CfbCD, and CfbE. Additionally, methods for the preparation of the *mcr* operon protein McrD are described. As noted, McrD is not required for the biosynthesis of coenzyme F430 but has been found to enhance its yield in CfbE reactions in vitro (Zheng et al., 2016). Each of these enzymes are prepared as His-tagged fusion proteins using *Escherichia coli* BL21(DE3) as the expression host and purified via immobilized metal affinity chromatography (IMAC). The only enzyme that requires removal of the His-tag to be active is CfbA, although in general removing affinity tags from recombinant proteins can enhance specific activity (e.g., such is the case with CfbB). The genes encoding HemC and HemD were cloned from *E. coli* BL21(DE3) genomic DNA, while those for SirA, SirC, CfbA, CfbB, CfbE, and McrD were amplified from *Methanosarcina acetivorans* C2A (Zheng et al., 2016). Each of these genes were inserted into the pET-28b(+) expression vector. In contrast, the *cfbC* and *cfbD* genes were cloned from *Methanosarcina thermophila* and were ligated into pET SUMO. In our hands, the thermophilic homolog of the CfbCD complex from *M. thermophila* was more stable/active then the corresponding enzyme from *M. acetivorans* C2A.

2.1 HemC

HemC, porphobilinogen deaminase or hydroxymethylbilane synthase, catalyzes the polymerization of four equivalents of porphobilinogen to form the first common tetrapyrrolic intermediate hydroxymethylbilane (Fig. 3). HemC utilizes an oxygen-sensitive dipyrromethane cofactor covalently linked to an active site cysteine residue to carry out this transformation (Jordan & Warren, 1987). Thus, the purification of HemC must be carried out under anaerobic conditions. The protocol for the expression and purification of HemC is as follows:

1. Inoculate 3×5 mL of sterile Luria–Bertani (LB) media containing 50 µg/mL kanamycin with *E. coli* BL21(DE3) cells transformed with the pET-28b (+)-*hemC* plasmid. Grow the starter cultures in a water bath shaker at 37 °C and 200 rpm until the OD_{600} reaches 0.6.
2. Inoculate 6 Erlenmeyer flasks containing 1 L of sterile LB media and 50 µg/mL kanamycin with 2 mL of the starter culture. Grow the cultures in an incubator shaker at 37 °C and 200 rpm until the OD_{600} reaches 0.6.
3. Lower the temperature to 15 °C and incubate the cultures for 1 h. Induce each culture with 60 µM isopropyl β-D-thiogalactopyranoside (IPTG) and incubate the cultures for an additional 12 h.
4. Harvest the cells via centrifugation at 6500 *g* and 4 °C for 15 min. Decant the supernatant and transfer the cell pellets to an anaerobic

Preparation of coenzyme F430 biosynthetic enzymes and intermediates 153

chamber. Resuspend the pellets in an equal volume of degassed lysis buffer (50 mM sodium phosphate, pH 8.0, 300 mM NaCl, 5 mM imidazole, and 20% glycerol).

5. Lyse the resuspended cells using a sonicator in the anaerobic chamber and transfer them to airtight ultracentrifuge tubes. Centrifuge at 70,000 g for 30 min at 4 °C.

6. Load the supernatant onto a column packed with 6 mL of Ni-charged IMAC resin in the anaerobic chamber. Wash the column with five column volumes of degassed wash buffer (50 mM sodium phosphate, pH 8.0, 300 mM NaCl, and 5 mM imidazole).

7. Elute HemC from the column using three column volumes of degassed elution buffer (50 mM sodium phosphate, pH 8.0, 300 mM NaCl, and 500 mM imidazole) and collect the eluate in 1 mL fractions.

8. Combine the HemC containing fractions and exchange the buffer with degassed storage buffer (100 mM Tris–HCl, pH 8.0, and 15% glycerol) using a spin concentrator (30 kDa MW cutoff) in the anaerobic chamber.

9. The yield and purity of HemC (MW = 36.0 kDa, $\varepsilon_{280} = 1.85 \times 10^4/(\text{M cm})$ with the His-tag) can be confirmed using UV–visible spectrophotometry and sodium dodecyl sulfate–polyacrylamide gel electrophoresis (SDS–PAGE) analysis.

2.2 HemD

HemD, uroporphyrinogen III synthase, effects the conversion of hydroxymethylbilane to uroporphyrinogen III, the last common precursor of all tetrapyrroles (Fig. 3). The HemD-catalyzed transformation is a dehydrative cyclization reaction that occurs with concomitant inversion of the D ring (Stamford, Capretta, & Battersby, 1995). In the absence of HemD, hydroxymethylbilane generated in the HemC reaction can spontaneously cyclize without D ring inversion to form uroporphyrinogen I (Frydman, Frydman, Valasinas, Levy, & Feinstein, 1976). Thus, HemD functions as a cosynthase with HemC to produce uroporphyrinogen III. The protocol for the expression and purification of HemD is as follows:

1. Inoculate 3×5 mL of sterile LB media containing 50 μg/mL kanamycin with *E. coli* BL21(DE3) cells transformed with the pET-28b(+)-*hemD* plasmid. Grow the starter cultures in a water bath shaker at 37 °C and 200 rpm until the OD_{600} reaches 0.6.

2. Inoculate 6 Erlenmeyer flasks containing 1 L of sterile LB media and 50 µg/mL kanamycin with 2 mL of the starter culture. Grow the cultures in an incubator shaker at 37 °C and 200 rpm until the OD_{600} reaches 0.6.

3. Lower the temperature to 18 °C and incubate the cultures for 1 h. Induce each culture with 75 µM IPTG and incubate the cultures for an additional 12 h.

4. Harvest the cells via centrifugation at 6500 g and 4 °C for 15 min. Decant the supernatant and resuspend the pellets in an equal volume of lysis buffer (50 mM sodium phosphate, pH 8.0, 300 mM NaCl, 5 mM imidazole, and 20% glycerol).

5. Lyse the resuspended cells via sonication and centrifuge at 16,000 g for 30 min at 4 °C.

6. Load the supernatant onto a column packed with 6 mL of Ni-charged IMAC resin. Wash the column with five column volumes of wash buffer (50 mM sodium phosphate, pH 8.0, 300 mM NaCl, and 5 mM imidazole).

7. Elute HemD from the column using three column volumes of elution buffer (50 mM sodium phosphate, pH 8.0, 300 mM NaCl, and 500 mM imidazole) and collect the eluate in 1 mL fractions.

8. Combine the HemD containing fractions and exchange the buffer with storage buffer (100 mM Tris–HCl, pH 8.0, and 15% glycerol) using a spin concentrator (10 kDa MW cutoff).

9. The yield and purity of HemD (MW = 30.0 kDa, $\varepsilon_{280} = 4.60 \times 10^4 /(M\ cm)$ with the His-tag) can be confirmed using UV–visible spectrophotometry and SDS–PAGE analysis.

2.3 SirA

SirA is a SAM-dependent uroporphyrinogen III C-methyltransferase (SUMT) that catalyzes the successive methylation of uroporphyrinogen III at C2 and C7 to form precorrin-2 (dihydrosirohydrochlorin) (Fig. 3) (Blanche, Debussche, Thibaut, Crouzet, & Cameron, 1989). As noted, formation of precorrin-2 is an important branch point in tetrapyrrole biosynthesis and the genes encoding SUMTs often cluster with those for late-stage tetrapyrrole biosynthesis and/or utilization where they can have an alternative annotation (e.g., $cobA$ in the cobalamin biosynthetic pathway (Crouzet et al., 1990) or $cysG$ in cysteine biosynthesis, which utilizes the siroheme-containing sulfite reductase (Tei, Murata, & Kimura, 1990)). Some $cysG$ homologs encode a multifunctional SUMT that combines functionalities with the next enzyme(s) in the pathway by including a

Preparation of coenzyme F430 biosynthetic enzymes and intermediates

domain with precorrin-2 dehydrogenase (SirC) and possibly ferrochelatase (SirB) activity (Spencer, Stolowich, Roessner, & Scott, 1993). The *sirA* gene from *M. acetivorans* C2A encodes a simple monofunctional SUMT (Zheng et al., 2016). The protocol for the expression and purification of this SirA is as follows:

1. Inoculate 3×5 mL of sterile LB media containing 50 µg/mL kanamycin with *E. coli* BL21(DE3) cells transformed with the pET-28b(+)-*sirA* plasmid. Grow the starter cultures in a water bath shaker at 37 °C and 200 rpm until the OD_{600} reaches 0.6.
2. Inoculate 6 Erlenmeyer flasks containing 1 L of sterile LB media and 50 µg/mL kanamycin with 2 mL of the starter culture. Grow the cultures in an incubator shaker at 37 °C and 200 rpm until the OD_{600} reaches 0.6.
3. Lower the temperature to 18 °C and incubate the cultures for 1 h. Induce each culture with 100 µM IPTG and incubate the cultures for an additional 14 h.
4. Harvest the cells via centrifugation at $6500\,g$ and 4 °C for 15 min. Decant the supernatant and resuspend the pellets in an equal volume of lysis buffer (50 mM sodium phosphate, pH 8.0, 300 mM NaCl, 5 mM imidazole, and 20% glycerol).
5. Lyse the resuspended cells via sonication and centrifuge at $16,000\,g$ for 30 min at 4 °C.
6. Load the supernatant onto a column packed with 6 mL of Ni-charged IMAC resin. Wash the column with five column volumes of wash buffer (50 mM sodium phosphate, pH 8.0, 300 mM NaCl, and 5 mM imidazole).
7. Elute SirA from the column using three column volumes of elution buffer (50 mM sodium phosphate, pH 8.0, 300 mM NaCl, and 500 mM imidazole) and collect the eluate in 1 mL fractions.
8. Combine the SirA containing fractions and exchange the buffer with storage buffer (100 mM Tris–HCl, pH 8.0, and 15% glycerol) using a spin concentrator (10 kDa MW cutoff).
9. The yield and purity of SirA (MW = 29.9 kDa, $\varepsilon_{280} = 2.14 \times 10^4/$ (M cm) with the His-tag) can be confirmed using UV–visible spectrophotometry and SDS–PAGE analysis.

2.4 SirC

SirC is a $NAD(P)^+$-dependent precorrin-2 dehydrogenase that catalyzes the formation of sirohydrochlorin (Fig. 3), an important branch point intermediate in the biosynthesis of all C2 and C7 methylated tetrapyrroles (Schubert et al., 2008). Certain homologs of SirC, for example Met8p from *Saccharomyces cerevisiae*, catalyze both dehydrogenation and ferrochelation (Schubert et al., 2002). However, SirC from *M. acetivorans* C2A is monofunctional and can catalyze the dehydrogenation of precorrin-2 using either NAD^+ or $NADP^+$ (Zheng et al., 2016). The protocol for the expression and purification of SirC is as follows:

1. Inoculate 3×5 mL of sterile LB media containing 50 µg/mL kanamycin with *E. coli* BL21(DE3) cells transformed with the pET-28b(+)-*sirC* plasmid. Grow the starter cultures in a water bath shaker at 37 °C and 200 rpm until the OD_{600} reaches 0.6.
2. Inoculate 6 Erlenmeyer flasks containing 1 L of sterile LB media and 50 µg/mL kanamycin with 2 mL of the starter culture. Grow the cultures in an incubator shaker at 37 °C and 200 rpm until the OD_{600} reaches 0.6.
3. Lower the temperature to 18 °C and incubate the cultures for 1 h. Induce each culture with 100 µM IPTG and incubate the cultures for an additional 14 h.
4. Harvest the cells via centrifugation at $6500\,g$ and 4 °C for 15 min. Decant the supernatant and resuspend the pellets in an equal volume of lysis buffer (50 mM sodium phosphate, pH 8.0, 300 mM NaCl, 5 mM imidazole, and 20% glycerol).
5. Lyse the resuspended cells via sonication and centrifuge at $16,000\,g$ for 30 min at 4 °C.
6. Load the supernatant onto a column packed with 6 mL of Ni-charged IMAC resin. Wash the column with five column volumes of wash buffer (50 mM sodium phosphate, pH 8.0, 300 mM NaCl, and 5 mM imidazole).
7. Elute SirC from the column using three column volumes of elution buffer (50 mM sodium phosphate, pH 8.0, 300 mM NaCl, and 500 mM imidazole) and collect the eluate in 1 mL fractions.
8. Combine the SirC containing fractions and exchange the buffer with storage buffer (100 mM Tris–HCl, pH 8.0, and 15% glycerol) using a spin concentrator (10 kDa MW cutoff).
9. The yield and purity of SirC (MW = 26.1 kDa, $\varepsilon_{280} = 1.85 \times 10^4/$ (M cm) with the His-tag) can be confirmed using UV–visible spectrophotometry and SDS–PAGE analysis.

2.5 CfbA

CfbA catalyzes nickelochelation of sirohydrochlorin in the first step unique to the coenzyme F430 biosynthetic pathway (Fig. 4) (Zheng et al., 2016). CfbA is a small, homodimeric class II chelatase (Fujishiro & Ogawa, 2021) that shares homology with the other branchpoint chelatases CbiX and SirB, from the cobalamin and siroheme biosynthetic pathways, respectively (Leech, Raux-Deery, Heathcote, & Warren, 2002). The protocol for the expression and purification of CfbA is as follows:

1. Inoculate $3 \times 5\,mL$ of sterile LB media containing $50\,\mu g/mL$ kanamycin with *E. coli* BL21(DE3) cells transformed with the pET-28b (+)-*cfbA* plasmid. Grow the starter cultures in a water bath shaker at $37\,°C$ and 200 rpm until the OD_{600} reaches 0.6.
2. Inoculate 6 Erlenmeyer flasks containing 1 L of sterile LB media and $50\,\mu g/mL$ kanamycin with $2\,mL$ of the starter culture. Grow the cultures in an incubator shaker at $37\,°C$ and 200 rpm until the OD_{600} reaches 0.6.
3. Lower the temperature to $18\,°C$ and incubate the cultures for 1 h. Induce each culture with $400\,\mu M$ IPTG and incubate the cultures for an additional 16 h.
4. Harvest the cells via centrifugation at $6500\,g$ and $4\,°C$ for 15 min. Decant the supernatant and resuspend the pellets in an equal volume of lysis buffer (100 mM sodium phosphate, pH 8.0, 300 mM NaCl, 5 mM imidazole, and 20% glycerol).
5. Lyse the resuspended cells via sonication and centrifuge at $16,000\,g$ for 30 min at $4\,°C$.
6. Load the supernatant onto a column packed with 6 mL of Ni-charged IMAC resin. Wash the column with five column volumes of wash buffer (100 mM sodium phosphate, pH 8.0, 300 mM NaCl, and 5 mM imidazole).
7. Add thrombin (80 units/mL of IMAC resin), cap the column, and incubate it overnight at $4\,°C$ on a nutating shaker.
8. Elute CfbA with 100 mM Tris–HCl buffer and apply the resulting eluate to a *p*-aminobenzamidine agarose column to remove the thrombin. Collect the eluate in 1 mL fractions.
9. Combine the CfbA containing fractions and exchange the buffer with storage buffer (100 mM Tris–HCl, pH 8.0, and 15% glycerol) using a spin concentrator (10 kDa MW cutoff).

10. The yield and purity of CfbA (MW = 13.7 kDa, $\varepsilon_{280} = 4.47 \times 10^3/$ (M cm) without the His-tag) can be confirmed using UV–visible spectrophotometry and SDS–PAGE analysis.

2.6 CfbB

CfbB catalyzes the sequential amidation of Ni-sirohydrochlorin to form Ni-sirohydrochlorin a,c-diamide (Fig. 4) (Zheng et al., 2016). CfbB is homologous to the glutamine amidotransferase cobyrinic acid a,c-diamide synthetase, CbiA, which catalyzes an analogous ATP-dependent amidation of the a- and c-acetic acid arms of cobyrinic acid in the cobalamin biosynthetic pathway (Fresquet, Williams, & Raushel, 2004). The protocol for the expression and purification of CfbB is as follows:

1. Inoculate 3×5 mL of sterile LB media containing 50 µg/mL kanamycin with *E. coli* BL21(DE3) cells transformed with the pET-28b(+)-*cfbB* plasmid. Grow the starter cultures in a water bath shaker at 37 °C and 200 rpm until the OD_{600} reaches 0.6.
2. Inoculate 6 Erlenmeyer flasks containing 1 L of sterile LB media and 50 µg/mL kanamycin with 2 mL of the starter culture. Grow the cultures in an incubator shaker at 37 °C and 200 rpm until the OD_{600} reaches 0.6.
3. Lower the temperature to 25 °C and incubate the cultures for 1 h. Induce each culture with 100 µM IPTG and incubate the cultures for an additional 12 h.
4. Harvest the cells via centrifugation at $6500\,g$ and 4 °C for 15 min. Decant the supernatant and resuspend the pellets in an equal volume of lysis buffer (100 mM sodium phosphate, pH 8.0, 300 mM NaCl, 5 mM imidazole, and 20% glycerol).
5. Lyse the resuspended cells via sonication and centrifuge at $16,000\,g$ for 30 min at 4 °C.
6. Load the supernatant onto a column packed with 6 mL of Ni-charged IMAC resin. Wash the column with five column volumes of wash buffer (100 mM sodium phosphate, pH 8.0, 300 mM NaCl, and 5 mM imidazole).
7. Elute CfbB from the column using three column volumes of elution buffer (50 mM sodium phosphate, pH 8.0, 300 mM NaCl, and 500 mM imidazole) and collect the eluate in 1 mL fractions.
8. Combine the CfbB containing fractions and exchange the buffer with storage buffer (100 mM Tris–HCl, pH 8.0, and 15% glycerol) using a spin concentrator (30 kDa MW cutoff).

Preparation of coenzyme F430 biosynthetic enzymes and intermediates 159

9. The yield and purity of CfbB (MW = 55.6 kDa, $\varepsilon_{280} = 3.13 \times 10^4/$ (M cm) with the His-tag) can be confirmed using UV–visible spectrophotometry and SDS–PAGE analysis.

2.7 CfbCD

The CfbCD complex converts Ni-sirohydrochlorin a,c-diamide to 15,17^3-seco-F430-17^3-acid in the key step of the coenzyme F430 biosynthetic pathway (Fig. 4) (Zheng et al., 2016). CfbC and CfbD are homologous to the Fe protein (NifH) and MoFe protein (NifD and NifK) components of nitrogenase, respectively (Ghebreamlak & Mansoorabadi, 2020). CfbC and CfbD can be purified together as a complex using a His-tag on the N-terminus of the CfbC subunit (Zheng et al., 2016) or individually as SUMO-tagged fusion proteins. Supplementation with iron and cysteine during induction is sufficient for the production of CfbCD with high iron sulfur cluster content. However, if necessary, the iron sulfur content can be further enhanced by coexpression with the iron sulfur cluster (*isc*) biosynthesis genes or via chemical reconstitution with iron and sulfide after purification (Zheng et al., 2016). The protocol for the expression and purification of SUMO-tagged CfbC and CfbD is as follows:

1. Inoculate 3×5 mL of sterile LB media containing 50 μg/mL kanamycin with *E. coli* BL21(DE3) cells transformed with either the pET SUMO-*cfbC* or pET SUMO-*cfbD* plasmids. Grow the starter cultures in a water bath shaker at 37 °C and 200 rpm until the OD_{600} reaches 0.6.
2. Inoculate 6 Erlenmeyer flasks containing 1 L of sterile LB media and 50 μg/mL kanamycin with 2 mL of the starter culture. Grow the cultures in an incubator shaker at 37 °C and 200 rpm until the OD_{600} reaches 0.6.
3. Lower the temperature to 25 °C and incubate the cultures for 1 h. Induce each culture with 300 μM IPTG, supplement with 3 mM $FeSO_4$ and l-cysteine, and incubate the cultures for an additional 14 h.
4. Harvest the cells via centrifugation at $6500\,g$ and 4 °C for 15 min. Decant the supernatant and transfer the cell pellets to an anaerobic chamber. Resuspend the pellets in an equal volume of degassed lysis buffer (50 mM sodium phosphate, pH 8.0, 300 mM NaCl, 5 mM imidazole, and 20% glycerol).
5. Lyse the resuspended cells using a sonicator in the anaerobic chamber and transfer them to airtight ultracentrifuge tubes. Centrifuge at $70,000\,g$ for 30 min at 4 °C.

6. Load the supernatant onto a column packed with 6 mL of Ni-charged IMAC resin in the anaerobic chamber. Wash the column with five column volumes of degassed wash buffer (50 mM sodium phosphate, pH 8.0, 300 mM NaCl, and 5 mM imidazole).

7. Elute CfbC or CfbD from the column using three column volumes of elution buffer (50 mM sodium phosphate, pH 8.0, 300 mM NaCl, and 500 mM imidazole) and collect the eluate in 1 mL fractions.

8. Combine the CfbC or CfbD containing fractions and exchange the buffer with degassed storage buffer (100 mM Tris–HCl, pH 8.0, and 15% glycerol) using a spin concentrator (30 kDa MW cutoff).

9. The yield and purity of CfbC (MW = 41.9 kDa, $\varepsilon_{280} = 1.34 \times 10^4$/(M cm) with the SUMO-tag) and CfbD (MW=54.3 kDa, $\varepsilon_{280} = 1.64 \times 10^4$/ (M cm) with the SUMO-tag) can be confirmed using UV–visible spectrophotometry and SDS–PAGE analysis.

2.8 CfbE

CfbE catalyzes the ATP-dependent conversion of 15,17^{3}seco-F430-17^3-acid to coenzyme F430 in the final step of the biosynthetic pathway (Fig. 4) (Zheng et al., 2016). CfbE is homologous to ATP-dependent Mur ligases, which facilitate nonribosomal peptide bond formation during the biosynthesis of peptidoglycan (Kouidmi, Levesque, & Paradis-Bleau, 2014). The protocol for the expression and purification of CfbE is as follows:

1. Inoculate 3×5 mL of sterile LB media containing 50 µg/mL kanamycin with *E. coli* BL21(DE3) cells transformed with the pET-28b(+)-*cfbE* plasmid. Grow the starter cultures in a water bath shaker at 37 °C and 200 rpm until the OD_{600} reaches 0.6.

2. Inoculate 6 Erlenmeyer flasks containing 1 L of sterile LB media and 50 µg/mL kanamycin with 2 mL of the starter culture. Grow the cultures in an incubator shaker at 37 °C and 200 rpm until the OD_{600} reaches 0.6.

3. Lower the temperature to 25 °C and incubate the cultures for 1 h. Induce each culture with 100 µM IPTG and incubate the cultures for an additional 14 h.

4. Harvest the cells via centrifugation at $6500\,g$ and 4 °C for 15 min. Decant the supernatant and resuspend the pellets in an equal volume of lysis buffer (50 mM sodium phosphate, pH 8.0, 300 mM NaCl, 5 mM imidazole, and 20% glycerol).

5. Lyse the resuspended cells via sonication and centrifuge at $16,000\,g$ for 30 min at 4 °C.

Preparation of coenzyme F430 biosynthetic enzymes and intermediates   161

6. Load the supernatant onto a column packed with 6 mL of Ni-charged IMAC resin. Wash the column with five column volumes of wash buffer (50 mM sodium phosphate, pH 8.0, 300 mM NaCl, and 5 mM imidazole).

7. Elute CfbE from the column using three column volumes of elution buffer (50 mM sodium phosphate, pH 8.0, 300 mM NaCl, and 500 mM imidazole) and collect the eluate in 1 mL fractions.

8. Combine the CfbE containing fractions and exchange the buffer with storage buffer (100 mM Tris–HCl, pH 8.0, and 15% glycerol) using a spin concentrator (30 kDa MW cutoff).

9. The yield and purity of CfbE (MW = 52.6 kDa, $\varepsilon_{280} = 1.89 \times 10^4/$ (M cm) with the His-tag) can be confirmed using UV–visible spectrophotometry and SDS–PAGE analysis.

2.9 McrD

The *mcr* operon protein McrD (Cram et al., 1987), while not required for the biosynthesis of coenzyme F430, was found to enhance the yield of the CfbE reaction (Zheng et al., 2016). This effect was thought to arise from the alleviation of product inhibition of the CfbE reaction via the binding of McrD to coenzyme F430 and/or CfbE (Zheng et al., 2016). McrD was previously found to physically associate with MCR (Sherf & Reeve, 1990) and more recently to bind asymmetrically to *apo* and semi-*apo* MCR (Chadwick et al., 2023). Thus, McrD likely serves as a chaperone that facilitates the insertion of coenzyme F430 during the maturation of MCR. The protocol for the expression and purification of McrD is as follows:

1. Inoculate 3×5 mL of sterile LB media containing 50 µg/mL kanamycin with *E. coli* BL21(DE3) cells transformed with the pET-28b(+)-*mcrD* plasmid. Grow the starter cultures in a water bath shaker at 37 °C and 200 rpm until the OD_{600} reaches 0.6.

2. Inoculate 6 Erlenmeyer flasks containing 1 L of sterile LB media and 50 µg/mL kanamycin with 2 mL of the starter culture. Grow the cultures in an incubator shaker at 37 °C and 200 rpm until the OD_{600} reaches 0.6.

3. Lower the temperature to 25 °C and incubate the cultures for 1 h. Induce each culture with 100 µM IPTG and incubate the cultures for an additional 16 h.

4. Harvest the cells via centrifugation at 6500 *g* and 4 °C for 15 min. Decant the supernatant and resuspend the pellets in an equal volume of

lysis buffer (50 mM sodium phosphate, pH 8.0, 300 mM NaCl, 5 mM imidazole, and 20% glycerol).
5. Lyse the resuspended cells via sonication and centrifuge at 16,000 g for 30 min at 4 °C.
6. Load the supernatant onto a column packed with 6 mL of Ni-charged IMAC resin. Wash the column with five column volumes of wash buffer (50 mM sodium phosphate, pH 8.0, 300 mM NaCl, and 5 mM imidazole).
7. Elute McrD from the column using three column volumes of elution buffer (50 mM sodium phosphate, pH 8.0, 300 mM NaCl, and 500 mM imidazole) and collect the eluate in 1 mL fractions.
8. Combine the McrD containing fractions and exchange the buffer with storage buffer (100 mM Tris–HCl, pH 8.0, and 15% glycerol) using a spin concentrator (10 kDa MW cutoff).
9. The yield and purity of McrD (MW = 21.5 kDa, ε_{280} = 2.98 × 10^3/ (M cm) with the His-tag) can be confirmed using UV–visible spectrophotometry and SDS–PAGE analysis.

3. Synthesis and purification of coenzyme F430 biosynthetic intermediates

The coenzyme F430 biosynthesis enzymes purified via the above protocols can readily be used for the in vitro synthesis of pathway intermediates on the milligram scale. Many of the enzymes (e.g., CfbCD) and intermediates (e.g., precorrin-2) are oxygen- and/or light-sensitive, so the reactions must be carried out using degassed components in an anoxic environment (e.g., in an anaerobic chamber or on a Schlenk line) in opaque or covered tubes. Reaction buffers should be degassed and passed through a Chelex 100 resin to remove trace divalent metal ions that could lead to undesired chelation into sirohydrochlorin. Sirohydrochlorin can be biosynthesized from commercially available reagents using purified HemC, HemD, SirA, and SirC (Fig. 3). Coenzyme F430 and its biosynthetic precursors can then be obtained via the subsequent addition of the corresponding Cfb enzymes and cosubstrates to the sirohydrochlorin reactions (Fig. 4). Inclusion of the next enzyme in the pathway without their cosubstrates can enhance the rate/yield of a particular Cfb enzyme-catalyzed reaction (presumably via the alleviation of product inhibition), as can the inclusion of an ATP-regeneration system for the ATP-dependent steps

Preparation of coenzyme F430 biosynthetic enzymes and intermediates

(Zheng et al., 2016). Additionally, in our hands, the yields from several small volume (e.g., 1 mL) reactions was better than from one large volume reaction. Each reaction product can be purified using either column (e.g., C18) or thin layer chromatography. The organic elution solvents must then be removed (e.g., by boiling or rotary evaporation) prior to use in subsequent applications.

3.1 Sirohydrochlorin

Sirohydrochlorin can be prepared from porphobilinogen, SAM, and NAD $(P)^+$ using HemC, HemD, SirA, and SirC purified as described above (Fig. 3). The protocol for the synthesis and purification of sirohydrochlorin is as follows:

1. Prepare 10×1 mL reaction mixtures in an anaerobic chamber by combining porphobilinogen (1 mM), SAM (1 mM), NADP$^+$ (1 mM), MgCl$_2$ (4 mM), HemC (2 μM), HemD (2 μM), SirA (4 μM), and SirC (14 μM) in degassed/demetallated 100 mM Tris–HCl buffer, pH 8.0. Incubate the reaction mixtures at 37 °C in a heat block for 6 h. The formation of sirohydrochlorin is accompanied by a color change in the solution from clear to a deep magenta (Fig. 4).
2. Quench the reaction mixtures by heating at 95 °C for 15 min. Allow the reaction mixtures to cool and centrifuge at $16,000\,g$ for 10 min.
3. Combine the supernatants and load them onto a column packed with 50 mL of C18 resin. Wash the column with two column volumes of degassed wash buffer (15% acetonitrile, 0.5% formic acid in water).
4. Elute sirohydrochlorin using two column volumes of degassed elution buffer (25% acetonitrile, 0.5% formic acid in water). Collect the magenta-colored fraction.
5. Heat the sirohydrochlorin-containing fraction at 85 °C for 45 min to remove the acetonitrile.
6. The yield and purity of sirohydrochlorin can be assessed using HPLC (Zheng et al., 2016).

3.2 Ni-sirohydrochlorin

Ni-sirohydrochlorin can be prepared from porphobilinogen, SAM, NAD $(P)^+$, and Ni^{2+} ions using HemC, HemD, SirA, SirC, and CfbA purified as described above (Fig. 3 and Fig. 4). Ni^{2+} should not be added until after the accumulation of sirohydrochlorin, as it interferes with the preceding reactions leading to low yields of Ni-sirohydrochlorin and the formation of shunt products such as uroporphyrins. Inclusion of CfbB without glutamine and ATP can enhance the yield of the Ni-sirohydrochlorin reaction by alleviating product inhibition of CfbA (Zheng et al., 2016). The protocol for the synthesis and purification of Ni-sirohydrochlorin is as follows:

1. Prepare 10×1 mL reaction mixtures in an anaerobic chamber by combining porphobilinogen (1 mM), SAM (1 mM), $NADP^+$ (1 mM), $MgCl_2$ (4 mM), HemC (2 µM), HemD (2 µM), SirA (4 µM), SirC (14 µM), and CfbA (14 µM) in degassed/demetallated 100 mM Tris–HCl buffer, pH 8.0. Incubate the reaction mixtures at 37 °C in a heat block for 4 h.
2. Add 20 µL of 15 mM $NiCl_2$ to each reaction mixture and incubate for an additional 4 h. The formation of Ni-sirohydrochlorin is accompanied by a color change in the solution to a deep burgundy (Fig. 4).
3. Quench the reaction mixtures by heating at 95 °C for 15 min. Allow the reaction mixtures to cool and centrifuge at $16,000\,g$ for 10 min.
4. Combine the supernatants and load them onto a column packed with 50 mL of C18 resin. Wash the column with two column volumes of degassed wash buffer (15% acetonitrile, 0.5% formic acid in water).
5. Elute Ni-sirohydrochlorin using two column volumes of degassed elution buffer (25% acetonitrile, 0.5% formic acid in water). Collect the burgundy-colored fraction.
6. Heat the Ni-sirohydrochlorin-containing fraction at 85 °C for 45 min to remove the acetonitrile.
7. The yield and purity of Ni-sirohydrochlorin can be assessed using HPLC (Zheng et al., 2016).

3.3 Ni-sirohydrochlorin *a,c*-diamide

Ni-sirohydrochlorin *a,c*-diamide can be prepared from porphobilinogen, SAM, $NAD(P)^+$, Ni^{2+} ions, glutamine, and ATP using HemC, HemD, SirA, SirC, CfbA, and CfbB purified as described above (Figs. 3 and 4). Ammonia can be used as a cosubstrate in lieu of glutamine. Inclusion of CfbCD without a source of reducing equivalents (e.g., sodium dithionite) can enhance the yield of the Ni-sirohydrochlorin *a,c*-diamide reaction by alleviating product inhibition of CfbB (Zheng et al., 2016). The reaction can also be driven forward by including an excess of glutamine and/or ATP or with an ATP-regeneration system (Zheng et al., 2016). The protocol for the synthesis and purification of Ni-sirohydrochlorin *a,c*-diamide is as follows:

1. Prepare 10×1 mL reaction mixtures in an anaerobic chamber by combining porphobilinogen (1 mM), SAM (1 mM), NADP$^+$ (1 mM), MgCl$_2$ (4 mM), HemC (2 μM), HemD (2 μM), SirA (4 μM), SirC (14 μM), CfbA (14 μM), and CfbB (14 μM) in degassed/demetallated 100 mM Tris–HCl buffer, pH 8.0. Incubate the reaction mixtures at 37 °C in a heat block for 4 h.
2. Add 20 μL of a solution containing 13 mM NiCl$_2$, 200 mM ʟ-glutamine, and 200 mM ATP to each reaction mixture and incubate for an additional 4 h. The formation of Ni-sirohydrochlorin a,c-diamide is accompanied by a color change in the solution to a deep burgundy (Fig. 4).
3. Quench the reaction mixtures by heating at 95 °C for 15 min. Allow the reaction mixtures to cool and centrifuge at $16,000\,g$ for 10 min.
4. Combine the supernatants and load them onto a column packed with 50 mL of C18 resin. Wash the column with two column volumes of degassed wash buffer (15% acetonitrile, 0.5% formic acid in water).
5. Elute Ni-sirohydrochlorin a,c-diamide using two column volumes of degassed elution buffer (25% acetonitrile, 0.5% formic acid in water). Collect the burgundy-colored fraction.
6. Heat the Ni-sirohydrochlorin a,c-diamide-containing fraction at 85 °C for 45 min to remove the acetonitrile.
7. The yield and purity of Ni-sirohydrochlorin a,c-diamide can be assessed using HPLC (Zheng et al., 2016).

3.4 15,17³-seco-F430-17³-acid

15,17^{3}Seco-F430-17^3-acid can be prepared from porphobilinogen, SAM, NAD(P)$^+$, Ni^{2+} ions, glutamine, ATP, and sodium dithionite using HemC, HemD, SirA, SirC, CfbA, CfbB, and CfbCD purified as described above (Fig. 3 and Fig. 4). NADPH and spinach ferredoxin/ferredoxin reductase can be used as the reducing system in lieu of sodium dithionite. An excess of ATP (Moore et al., 2017) or an ATP regeneration system (e.g., phosphoenolpyruvate and pyruvate kinase) (Zheng et al., 2016) can be used to drive the CfbCD reaction forward. The protocol for the synthesis and purification of 15,17^3-seco-F430-17^3-acid is as follows:

1. Prepare 10×1 mL reaction mixtures in an anaerobic chamber by combining porphobilinogen (1 mM), SAM (1 mM), NADP$^+$ (1 mM), MgCl$_2$ (4 mM), HemC (2 μM), HemD (2 μM), SirA (4 μM), SirC (14 μM), CfbA (14 μM), and CfbB (14 μM) in degassed/demetallated

100 mM Tris–HCl buffer, pH 8.0. Incubate the reaction mixtures at 37 °C in a heat block for 4 h.

2. Add 20 μL of a solution containing 13 mM $NiCl_2$, 200 mM L-glutamine, and 200 mM ATP to each reaction mixture and incubate for an additional 4 h.

3. Add 15 nmol of the CfbCD complex and 20 μL of a solution containing 300 mM $MgCl_2$, 300 mM ATP, and 300 mM sodium dithionite. Incubate the reaction mixtures for an additional 6 h. The formation of $15,17^3$-seco-F430-17^3-acid is accompanied by a color change in the solution to a pale yellow (Fig. 4).

4. Quench the reaction mixtures by heating at 95 °C for 15 min. Allow the reaction mixtures to cool and centrifuge at 16,000 g for 10 min.

5. Combine the supernatants and load them onto a column packed with 50 mL of C18 resin. Wash the column with two column volumes of degassed wash buffer (15% acetonitrile, 0.5% formic acid in water).

6. Elute $15,17^3$-seco-F430-17^3-acid using two column volumes of degassed elution buffer (25% acetonitrile, 0.5% formic acid in water). Collect the yellow-colored fraction.

7. Heat the $15,17^3$-seco-F430-17^3-acid-containing fraction at 85 °C for 45 min to remove the acetonitrile.

8. The yield and purity of $15,17^3$-seco-F430-17^3-acid can be assessed using HPLC (Zheng et al., 2016).

3.5 Coenzyme F430

Coenzyme F430 can be prepared from porphobilinogen, SAM, $NAD(P)^+$, Ni^{2+} ions, glutamine, ATP, and sodium dithionite using HemC, HemD, SirA, SirC, CfbA, CfbB, CfbCD, and CfbE purified as described above (Fig. 3 and Fig. 4). Inclusion of McrD can enhance the yield of the coenzyme F430 reaction by alleviating product inhibition of CfbE (Zheng et al., 2016). The protocol for the synthesis and purification of coenzyme F430 is as follows:

1. Prepare 10×1 mL reaction mixtures in an anaerobic chamber by combining porphobilinogen (1 mM), SAM (1 mM), $NADP^+$ (1 mM), $MgCl_2$ (4 mM), HemC (2 μM), HemD (2 μM), SirA (4 μM), SirC (14 μM), CfbA (14 μM), and CfbB (14 μM) in degassed/demetallated 100 mM Tris–HCl buffer, pH 8.0. Incubate the reaction mixtures at 37 °C in a heat block for 4 h.

Preparation of coenzyme F430 biosynthetic enzymes and intermediates 167

2. Add 20 μL of a solution containing 13 mM $NiCl_2$, 200 mM L-glutamine, and 200 mM ATP to each reaction mixture and incubate for an additional 4 h.

3. Add 15 nmol of the CfbCD complex, 15 nmol CfbE, 15 nmol McrD, and 20 μL of a solution containing 300 mM $MgCl_2$, 300 mM ATP, and 300 mM sodium dithionite. Incubate the reaction mixtures for an additional 6 h.

4. Quench the reaction mixtures by heating at 95 °C for 15 min. Allow the reaction mixtures to cool and centrifuge at 16,000 g for 10 min.

5. Combine the supernatants and load them onto a column packed with 50 mL of C18 resin. Wash the column with two column volumes of degassed wash buffer (15% acetonitrile, 0.5% formic acid in water).

6. Elute coenzyme F430 using two column volumes of degassed elution buffer (25% acetonitrile, 0.5% formic acid in water). Collect the yellow-colored fraction.

7. Heat the coenzyme F430-containing fraction at 85 °C for 45 min to remove the acetonitrile.

8. The yield and purity of coenzyme F430 can be assessed using HPLC (Zheng et al., 2016).

4. Concluding remarks

Coenzyme F430 is the nickel-containing prosthetic group of MCR, the key enzyme in pathways for the biological formation and anaerobic oxidation of methane. This unique coenzyme is the most highly reduced tetrapyrrole found in nature and is biosynthesized from the common intermediate porphobilinogen in eight enzymatic steps. Methods for the preparation of these enzymes and of their respective substrates/products in high yields and purity are described herein. As methane is a potent greenhouse gas and biofuel, it is our hope these methods will help to enable mechanistic studies of the coenzyme F430 biosynthesis enzymes, which have potential applications in the development of novel inhibitors of methanogenesis and in the heterologous assembly of MCR for use in natural gas conversion strategies.

Acknowledgments

This work was supported by a grant from the U.S. Department of Energy, Office of Science, Basic Energy Sciences (DE-SC0023451).

References

Allen, K. D., Wegener, G., & White, R. H. (2014). White discovery of multiple modified F430 coenzymes in methanogens and anaerobic methanotrophic archaea suggests possible new roles for F430 in nature. *Applied and Environmental Microbiology, 80,* 6403–6411.

Battersby, A. R. (2000). Tetrapyrroles: The pigments of life. *Natural Product Reports, 17,* 507–526.

Blanche, F., Debussche, L., Thibaut, D., Crouzet, J., & Cameron, B. (1989). Purification and characterization of S-adenosyl-L-methionine:uroporphyrinogen III methyltransferase from *Pseudomonas denitrificans. Journal of Bacteriology, 171,* 4222–4231.

Bryant, D. A., Hunter, C. N., & Warren, M. J. (2020). Biosynthesis of the modified tetrapyrroles – The pigments of life. *The Journal of Biological Chemistry, 295,* 6888–6925.

Chadwick, G. L., Joiner, A. M. N., Ramesh, S., Mitchell, D. A., & Nayak, D. D. (2023). McrD binds asymmetrically to methyl-coenzyme M reductase improving active-site accessibility during assembly. *Proceedings of the National Academy of Sciences of the United States of America, 120,* e2302815120.

Conrad, R. (2009). The global methane cycle: Recent advances in understanding the microbial processes involved. *Environmental Microbiology Reports, 1,* 285–292.

Cram, D. S., Sherf, B. A., Libby, R. T., Mattaliano, R. J., Ramachandran, K. L., & Reeve, J. N. (1987). Structure and expression of the genes, mcrBDCGA, which encode the subunits of component C of methyl coenzyme M reductase in *Methanococcus vannielii. Proceedings of the National Academy of Sciences of the United States of America, 84,* 3992–3996.

Crouzet, J., Cauchois, L., Blanche, F., Debussche, L., Thibaut, D., Rouyez, M. C., et al. (1990). Nucleotide sequence of a *Pseudomonas denitrificans* 5.4-kilobase DNA fragment containing five cob genes and identification of structural genes encoding S-adenosyl-L-methionine:uroporphyrinogen III methyltransferase and cobyrinic acid a,c-diamide synthase. *Journal of Bacteriology, 172,* 5968–5979.

Dailey, H. A., Dailey, T. A., Gerdes, S., Jahn, D., Jahn, H. A., O'Brian, M. R., et al. (2017). Prokaryotic heme biosynthesis: Multiple pathways to a common essential product. *Microbiology and Molecular Biology Reviews: MMBR, 81,* e00048-16.

Ermler, U., & Grabarse, W. (1997). Crystal structure of methyl-coenzyme M reductase: The key enzyme of biological methane formation. *Science (New York, N. Y.), 278,* 1457–1462.

Fresquet, V., Williams, L., & Raushel, F. M. (2004). Mechanism of cobyrinic acid a,c-diamide synthetase from *Salmonella typhimurium* LT2. *Biochemistry, 43,* 10619–10627.

Frydman, B., Frydman, R. B., Valasinas, A., Levy, E. S., & Feinstein, G. (1976). Biosynthesis of uroporphyrinogens from porphobilinogen: Mechanism and the nature of the process. *Philosophical Transactions of the Royal Society of London. Series B, Biological Sciences, 273,* 137–160.

Fujishiro, T., & Ogawa, S. (2021). The nickel-sirohydrochlorin formation mechanism of the ancestral class II chelatase CfbA in coenzyme F430 biosynthesis. *Chemical Science, 12,* 2172–2180.

Garcia, P. S., Gribaldo, S., & Borrel, G. (2022). Diversity and evolution of methane-related pathways in archaea. *Annual Review of Microbiology, 76,* 727–755.

Ghebreamlak, S. M., & Mansoorabadi, S. O. (2020). Divergent members of the nitrogenase superfamily: Tetrapyrrole biosynthesis and beyond. *Chembiochem: a European Journal of Chemical Biology, 21,* 1723–1728.

Gilles, H., & Thauer, R. K. (1983). Uroporphyrinogen III, an intermediate in the biosynthesis of the nickel-containing factor F430 in *Methanobacterium thermoautotrophicum. European Journal of Biochemistry/FEBS, 135,* 109–112.

Hahn, C. J., Lemaire, O. N., Kahnt, J., Engilberge, S., Wegener, G., & Wagner, T. (2021). Crystal structure of a key enzyme for anaerobic ethane activation. *Science (New York, N. Y.)*, *373*, 118–121.

Jordan, P. M., & Warren, M. J. (1987). Evidence for a dipyrromethane cofactor at the catalytic site of *E. coli* porphobilinogen deaminase. *FEBS Letters*, *225*, 87–92.

Kouidmi, I., Levesque, R. C., & Paradis-Bleau, C. (2014). The biology of Mur ligases as an antibacterial target. *Molecular Microbiology*, *94*, 242–253.

Leech, H. K., Raux-Deery, E., Heathcote, P., & Warren, M. J. (2002). Production of cobalamin and sirohaem in *Bacillus megaterium*: An investigation into the role of the branchpoint chelatases sirohydrochlorin ferrochelatase (SirB) and sirohydrochlorin cobalt chelatase (CbiX). *Biochemical Society Transactions*, *30*, 610–613.

Lemaire, O. N., & Wagner, T. (2022). A structural view of alkyl-coenzyme M reductases, the first step of alkane anaerobic oxidation catalyzed by archaea. *Biochemistry*, *61*, 805–821.

Mansoorabadi, S. O., Zheng, K., & Ngo, P. D. (2017). Biosynthesis of coenzyme F430 and the post-translational modification of the active site region of methyl-coenzyme M reductase. In M. K. Johnson, & R. A. Scott (Eds.). *Metalloprotein active site assembly*. West Sussex, UK: John Wiley & Sons, Ltd.

Moore, S. J., Sowa, S. T., Schuchardt, C., Deery, E., Lawrence, A. D., Ramos, J. V., et al. (2017). Elucidation of the biosynthesis of the methane catalyst coenzyme F430. *Nature*, *543*, 78–82.

Moore, C. W., Zielinska, B., Pétron, G., & Jackson, R. B. (2014). Air impacts of increased natural gas acquisition, processing, and use: A critical review. *Environmental Science & Technology*, *48*, 8349–8359.

Oh-hama, T. (1997). Evolutionary consideration on 5-aminolevulinate synthase in nature. *Origins of Life and Evolution of the Biosphere: the Journal of the International Society for the Study of the Origin of Life*, *27*, 405–412.

Pfaltz, A., Jaun, B., Fassler, A., Eschenmoser, A., Jaenchen, R., Gilles, H. H., et al. (1982). Factor F430 from methanogenic bacteria: Structure of the porphninoid ligand system. *Helvetica Chimica Acta*, *65*, 828–865.

Schubert, H. L., Raux, E., Brindley, A. A., Leech, H. K., Wilson, K. S., Hill, C. P., et al. (2002). The structure of *Saccharomyces cerevisiae* Met8p, a bifunctional dehydrogenase and ferrochelatase. *The EMBO Journal*, *21*, 2068–2075.

Schubert, H. L., Rose, R. S., Leech, H. K., Brindley, A. A., Hill, C. P., Rigby, S. E., et al. (2008). Structure and function of SirC from *Bacillus megaterium*: A metal-binding precorrin-2 dehydrogenase. *The Biochemical Journal*, *415*, 257–263.

Sherf, B. A., & Reeve, J. N. (1990). Identification of the *mcrD* gene product and its association with component C of methyl coenzyme M reductase in *Methanococcus vannielii*. *Journal of Bacteriology*, *172*, 1828–1833.

Shima, S., Krueger, M., Weinert, T., Demmer, U., Kahnt, J., Thauer, R. K., et al. (2011). Structure of a methyl-coenzyme M reductase from Black Sea mats that oxidize methane anaerobically. *Nature*, *481*, 98–101.

Spencer, J. B., Stolowich, N. J., Roessner, C. A., & Scott, A. I. (1993). The *Escherichia coli* *cysG* gene encodes the multifunctional protein, siroheme synthase. *FEBS Letters*, *335*, 57–60.

Stamford, N. P., Capretta, A., & Battersby, A. R. (1995). Expression, purification and characterisation of the product from the *Bacillus subtilis* hemD gene, uroporphyrinogen III synthase. *European Journal of Biochemistry / FEBS*, *231*, 236–241.

Tei, H., Murata, K., & Kimura, A. (1990). Molecular cloning of the *cys* genes (*cysC*, *cysD*, *cysH*, *cysI*, *cysJ*, and *cysG*) responsible for cysteine biosynthesis in *Escherichia coli* K-12. *Biotechnology and Applied Biochemistry*, *12*, 212–216.

Thauer, R. K. (2019). Methyl (alkyl)-coenzyme M reductases: Nickel F-430-containing enzymes involved in anaerobic methane formation and in anaerobic oxidation of methane or of short chain alkanes. *Biochemistry, 58*, 5198–5220.

Zheng, K., Ngo, P. D., Owens, V. L., Yang, X. P., & Mansoorabadi, S. O. (2016). The biosynthetic pathway of coenzyme F430 in methanogenic and methanotrophic archaea. *Science (New York, N. Y.), 354*, 339–342.

CHAPTER EIGHT

Purification and biochemical characterization of methanobactin biosynthetic enzymes

Reyvin M. Reyes and Amy C. Rosenzweig*

Departments of Molecular Biosciences and of Chemistry, Northwestern University, Evanston, IL, United States
*Corresponding author. e-mail address: amyr@northwestern.edu

Contents

1. Introduction	172
2. Expression and purification of *M. trichosporium* OB3b MbnBC complexes	174
2.1 Plasmid construction and transformation into *Escherichia coli*	174
2.2 Large-scale growth and induction of MbnBC expression	175
2.3 Purification of MbnBC for biochemical studies	176
2.4 Purification of MbnBC for crystallization	177
3. *In vitro* modification of MbnA using purified MbnBC	179
4. Crystallization and structure determination of *M. trichosporium* OB3b MbnBC	181
5. Expression and purification of *M. trichosporium* OB3b MbnN	182
6. Conclusions	185
Acknowledgments	185
References	185

Abstract

Methanobactin (Mbn) is a ribosomally synthesized and post-translationally modified peptide (RiPP) natural product that binds Cu(I) with high affinity. The copper-chelating thioamide/oxazolone groups in Mbn are installed on the precursor peptide MbnA by the core enzyme complex, MbnBC, which includes the multinuclear non-heme iron-dependent oxidase (MNIO) MbnB and its RiPP recognition element-containing partner protein MbnC. For the extensively characterized Mbn biosynthetic gene cluster (BGC) from the methanotroph *Methylosinus trichosporium* OB3b, the tailoring aminotransferase MbnN further modifies MbnA after leader sequence cleavage by an unknown mechanism. Here we detail methods to express and purify *M. trichosporium* OB3b MbnBC and MbnN along with protocols for assessing MbnA modification by MbnBC and MbnN aminotransferase activity. In addition, we describe crystallization and structure determination of MbnBC. These procedures can be adapted for other MNIOs and partner proteins encoded in Mbn and Mbn-like BGCs. Furthermore, these methods provide a first step toward in vitro biosynthesis of Mbns and related natural products as potential therapeutics.

Methods in Enzymology, Volume 702
ISSN 0076-6879, https://doi.org/10.1016/bs.mie.2024.06.011
Copyright © 2024 Elsevier Inc. All rights are reserved, including those for text and data mining, AI training, and similar technologies.

1. Introduction

Ribosomally synthesized and post-translationally modified peptides (RiPPs) are a vast group of bioactive natural products (Arnison et al., 2013; Montalbán-López et al., 2021) with significant potential for drug discovery and development (Pfeiffer, Schröder, & Mordhorst, 2024). RiPPs are biosynthesized from precursor peptides composed of a leader sequence and a core sequence. The leader sequence, which is ultimately removed, facilitates recognition by biosynthetic enzymes that post-translationally modify the core sequence. There are more than 40 classes of RiPPs, defined on the basis of core PTMs, along with numerous in class variations installed by tailoring enzymes (Nguyen, Mitchell, & van der Donk, 2024). Because RiPP biosynthetic pathways include multiple transformations catalyzed by enzymes encoded in a biosynthetic gene cluster (BGC), they are fertile ground for the discovery of new reactions and classes of enzymes (Nguyen et al., 2024). One prime example of enzyme discovery via RiPP characterization is the identification of the multinuclear non-heme iron-dependent oxidases (MNIOs), formerly known as the domain of unknown function 692 (DUF692) family. These enzymes, which adopt a TIM barrel fold, utilize their multi-iron cofactors to perform a range of reactions, including heterocycle and macrocycle formation (Ayikpoe, Zhu, Chen, Ting, & van der Donk, 2023; Kenney et al., 2018), carbon excision (Ting et al., 2019; Yu & van der Donk, 2022), and amino acid cleavage (Chioti, Clark, Ganley, Han, & Seyedsayamdost, 2024; Nguyen et al., 2024), often targeting cysteine residues. For each system, recognition of the substrate precursor peptides is facilitated by a partner protein or domain containing a RiPP recognition element (RRE).

The founding member of the MNIO family is *Methylosinus trichosporium* OB3b MbnB, which, in complex with its partner protein MbnC, performs the key processing step in the formation of methanobactin (Mbn) from its precursor peptide MbnA (Jodts et al., 2024; Kenney et al., 2018; Park et al., 2022). Methanobactin is a copper-chelating natural product, also called a "chalkophore," produced and secreted by some methanotrophic (methane-oxidizing) bacteria under copper-starved conditions (Kenney & Rosenzweig, 2018a, 2018b; Semrau, DiSpirito, Obulisamy, & Kang-Yun, 2020). Copper-bound Mbn is then imported back into the cell (Dassama, Kenney, Ro, Zielazinski, & Rosenzweig, 2016; Gu et al., 2016) where the copper is hypothesized to be released and used for physiological processes, including methane oxidation by the copper-dependent particulate methane monooxygenase (pMMO) (Tucci & Rosenzweig, 2024). MbnB performs four-electron oxidations of two cysteine

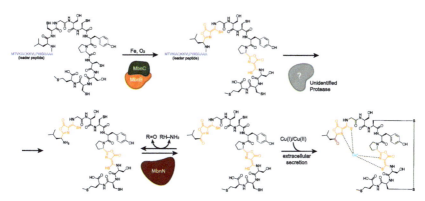

Fig. 1 Biosynthetic pathway of *M. trichosporium* OB3b Mbn. The post-translational modifications installed by MbnBC and MbnN are shown in orange and maroon, respectively. The two nitrogen and two sulfur atoms from the oxazolone/thioamide pairs form a four-coordinate binding site for Cu(I).

residues in MbnA to oxazolone/thioamide pairs (Fig. 1) (Kenney et al., 2018), which together coordinate Cu(I) with high affinity (10^{19-21} M^{-1}) (Behling et al., 2008; El Ghazouani et al., 2012; Kenney et al., 2016; Krentz et al., 2010; Park et al., 2021). The catalytically active species in MbnB is a mixed-valent Fe(II)/Fe(III) cluster (Park et al., 2022), and the cysteine sulfur from MbnA has been demonstrated to bind to the Fe(III) in the initial steps of the reaction (Jodts et al., 2024). Crystal structures of MbnB in complex with MbnC or both MbnC and MbnA show that the MbnC RRE recognizes the leader peptide of MbnA (Dou et al., 2022; Park et al., 2022). MbnB and MbnC are encoded in all Mbn BGCs, including those from non-methanotrophic bacteria (Kenney & Rosenzweig, 2013), and are thus the central processing enzymes in the Mbn biosynthetic pathway.

A number of tailoring enzymes confer additional modifications on Mbns and Mbn-like natural products in different species. In the case of *M. trichosporium* OB3b Mbn, the pyridoxal 5′-phosphate (PLP)-dependent aminotransferase MbnN replaces the *N*-terminal amino group of the core peptide with a carbonyl group, stabilizing mature Mbn (Fig. 1) (Park, Kenney, Schachner, Kelleher, & Rosenzweig, 2018). In several Mbns, one of the oxazolone/thioamide groups is replaced with a pyrazinedione ring, hypothesized to be installed by an FAD-dependent oxidoreductase, MbnF, found in some Mbn BGCs. Similarly, some Mbns contain a sulfonated threonine, consistent with the presence of a gene encoding the sulfotransferase MbnS in their BGCs. The functions of other tailoring enzymes, such as the putative dioxygenase MbnD, remain unclear. Notably, the Mbn BGC in *Vibrio fluvialis* encodes MbnB as well as an additional MNIO, MbnX, which performs a tailoring N-Cα bond cleavage reaction (Chioti et al., 2024). Many of these tailoring reactions likely occur after cleavage of the leader peptide by a yet-to-be-identified protease; there are no proteases encoded in the Mbn BGCs (Kenney & Rosenzweig, 2013).

In this chapter, we describe the heterologous expression and purification of *M. trichosporium* OB3b MbnBC and *M. trichosporium* OB3b MbnN, including MbnA modification assays and crystallization protocols for MbnBC (Jodts et al., 2024; Kenney et al., 2018; Park et al., 2018, 2022). Since the MbnN substrate can only be obtained from a Δ*mbnN* strain of *M. trichosporium* OB3b (Kenney et al., 2018; Park et al., 2018), activity assays using a commercial aminotransferase activity kit are outlined. These procedures can serve as a starting point for studying other Mbn or Mbn-like BGCs implicated in different RiPP biosynthetic pathways. Moreover, Mbn has shown promise as a therapeutic for Wilson disease (Einer et al., 2023; Lichtmannegger et al., 2016; Müller, Lichtmannegger, Zischka, Sperling, & Karst, 2018), and these methods could be deployed in drug development workflows.

2. Expression and purification of *M. trichosporium* OB3b MbnBC complexes

2.1 Plasmid construction and transformation into *Escherichia coli*

High-yield expression of active MbnBC complexes depends on construct design and expression systems, and may vary among homologs. Optimization of these parameters should be conducted. The optimized constructs for MbnBC from *M. trichosporium* OB3b include an *N*-terminal His$_6$ tag on the MbnC subunit and a *C*-terminal Strep II or S tag on the MbnB subunit in a pACYC-Duet1 vector (Jodts et al., 2024; Kenney et al., 2018; Park et al., 2022). MbnBC variants with site-specific mutations and with different combinations of affinity tags can also be expressed and purified and have been useful for crystallography (Park et al., 2022). Plasmids encoding both MbnB and MbnC can be constructed via gene synthesis (GenScript) and then transformed into expression cell lines. The procedures described below yield the most active MbnBC as determined by the ~335 nm absorbance resulting from MbnBC modification of MbnA (Section 3) unless stated otherwise.

Purification and biochemical characterization of methanobactin biosynthetic enzymes **175**

1. Obtain the pACYC-Duet1 plasmid encoding the MbnBC subunits via gene synthesis and prepare a 10 ng/μL working stock by dissolving or diluting the DNA in ddH$_2$O. Add 2 μL of the MbnBC plasmid to 20 μL of *E. coli* NiCo21 (DE3) (New England Biolabs, C2529H) in a pre-chilled microcentrifuge tube.

2. After incubating the cells on ice for 20–30 min, heat shock the cells by placing the microcentrifuge tube on a heat block pre-warmed to 42 °C for 45 s, and then chill it back on ice for 2 min.

3. Add 200 μL of SOC outgrowth medium (New England Biolabs, B9020S) to the cells and incubate the cell culture in a shaking incubator for 45 min to 1 h at 37 °C.

4. Plate 75–100 μL of cell culture on a pre-warmed LB agar plate containing 34 μg/mL chloramphenicol and incubate the plate overnight at 37 °C to allow colony formation.

5. To create a glycerol stock of a single colony, inoculate 5 mL of LB media containing 34 μg/mL chloramphenicol and incubate while shaking for 14–18 h at 37 °C. Add 500 μL of the culture to 500 μL of sterile 80% glycerol, then freeze and store at −80 °C until further use.

2.2 Large-scale growth and induction of MbnBC expression

Typically, MbnBC-expressing *E. coli* NiCo21 (DE3) cells are grown in autoinduction medium (Studier, 2014), which contains concentrations of glucose that enable cell growth to mid-to-late log phase without protein induction. Upon glucose depletion, inducing sugars such as lactose present in the medium are internalized and used by the cells after glucose depletion, initiating protein expression. Optimized expression of MbnBC involves overnight induction of protein expression at a lower temperature, after which the cells are harvested and frozen until purification. Expression induction using IPTG can also be used to produce active MbnBC, but further optimization will be necessary.

1. One day before the growth, prepare the desired amount of defined autoinduction media (Studier, 2014) in 2 L baffled flasks with the omission of FeCl$_3$ in the 1000× metal solution. The typical large-scale growth is 6–12 L of media to yield ~5 mg/L purified MbnBC. In addition, inoculate an overnight culture of LB media containing 34 μg/mL chloramphenicol with the glycerol stock of transformed *E. coli* NiCo21 (DE3) cells ensuring that there is greater than 10 mL allotted for every 1 L of autoinduction media.

2. On the day of the growth, add 34 µg/mL chloramphenicol to each liter of culture as well as ferrous ammonium sulfate to a final concentration of 160 µM. Iron supplementation of the media is necessary for the expression of catalytically active MbnBC. Differential iron supplementation leads to varying degrees of MbnB iron loading and activity with 160 µM resulting in the MbnBC species with the highest MbnA modification activity (Park et al., 2022). The optimal iron supplementation level should be determined for each MbnBC homolog.

3. Inoculate each liter of media with 10 mL of overnight culture and incubate while shaking at 37 °C. During the growth, monitor the OD_{600} of the cultures until it reaches a value of 0.4–0.6, which should take 3–4 h. To induce protein expression, reduce the temperature to 18 °C and continue to incubate for 20–24 h.

4. Separate the cells from the media by centrifuging the cultures at $6000 \times g$ for 10–15 min at 4 °C. The cell mass can be harvested and transferred to a sample bag using a spatula to be weighed and flash frozen in liquid N_2 until lysis and purification. Typical cell yields are 5–10 g/L cell culture.

2.3 Purification of MbnBC for biochemical studies

MbnBC from *M. trichosporium* OB3b is catalytically active even when fused to affinity tags (Jodts et al., 2024; Kenney et al., 2018; Park et al., 2022). As such, isolation of the complex only requires a two-step aerobic purification of nickel affinity chromatography followed by size exclusion chromatography without an affinity tag cleavage step. The resultant protein is sufficiently pure for functional and spectroscopic studies. Affinity tag cleavage for crystallization is described in Section 2.4.

1. Prepare 1 L of each of the following buffers, making sure to sterile filter and chill to 4 °C before use:
   a. Buffer A – 250 mM NaCl, 25 mM MOPS pH 7.2, 10% glycerol
   b. Buffer B – 250 mM NaCl, 25 mM MOPS pH 7.2, 10% glycerol, 1 M imidazole

2. Thaw and resuspend the cell pellet in Buffer A by stirring at 4 °C in the presence of EDTA-free cOmplete protease inhibitor (Roche) and DNAse.

3. Lyse the cells on ice via sonication at 4 °C for 5–10 min depending on the total mass of cell pellet. Typically, ~30 g of cells can be lysed with 5 min of sonication while > 60 g of cells will require closer to 10 min. To maintain the temperature of the cell slurry, sonicate the mixture for 1 s followed by a rest of 3 s.

Purification and biochemical characterization of methanobactin biosynthetic enzymes — **177**

4. Separate soluble protein from cell debris and other insoluble proteins by centrifuging the lysate at 22,000 × g for 1 h at 4 °C. Isolate the clarified lysate from the pellet by pipetting it into a pre-chilled container.

5. Equilibrate a pre-charged Cytiva HP-His Trap chelating column with Buffer A and use a peristaltic pump to load the column with the clarified lysate.

6. Elute any proteins that do not bind nickel by washing the column with 10 column volumes (CVs) of Buffer A while collecting 5 mL fractions. Elute the remaining nickel binding proteins via a 15 CV gradient of 0–50% Buffer B while collecting 2 mL fractions. Fully wash the column with 2 CVs of Buffer B and re-equilibrate with 5 CVs of Buffer A. To detect and visualize any protein-containing fractions, measure the absorbance of each fraction and/or run SDS-PAGE. MbnB and MbnC have molecular masses of ~32 and ~22 kDa, respectively.

7. Use an Amicon Ultra centrifugal filter with a molecular weight cutoff (MWCO) of 30 kDa to concentrate any MbnBC-containing fractions to ≤2 mL. Centrifuge the resulting protein sample at 20,000 × g for at least 5 min to pellet and separate any aggregated particulates.

8. Inject the MbnBC sample onto a Cytiva Superdex 75 size exclusion chromatography column that has been pre-equilibrated with Buffer A and run a 1 CV isocratic gradient of Buffer A while collecting 2 mL fractions. Detect protein-containing fractions as described above. A representative SDS-PAGE gel of purified MbnBC is shown in Fig. 2.

9. Concentrate MbnBC-containing fractions using a 30 kDa MWCO Amicon Ultra centrifugal filter and measure the concentration using the sample's absorbance at 280 nm (MbnBC $\varepsilon = 80{,}705\ \mathrm{M}^{-1}\ \mathrm{cm}^{-1}$, 1.478 g/L). MbnBC samples are typically concentrated to 1 mM for storage and subsequent MbnA modification assays. Aliquot samples of protein and flash freeze in liquid N_2 until further use.

2.4 Purification of MbnBC for crystallization

The affinity-tagged construct for MbnBC expression and purification described above did not crystallize. To obtain a crystallizable sample, multiple alterations to the sequence and purification protocol were necessary (Park et al., 2022). Only *M. trichosporium* OB3b MbnBC without affinity tags crystallized. In addition, the surface entropy reduction prediction (SERp) server (Goldschmidt, Cooper, Derewenda, & Eisenberg, 2007), which is no longer available, was used to design variants with multiple surface mutations on both subunits, detailed in (Park et al., 2022). This section details an extended purification process involving

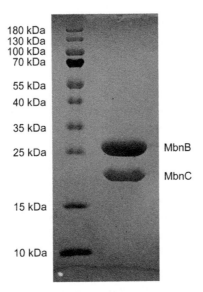

Fig. 2 Denaturing SDS-PAGE (16% Tris-glycine) of purified *M. trichosporium* OB3b MbnBC (~85 μM).

the cleavage of an N-terminal His$_6$+SUMO tag fused to the MbnC subunit using SUMO protease. It is important to test the activity of each complex variant before crystallization, described in Section 3.1.

1. Follow Sections 2.1 and 2.2 to transform and express the MbnBC complex containing an N-terminal His$_6$+SUMO tag on MbnC and no tag on MbnB in *E. coli* NiCo21 (DE3) cells. This construct can be obtained from gene synthesis services or DNA subcloning from the tagged MbnBC plasmid described in Section 2.1. Variants designed using the SERp server were generated by either gene synthesis services or by using the Quikchange Lightning Site-Directed Mutagenesis kits (Agilent).
2. In addition to sterile and chilled Buffers A and B from Section 2.3, prepare Buffers C and D:
    a. Buffer C – 250 mM NaCl, 25 mM MOPS pH 7.2, 10% glycerol, 10 mM imidazole
    b. Buffer D – 250 mM NaCl, 25 mM MOPS pH 7.2
3. Follow steps 2–6 in Section 2.3 to isolate nickel-binding proteins, including MbnBC, from the cell lysate. As described in step 7 of Section 2.3, concentrate MbnBC-containing fractions identified by SDS-PAGE by using an Amicon Ultra centrifugal filter with a MWCO of 30 kDa. Once the fractions have been concentrated to ≤1 mL, conduct a buffer exchange to remove the

imidazole by diluting the sample to 10 mL with Buffer A and concentrating again to ≤1 mL. Repeat this cycle once more.

4. In a 15 mL conical tube, dilute the protein sample to 10 mL using Buffer A. Add an appropriate amount of SUMO protease, which can be determined by conducting cleavage optimization studies. For a starting cell pellet of 30 g, an excess amount of 4 mg of SUMO protease is typically added. Incubate the reaction while shaking overnight at 4 °C.

5. On the next day, centrifuge the reaction at 20,000 × g for at least 10 min to isolate any aggregated particulates. Concentrate the soluble fraction to ≤2 mL using a 30 kDa MWCO Amicon Ultra centrifugal filter. Spin the sample one more time for at least 5 min to pellet any aggregates.

6. While collecting 5 mL fractions, inject the sample onto a pre-charged Cytiva HP-His Trap chelating column equilibrated in Buffer C and wash with 10 CVs of Buffer C. The cleaved and untagged MbnBC should elute within the first 1–5 fractions. Continue to elute the rest of the uncleaved MbnBC and other remaining nickel-binding proteins by running a 15 CV gradient from 0 to 50% Buffer B while collecting 2 mL fractions. Fully wash the column with 2 CVs of Buffer B and re-equilibrate with 5 CVs of Buffer C. Detect and visualize relevant protein fractions by SDS-PAGE as described above (Fig. 2).

7. Follow steps 7–8 in Section 2.3 for size exclusion chromatographic separation and detection of MbnBC. Concentrate relevant fractions using a 30 kDa MWCO Amicon Ultra centrifugal filter to ≤1 mL, and conduct a buffer exchange to remove the glycerol by diluting the sample to 10 mL with Buffer D and concentrating it to ≤1 mL. Repeat twice more. Measure the concentration using the sample's absorbance at 280 nm (MbnBC $\varepsilon = 80{,}705 \, \text{M}^{-1} \, \text{cm}^{-1}$, 1.478 g/L). Flash freeze aliquots of the sample in liquid N_2 and store at −80 °C until further use.

3. *In vitro* modification of MbnA using purified MbnBC

Modification of the MbnA precursor peptide using purified MbnBC requires the prior reduction of the MbnB iron cofactor in an anaerobic chamber. Once reduced to the active mixed-valent Fe(II)/Fe(III) species, addition of precursor peptide and O_2 will result in the modification of specific MbnA core peptide cysteine residues to oxazolone/thioamide pairs, a reaction that can be monitored using UV-Vis spectroscopy. The following procedure describes conducting reactions at an appropriate scale for monitoring MbnBC activity.

Reaction scales can be increased to synthesize sufficient concentrations of intermediate peptides for downstream reactions, although full synthesis of Mbn has not been achieved in vitro due to the unknown identity of the leader peptide protease and the instability of biosynthetic intermediates (Park et al., 2018).

1. Obtain the MbnA precursor peptide via peptide synthesis services. A heterologous expression system for MbnA has not been reported. Lyophilized samples should be dissolved in a degassed 50% (v/v) acetonitrile solution to a final concentration of 20 mM in a Coy or other anaerobic chamber to prevent introduction of O_2, which could initiate the MbnBC reaction prematurely. Flash freeze aliquots of MbnA in liquid N_2 and store at $-80\,°C$ until further use.

2. Before conducting modification reactions, degas a stock of Buffer A (Section 2.3) using a Schlenk line or allowing it to stir overnight inside the anaerobic chamber.

3. Degas an aliquot of MbnBC by allowing it to incubate uncapped inside the anaerobic chamber for 1 h at 4 °C. Prepare a 150 μL dilution of purified MbnBC to a final concentration of 100 μM using pre-chilled degassed Buffer A. Reduce the complex by adding 3–5 equivalents of L-ascorbic acid sodium salt and incubate for at least 1 h at 4 °C. This procedure has been shown by EPR spectroscopy to maximize the amount of active Fe(II)/Fe(III) species (Park et al., 2022).

4. Set-up the modification reaction in the anaerobic chamber by transferring the reduced MbnBC sample into a quartz cuvette with a minimum 150 μL sample volume and cap with a rubber stopper. For the blank, add Buffer A to another quartz cuvette. In the anaerobic chamber, aliquot 3-5 equivalents of MbnA peptide in a microcentrifuge tube to be added to the reaction in step 5. Set up the spectrophotometer to conduct a kinetic study, measuring the absorbance at 300–750 nm for at least 30 min.

5. Remove both quartz cuvettes and the MbnA aliquot from the anaerobic chamber. Blank the spectrophotometer with the quartz cuvette containing Buffer A. In the other cuvette containing the reduced MbnBC, quickly initiate the modification reaction at room temperature (~23 °C) by adding the MbnA and pipetting up and down repeatedly to introduce O_2 into the solution while taking absorbance measurements. Alternatively, MbnA can be initially diluted with 75 μL O_2 saturated Buffer A, which can then be added to reduced MbnBC that has been diluted with 75 μL degassed Buffer A from step 4. If MbnBC modifies MbnA, an increase in absorbance at 330 nm will be observed (Fig. 3) (Kenney et al., 2018).

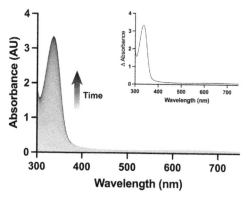

Fig. 3 Reaction of 100 μM ascorbate-reduced MbnBC with 300 μM MbnA over 30 min. Inset depicts the difference spectrum between the first and last spectrum.

4. Crystallization and structure determination of *M. trichosporium* OB3b MbnBC

MbnBC from *M. trichosporium* OB3b crystallizes overnight at 10 °C using a precipitant solution containing Li_2SO_4, bis-tris propane, and PEG 3350 with the sitting drop diffusion method. Crystals will form into rectangular prisms and plates (Fig. 4A) and can be harvested and frozen in a cryoprotectant solution of the precipitant with the addition of 10% ethylene glycol. MbnBC crystals belong to space group $C222_1$ with unit cell dimensions of $a = 49.78$ Å, $b = 215.64$ Å, $c = 214.15$ Å, and typically diffract to better than 3 Å resolution (Park et al., 2022). The asymmetric unit contains two copies of the MbnBC complex (Fig. 4B). Phasing and identification of Fe in the electron density map can be performed using anomalous data collected at the Fe absorption edge (7.1120 keV or 1.7433 Å), and additional structures can be determined by molecular replacement.

1. Optimal crystal formation occurs with reservoir solutions containing a range of PEG 3350 concentrations. Prepare 6 stock reservoir solutions with the following concentrations: 0.2 M Li_2SO_4, 0.1 M bis-tris propane pH 7.0, 20–26% PEG 3350, with each solution increasing by 1% PEG 3350 concentration. Prepare enough stock solutions to have ≥125 μL per well. Dilute MbnBC to 10 mg/mL using Buffer D (Section 2.4), preparing enough for 1 μL per drop to be set up.
2. In a CombiClover sitting drop vapor diffusion tray (MiTeGen, M-R-100954X), add 125 μL of reservoir solution into each large center well, with the PEG 3350 concentration increasing down the row. Add 1 μL of

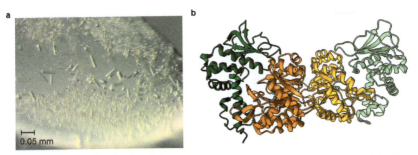

Fig. 4 Crystallization and structure of MbnBC. (A) MbnBC crystals visualized by light microscope. (B) Asymmetric unit of the MbnBC crystal lattice containing 2 copies of MbnBC colored in two shades of orange (MbnB) and green (MbnC) (PDB: 7TCR). The Fe ions are colored in teal.

10 mg/mL MbnBC and 1 μL of the corresponding reservoir solution to create 1:1 2 μL crystallization drops in each smaller well surrounding the reservoir well. Cover the tray with clear tape and incubate overnight at 10 °C.

3. Crystals should form and be visible using a light microscope overnight (Fig. 4A). To harvest and freeze for data collection, prepare cryoprotectant solutions containing identical concentrations of each reagent corresponding to the specific well conditions with 10% ethylene glycol. Pick a crystal using a loop and dip it into a small drop of the cryoprotectant solution. Flash freeze the crystal in liquid N_2 and store until data collection.

4. After diffraction data are collected, including some at the iron absorption edge, XDS (Kabsch, 2010) can be used to process data and initial structure determination using Fe single-wavelength anomalous dispersion (SAD) can be performed with autoSHARP (Vonrhein, Blanc, Roversi, & Bricogne, 2007). Model building and structure refinement can be done using Phenix (Liebschner et al., 2019) and Coot (Emsley, Lohkamp, Scott, & Cowtan, 2010).

5. Expression and purification of *M. trichosporium* OB3b MbnN

The expression and purification of active MbnN dimeric complexes from *M. trichosporium* OB3b required multiple optimizations as conditions such as buffer identity and pH can affect its aminotransferase activity (Park et al., 2018). Starting with a pNYCOMPS plasmid encoding MbnN with a

Purification and biochemical characterization of methanobactin biosynthetic enzymes **183**

C-terminal His$_6$ tag that is cleavable with TEV protease, the expression procedure is almost identical to that for MbnBC with minor changes. The purification protocol is similar to that for MbnBC, and includes an initial nickel affinity chromatography step, His$_6$ tag cleavage, and a final subtractive nickel affinity chromatography step. Use of the native substrate requires cultivating a $\Delta mbnN$ strain of *M. trichosporium* OB3b (Kenney et al., 2018; Park et al., 2018) so this procedure instead includes a test for general aminotransferase activity using a commercial kit.

1. Obtain a pNYCOMPS or other suitable plasmid encoding MbnN from *M. trichosporium* OB3b with a C-terminal His$_6$ tag and transform *E. coli* NiCo21 (DE3) as described in Section 2.1.

2. Follow the procedures in Section 2.2. to express MbnN-His$_6$ using autoinduction media. Instead of supplementing each liter of media with $160\,\mu M$ ferrous ammonium sulfate before inoculation, only supplement to a final concentration of $50\,\mu M$ as MbnN is not an iron-dependent enzyme.

3. Prepare 1 L of each of the following buffers, making sure to sterile filter and chill to $4\,°C$ before use:
   a. Buffer A – $250\,mM$ NaCl, $25\,mM$ phosphate buffer pH 8.0, 10% glycerol
   b. Buffer B – $250\,mM$ NaCl, $25\,mM$ phosphate buffer pH 8.0, 10% glycerol, 1 M imidazole
   c. Buffer C – $250\,mM$ NaCl, $25\,mM$ phosphate buffer pH 8.0, 10% glycerol, $10\,mM$ imidazole
   d. Dialysis buffer – $250\,mM$ NaCl, $25\,mM$ phosphate buffer pH 8.0, 10% glycerol

4. Thaw and resuspend the cell pellet in Buffer A by stirring at $4\,°C$ in the presence of EDTA-free cOmplete protease inhibitor and DNAse. Optional supplementation of the lysis mixture with $250\,\mu M$ PLP can be done to increase PLP loading and activity of resulting MbnN complexes, but is not absolutely necessary.

5. Lyse the cells on ice by sonication at $4\,°C$ for 5–10 min depending on the total mass of cell pellet. To maintain the temperature of the cell slurry, sonicate the mixture for 1 s followed by a rest for 5 s.

6. Follow steps 4–6 in Section 2.3 to conduct initial nickel affinity chromatography purification and detection of MbnN-containing fractions. MbnN has an expected molecular mass of $41.4\,kDa$ and an additional absorbance at 333 nm from the PLP cofactor. Concentrate MbnN-containing fractions using an Amicon Ultra centrifugal filter

with a MWCO of 30 kDa. Once the fractions have been concentrated to ≤1 mL, conduct a buffer exchange to remove imidazole by diluting the sample to 10 mL with Buffer A and concentrate again to ≤1 mL. Repeat this cycle once more.

7. Transfer the protein sample to a 15 mL conical tube and dilute to 5 mL using Buffer A supplemented with 1 mM DTT and 0.5 mM EDTA. Add an appropriate amount of TEV protease, which can be determined by conducting cleavage optimization studies. Typically, 1 mg of TEV protease is added for 30 g starting cell pellet. Inject the sample into a Slide-a-Lyzer dialysis cassette with a 10 kDa MWCO and conduct an overnight dialysis against the dialysis buffer at 4 °C.

8. Follow steps 5–6 in Section 2.4 to conduct subtractive nickel affinity chromatography to isolate the cleaved MbnN proteins. In addition to SDS-PAGE (Fig. 5A) and UV-Vis absorbance (Fig. 5B), running native PAGE is also very beneficial to determining the multimeric state of the resulting MbnN samples. Active MbnN forms a dimeric complex and should run at ~82 kDa on a native PAGE gel (Fig. 5C).

9. Concentrate purified MbnN fractions using a 30 kDa MWCO Amicon Ultra centrifugal filter until the sample is ≤1 mL and measure the concentration using the absorbance at 280 nm (MbnN $\varepsilon = 24,785$ M^{-1} cm^{-1}, 0.598 g/L). Flash freeze aliquots of the sample in liquid N$_2$ and store at −80 °C until further use.

10. Measure aminotransferase activity of the purified MbnN using a commercial kit such as the aspartate aminotransferase activity assay kit (Abcam, ab105135).

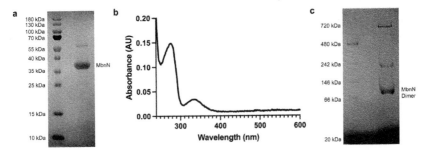

Fig. 5 Purification of *M. trichosporium* OB3b MbnN. (A) Denaturing 16% Tris-glycine SDS-PAGE of ~20 μM MbnN. (B) UV-Vis spectrum of ~3 μM MbnN. (C) Native PAGE (4–16% Bis-Tris) of ~20 μM MbnN.

6. Conclusions

The methods described here for heterologous expression, purification, and characterization of the Mbn biosynthetic enzyme complex MbnBC provide a roadmap for studying other Mbn or Mbn-like BGCs. In particular, MbnB-like MNIOs and their associated partner proteins are widespread and likely to catalyze a range of yet-to-be-discovered reactions (Chen & van der Donk, 2024). Tailoring enzymes such as MbnN are amenable to purification and biochemical characterization, as outlined here. However, functional assays remain challenging since the appropriate substrate requires leader peptide cleavage, of which the mechanism is not known. Identifying the elusive protease is thus an important goal for future studies.

Acknowledgments

This work was supported by NIH grants GM118035 (A.C.R.) and T32GM008449 (R. M. R.).

References

Arnison, P. G., Bibb, M. J., Bierbaum, G., Bowers, A. A., Bugni, T. S., Bulaj, G., ... van der Donk, W. A. (2013). Ribosomally synthesized and post-translationally modified peptide natural products: Overview and recommendations for a universal nomenclature. *Natural Product Reports, 30*, 108–160. https://doi.org/10.1039/C2NP20085F.

Ayikpoe, R. S., Zhu, L. Y., Chen, J. Y., Ting, C. P., & van der Donk, W. A. (2023). Macrocyclization and backbone rearrangement during RiPP biosynthesis by a SAM-dependent domain-of-unknown-function 692. *ACS Central Science, 9*(5), 1008–1018. https://doi.org/10.1021/acscentsci.3c00160.

Behling, L. A., Hartsel, S. C., Lewis, D. E., Dispirito, A. A., Choi, D. W., Masterson, L. R., ... Gallagher, W. H. (2008). NMR, mass spectrometry and chemical evidence reveal a different chemical structure for methanobactin that contains oxazolone rings. *Journal of the American Chemical Society, 130*, 12604–12605.

Chen, J. Y., & van der Donk, W. A. (2024). Multinuclear non-heme iron dependent oxidative enzymes (MNIOs) involved in unusual peptide modifications. *Current Opinion in Chemical Biology, 80*, 102467. https://doi.org/10.1016/j.cbpa.2024.102467.

Chioti, V. T., Clark, K. A., Ganley, J. G., Han, E. J., & Seyedsayamdost, M. R. (2024). N-Cα bond cleavage catalyzed by a multinuclear iron oxygenase from a divergent methanobactin-like RiPP gene cluster. *Journal of the American Chemical Society, 146*(11), 7313–7323. https://doi.org/10.1021/jacs.3c11740.

Dassama, L. M., Kenney, G. E., Ro, S. Y., Zielazinski, E. L., & Rosenzweig, A. C. (2016). Methanobactin transport machinery. *Proceedings of the National Academy of Sciences USA, 113*(46), 13027–13032. https://doi.org/10.1073/pnas.1603578113.

Dou, C., Long, Z., Li, S., Zhou, D., Jin, Y., Zhang, L., ... Cheng, W. (2022). Crystal structure and catalytic mechanism of the MbnBC holoenzyme required for methanobactin biosynthesis. *Cell Research*. https://doi.org/10.1038/s41422-022-00620-2.

Einer, C., Munk, D. E., Park, E., Akdogan, B., Nagel, J., Lichtmannegger, J., ... Zischka, H. (2023). ARBM101 (methanobactin SB2) drains excess liver copper via biliary excretion in Wilson's disease rats. *Gastroenterology*. https://doi.org/10.1053/j.gastro.2023.03.216.

El Ghazouani, A., Basle, A., Gray, J., Graham, D. W., Firbank, S. J., & Dennison, C. (2012). Variations in methanobactin structure influences copper utilization by methane-oxidizing bacteria. *Proceedings of the National Academy of Sciences USA, 109*(22), 8400–8404. https://doi.org/10.1073/pnas.1112921109.

Emsley, P., Lohkamp, B., Scott, W. G., & Cowtan, K. (2010). Features and development of Coot [Article]. *Acta Crystallographica, D66*, 486–501. https://doi.org/10.1107/s0907444910007493.

Goldschmidt, L., Cooper, D. R., Derewenda, Z. S., & Eisenberg, D. (2007). Toward rational protein crystallization: A Web server for the design of crystallizable protein variants. *Protein Science: A Publication of the Protein Society, 16*(8), 1569–1576. https://doi.org/10.1110/ps.072914007.

Gu, W., Farhan Ul Haque, M., Baral, B. S., Turpin, E. A., Bandow, N. L., Kremmer, E., ... Semrau, J. D. (2016). A TonB-dependent transporter is responsible for methanobactin uptake by *Methylosinus trichosporium* OB3b. *Applied and Environmental Microbiology, 82*(6), 1917–1923. https://doi.org/10.1128/aem.03884-15.

Jodts, R. J., Ho, M. B., Reyes, R. M., Park, Y. J., Doan, P. E., Rosenzweig, A. C., & Hoffman, B. M. (2024). Initial steps in methanobactin biosynthesis: Substrate binding by the mixed-valent diiron enzyme MbnBC. *Biochemistry*. https://doi.org/10.1021/acs.biochem.4c00011.

Kabsch, W. (2010). Integration, scaling, space-group assignment and post-refinement [Article]. *Acta Crystallographica, D66*, 133–144. https://doi.org/10.1107/s0907444909047374.

Kenney, G. E., Dassama, L. M. K., Pandelia, M. E., Gizzi, A. S., Martinie, R. J., Gao, P., ... Rosenzweig, A. C. (2018). The biosynthesis of methanobactin. *Science (New York, N. Y.), 359*(6382), 1411–1416. https://doi.org/10.1126/science.aap9437.

Kenney, G. E., Goering, A. W., Ross, M. O., DeHart, C. J., Thomas, P. M., Hoffman, B. M., ... Rosenzweig, A. C. (2016). Characterization of methanobactin from *Methylosinus* sp. LW4. *Journal of the American Chemical Society, 138*(35), 11124–11127. https://doi.org/10.1021/jacs.6b06821.

Kenney, G. E., & Rosenzweig, A. C. (2013). Genome mining for methanobactins. *BMC Biology, 11*, 17.

Kenney, G. E., & Rosenzweig, A. C. (2018a). Chalkophores. *Annual Review of Biochemistry, 87*, 645–676. https://doi.org/10.1146/annurev-biochem-062917-012300.

Kenney, G. E., & Rosenzweig, A. C. (2018b). Methanobactins: Maintaining copper homeostasis in methanotrophs and beyond. *The Journal of Biological Chemistry, 293*(13), 4606–4615. https://doi.org/10.1074/jbc.TM117.000185.

Krentz, B. D., Mulheron, H. J., Semrau, J. D., DiSpirito, A. A., Bandow, N. L., Haft, D. H., ... Gallagher, W. H. (2010). A comparison of methanobactins from *Methylosinus trichosporium* OB3b and *Methylocystis* strain SB2 predicts methanobactins are synthesized from diverse peptide precursors modified to create a common core for binding and reducing copper ions [Article]. *Biochemistry, 49*(47), 10117–10130. https://doi.org/10.1021/bi1014375.

Lichtmannegger, J., Leitzinger, C., Wimmer, R., Schmitt, S., Schulz, S., Kabiri, Y., ... Zischka, H. (2016). Methanobactin reverses acute liver failure in a rat model of Wilson disease. *The Journal of Clinical Investigation, 126*(7), 2721–2735. https://doi.org/10.1172/jci85226.

Liebschner, D., Afonine, P. V., Baker, M. L., Bunkóczi, G., Chen, V. B., Croll, T. I., ... Adams, P. D. (2019). Macromolecular structure determination using X-rays, neutrons and electrons: Recent developments in Phenix. *Acta Crystallographica Section D Structural Biology, 75*(Pt 10), 861–877. https://doi.org/10.1107/s2059798319011471.

Montalbán-López, M., Scott, T. A., Ramesh, S., Rahman, I. R., van Heel, A. J., Viel, J. H., ... van der Donk, W. A. (2021). New developments in RiPP discovery, enzymology and engineering. *Natural Product Reports, 38*(1), 130–239. https://doi.org/10.1039/d0np00027b.

Müller, J. C., Lichtmannegger, J., Zischka, H., Sperling, M., & Karst, U. (2018). High spatial resolution LA-ICP-MS demonstrates massive liver copper depletion in Wilson disease rats upon Methanobactin treatment. *Journal of Trace Elements in Medicine and Biology: Organ of the Society for Minerals and Trace Elements (GMS), 49*, 119–127. https://doi.org/10.1016/j.jtemb.2018.05.009.

Nguyen, D. T., Mitchell, D. A., & van der Donk, W. A. (2024). Genome mining for new enzyme chemistry. *ACS Catalysis, 14*(7), 4536–4553. https://doi.org/10.1021/acscatal.3c06322.

Nguyen, D. T., Zhu, L., Gray, D. L., Woods, T. J., Padhi, C., Flatt, K. M., ... van der Donk, W. A. (2024). Biosynthesis of macrocyclic peptides with C-terminal β-amino-α-keto acid groups by three different metalloenzymes. *bioRxiv*. https://doi.org/10.1101/2023.10.30.564719.

Park, Y. J., Jodts, R. J., Slater, J. W., Reyes, R. M., Winton, V. J., Montaser, R. A., ... Rosenzweig, A. C. (2022). A mixed-valent Fe(II)Fe(III) species converts cysteine to an oxazolone/thioamide pair in methanobactin biosynthesis. *Proceedings of the National Academy of Sciences USA, 119*(13), e2123566119. https://doi.org/10.1073/pnas.2123566119.

Park, Y. J., Kenney, G. E., Schachner, L. F., Kelleher, N. L., & Rosenzweig, A. C. (2018). Repurposed HisC aminotransferases complete the biosynthesis of some methanobactins. *Biochemistry, 57*(25), 3515–3523. https://doi.org/10.1021/acs.biochem.8b00296.

Park, Y. J., Roberts, G. M., Montaser, R., Kenney, G. E., Thomas, P. M., Kelleher, N. L., & Rosenzweig, A. C. (2021). Characterization of a copper-chelating natural product from the methanotroph *Methylosinus* sp. LW3. *Biochemistry, 60*(38), 2845–2850. https://doi.org/10.1021/acs.biochem.1c00443.

Pfeiffer, I. P. M., Schröder, M. P., & Mordhorst, S. (2024). Opportunities and challenges of RiPP-based therapeutics. *Natural Product Reports*. https://doi.org/10.1039/d3np00057e.

Semrau, J. D., DiSpirito, A. A., Obulisamy, P. K., & Kang-Yun, C. S. (2020). Methanobactin from methanotrophs: Genetics, structure, function and potential applications. *FEMS Microbiology Letters, 367*(5), fnaa045. https://doi.org/10.1093/femsle/fnaa045.

Studier, F. W. (2014). Stable expression clones and auto-induction for protein production in *E. coli. Methods in Molecular Biology, 1091*, 17–32. https://doi.org/10.1007/978-1-62703-691-7_2.

Ting, C. P., Funk, M. A., Halaby, S. L., Zhang, Z. G., Gonen, T., & van der Donk, W. A. (2019). Use of a scaffold peptide in the biosynthesis of amino acid-derived natural products. *Science (New York, N. Y.), 365*(6450), 280–284. https://doi.org/10.1126/science.aau6232.

Tucci, F. J., & Rosenzweig, A. C. (2024). Direct methane oxidation by copper- and iron-dependent methane monooxygenases. *Chemical Reviews, 124*, 1288–1320.

Vonrhein, C., Blanc, E., Roversi, P., & Bricogne, G. (2007). Automated structure solution with autoSHARP. *Methods in Molecular Biology, 364*, 215–230. http://www.ncbi.nlm.nih.gov/entrez/query.fcgi?cmd=Retrieve&db=PubMed&dopt=Citation&list_uids=17172768.

Yu, Y., & van der Donk, W. A. (2022). Biosynthesis of 3-thia-α-amino acids on a carrier peptide. *Proceedings of the National Academy of Sciences USA, 119*(29), e2205285119. https://doi.org/10.1073/pnas.2205285119.

CHAPTER NINE

Discovery, isolation, and characterization of diazeniumdiolate siderophores

Melanie Susman[1], Jin Yan[1], Christina Makris, and Alison Butler*

Department of Chemistry & Biochemistry, University of California, Santa Barbara, CA, United States
*Corresponding author. e-mail address: butler@chem.ucsb.edu

Contents

1. Introduction	190
2. Bioinformatics and genome mining to predict bacteria producing graminine-containing siderophores	192
2.1 Materials	192
2.2 Constructing an SSN	192
2.3 Identifying bacterial strains of interest	194
3. Bacterial growth conditions for production of siderophores	195
3.1 Materials	195
3.2 Protocols for bacterial culturing	196
4. Detection and isolation of the siderophores	198
4.1 Materials	198
4.2 CAS assay for general siderophore detection	199
4.3 Extraction of siderophores	200
4.4 Purification of siderophores	202
5. Identification and characterization of graminine-containing siderophores	203
5.1 Materials	204
5.2 Mass spectrometry and stable isotope labeling	205
5.3 Marfey's amino acid analysis	206
5.4 Photoreactivity of the diazeniumdiolate group	208
5.5 NMR spectroscopy characterization of the C-diazeniumdiolate and the photoproducts	211
6. Conclusions and future outlook	212
Acknowledgements	212
References	213

Abstract

The C-diazeniumdiolate (N-nitrosohydroxylamine) group in the amino acid graminine (Gra) is a newly discovered Fe(III) ligand in microbial siderophores. Graminine was first identified in the siderophore gramibactin, and since this discovery, other Gra-containing

[1] Co-first authors.

Methods in Enzymology, Volume 702
ISSN 0076-6879, https://doi.org/10.1016/bs.mie.2024.06.006
Copyright © 2024 Elsevier Inc. All rights reserved, including those for text and data
mining, AI training, and similar technologies.

189

siderophores have been identified, including megapolibactins, plantaribactin, gladiobactin, trinickiabactin (gramibactin B), and tistrellabactins. The C-diazeniumdiolate is photoreactive in UV light which provides a convenient characterization tool for this type of siderophore. This report details the process of genomics-driven identification of bacteria producing Gra-containing siderophores based on selected biosynthetic enzymes, as well as bacterial culturing, isolation and characterization of the C-diazeniumdiolate siderophores containing Gra.

1. Introduction

Cells of nearly all forms of life require iron to function. Bacteria and fungi often produce small organic molecules called "siderophores" which have a high affinity for Fe(III). The Fe(III)-siderophore complex is taken up by the cell for metabolic use (Sandy & Butler, 2009). Until recently, the ligands within characterized siderophores fell into the categories of catecholate, hydroxamate, α-hydroxycarboxylate, oxazoline and thiazoline, most commonly. Part of the non-proteogenic amino acid graminine (Gra), the C-type diazeniumdiolate functional group (also known as N-nitrosohydroxylamine) (Fig. 1) is a newly discovered siderophore ligand, first characterized in gramibactin (Gbt), a siderophore produced by the rhizospheric bacterium, *Paraburkholderia graminis* (Hermenau et al., 2018). Only a few other examples of C-type diazeniumdiolates are found in natural products (Dolak, Castle, Hannon, Argoudelis, & Reusser, 1983; Iimura, Takeuchi, Kondo, Matsuzaki, & Umezawa, 1972; Jenul et al., 2018; Murthy, Thiemann, Coronelli, & Sensi, 1966; Natori, Kataoka, Kato, Kawai, & Fusetani, 1997; Nishio et al., 1993; Sieber et al., 2021; Tamura, Murayama, & Hata, 1967). Since the discovery of gramibactin, several diazeniumdiolate siderophores have been characterized, including gramibactin, trinickiabactin (gramibactin B), megapolibactins, plantaribactin, gladiobactin, and tistrellabactins (Fig. 1). Most of the bacteria producing these siderophores are found in soil, with the exception of *Tistrella mobilis* KA081020-065, a marine strain which produces the tistrellabactins (Hermenau et al., 2018, 2019; Jiao, Du, Frediansyah, Jahanshah, & Gross, 2020; Makris, Leckrone, & Butler, 2023).

In addition to the ability to coordinate Fe(III), diazeniumdiolate siderophores have also been found to produce nitric oxide (NO) both in vivo and in vitro (Hermenau et al., 2019; Makris, Carmichael, Zhou, & Butler, 2022). Nitric oxide is a versatile signaling molecule in bacteria and higher eukaryotes (Tuteja, Chandra, Tuteja, & Misra, 2004). While synthetic N-diazeniumdiolate compounds are well-known NO donors in various chemical systems, little is

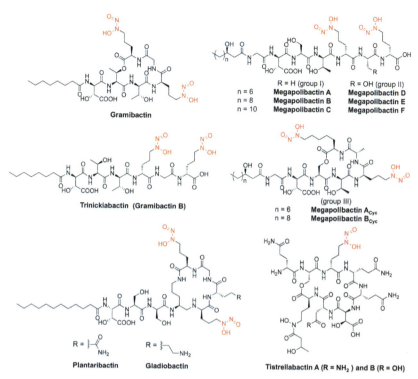

Fig. 1 Characterized C-diazeniumdiolate siderophores. The diazeniumdiolate group is shown in red.

known about the chemistry of C-diazeniumdiolate natural products and their ability to donate NO (Hrabie & Keefer, 2002; Li et al., 2020). It was recently established that upon UV photolysis (e.g., Hg(Ar) pen lamp or sunlight), C-diazeniumdiolate siderophores release the equivalent of NO and H, forming an oxime photoproduct, thus providing a pathway for bacteria to produce NO. The photoproduct loses its ability to bind Fe(III), which hints at a potential greater ecological role for these siderophores, and also provides an attractive rationale for the development of related compounds with pharmaceutical applications (Makris et al., 2022, 2023).

Through microbial gene knock-out studies on the biosynthetic gene cluster (BGC) of gramibactin, two enzymes, GrbD and GrbE, were found to be responsible for the biosynthesis of Gra, the non-canonical amino acid that bears the diazeniumdiolate functional group in diazeniumdiolate siderophores (Hermenau et al., 2019). L-Arg is the precursor of L-Gra based on the results of isotopic labeling studies (Makris et al., 2022). The biosynthesis of *C*-diazeniumdiolate compounds, including that of Gra, is a topic of much current interest (Jenul et al., 2018; Morgan & Li, 2020; Sieber et al., 2020, 2021; Wang & Ryan, 2023; Wang, Niikura, He, Daniel-Ivad, & Ryan, 2020).

In this chapter, we detail the process of genomics-driven discovery, bacterial culturing, and the isolation and characterization of the *C*-diazeniumdiolate siderophores containing Gra, to provide researchers with guidance to carry out experiments related to this novel class of siderophores.

2. Bioinformatics and genome mining to predict bacteria producing graminine-containing siderophores

Targeted discovery of new *C*-diazeniumdiolate siderophores is achieved by using multiple bioinformatics tools. Sequence similarity network (SSN) helps visualize the relationships among protein sequences (Atkinson, Morris, Ferrin, & Babbitt, 2009). The SSN of GrbD (or GrbE) is created and parameters are adjusted to select the entries which are likely in the BGCs of diazeniumdiolate siderophores. Accession codes of the genomes of certain bacterial strains are gained from NCBI, and the possible secondary metabolites BGCs are predicted using antiSMASH (antibiotics & Secondary Metabolite Analysis Shell) (Blin et al., 2023). By searching for GrbE (or GrbD) homologs and nonribosomal peptide synthetase (NRPS) adenylation (A) domains with the specificity code for Gra, candidate microbe-producing diazeniumdiolate siderophores can be predicted.

2.1 Materials

2.1.1 Equipment (software)
- NCBI (https://www.ncbi.nlm.nih.gov/)
- EFI—Enzyme similarity tool (EST) (https://efi.igb.illinois.edu/efi-est/)
- Cytoscape (https://cytoscape.org/)
- yFiles Layout Algorithms for Cytoscape (https://www.yworks.com/products/yfiles-layout-algorithms-for-cytoscape)
- antiSMASH (https://antismash.secondarymetabolites.org/)

2.2 Constructing an SSN
1. Obtain the sequence of GrbD (Accession code WP_006051176.1) or GrbE (Accession code WP_006051175.1) on NCBI website.
2. To create an SSN, first go to EFI–EST website.

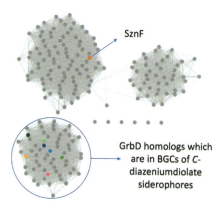

Fig. 2 SSN of GrbD (e-value = 5, alignment score threshold = 80).

3. Copy and paste the protein sequence into "Query Sequence" under "Sequence BLAST".
4. Edit UniProt BLAST query e-value under "BLAST Retrieval Options." The default value is set to 5. Increasing the e-value can help to retrieve more divergent homologs; decreasing the e-value gives more similar homologs. The value was set to 5 for creating SSN for GrbD (Fig. 2).
5. Edit other parameters (e.g., maximum number of sequences retrieved, sequence database, etc), if necessary.
6. Enter job name and email address to receive job updates by email.
7. There will be one email notification for successfully submitting the job and another email after the initial calculation is complete. Click on the link to finalize the SSN.
8. Go to the "Dataset analysis" tab and look for "percent identity vs. alignment score" graph. The alignment score corresponding to 35% identity is a good starting point to enter in "Alignment Score Threshold" under "SSN Finalization." The higher the alignment score, more smaller clusters will be obtained; when the alignment score is lower, there will be less clusters and the clusters will be bigger. For finalizing SSN of GrbD, 80 was entered.
9. An email is sent when the SSN is completed and available for download. Click on the link in the email and go to "Network Files".
10. Download the full network or representative node networks. In the representative node networks, the sequences with higher than a certain percent identity are put into the same node. The computer will process the SSN faster with fewer nodes in it.

11. Unzip the file and drag the.xgmml file into Cytoscape, which is an open-source software platform to visualize various networks.
12. Under "Layout", choose "yFiles Organic Layout".
13. Analyze the clusters; look for the cluster of which the sequences in BGCs of characterized diazeniumdiolate siderophores are in. Adjust the parameters to create SSN if necessary.
14. Taking the SSN of GrbD as an example (Fig. 2), the GrbD homologs which are in BGCs of characterized diazeniumdiolate siderophores are in the same cluster, and SznF, the N–N bond forming enzyme of streptozotocin, which does not have a diazeniumdiolate group, is in a separate cluster. Thus, the sequences in the former cluster are candidate GrbD homologs involved in the biosynthesis of diazeniumdiolate siderophores.

2.3 Identifying bacterial strains of interest

1. Find the NCBI IDs of the sequence of interest in SSN.
2. Search for the NCBI ID on NCBI website.
3. Choose "Identical Protein Groups"; look for the bacteria strain name.
4. Search for the bacterial strain name on NCBI website; choose "Nucleotide."
5. Acquire the accession code ("Accession:") under the entry name containing "whole genome shotgun sequencing project." Sort the results by sequence length at the bottom of this page.
6. In antiSMASH, select "Get from NCBI"; enter the accession code. Change parameters if necessary and submit the job.
7. After the job is completed, download the log file by clicking "Download" at the top right of the page and recheck results in a certain period of time without re-running the job. To recheck results, enter the log file name under "Results for existing job" at the start page.
8. Go through the predicted regions and look for the BGC candidates which may produce new diazeniumdiolate siderophores. The predicted type of the region is usually "NRPS" or "NRP-metallophore", and sometimes the name of the most similar known cluster contains the name of a characterized diazeniumdiolate siderophore. Click on regions of interest.
9. Look for the GrbD homolog—the homolog of GrbD can usually be found since the strain information is obtained from the SSN created for GrbD. At the bottom half of the page, check through the list of genes. GrbD homolog is usually labeled as "hypothetical protein" or "peptide synthetase", the function is "biosynthetic", and the amino acid sequence

length is between 450 to 500. Click on the possible gene, the "Gene details" at the top right of the page contains more information on this gene, in which "GrbD" is usually found.
10. Look for GrbE homolog—the homolog of GrbE is usually next to the GrbD homolog or very close to it. It is sometimes a stand-alone gene, predicted to be either "YqcI/YcgG protein" or "hypothetical protein", and in other cases fuses with a NRPS. The amino acid sequence length is around 290. If it is fused with a NRPS, "GrbE" will be listed in the "Gene details" of that NRPS module.
11. To check the prediction of the amino acid building blocks of this possible siderophore, click on "NRPS/PSK" modules at the bottom half of the page. Compare the predictions to the characterized siderophores to check its novelty. A "CAL" (CoA ligase) module suggests that this possible siderophore may have a fatty acid tail besides the peptide backbone. If a module is predicted to be "X" (or "D-X"), click on "NRPS/PKS substrates" at the middle-right of the page, and click on "+" and check the prediction of this amino acid. If the prediction is Gra (specificity code DVHRTGLVAK), this may be a new diazeniumdiolate siderophore.
12. Some bacterial strains can be obtained from sources like DSMZ (https://www.dsmz.de/), ATCC (https://www.atcc.org/), etc.

3. Bacterial growth conditions for production of siderophores

3.1 Materials

3.1.1 Equipment
- Freezer (−80 °C)
- Autoclave
- Orbital shaking incubator
- UV-Visible spectrophotometer

3.1.2 Reagents and supplies
- Large Erlenmeyer or Fernbach flask for bacterial culturing
- Hydrochloric acid (4 M)
- Ultrapure (e.g., MilliQ, 18 Ω) water
- Growth nutrients and salts
- Magnetic stirrer

- Stir bar
- Aluminum foil
- Ethanol to sterilize surfaces and gloves
- Bunsen burner to maintain sterile conditions
- Large petri dishes
- Agar
- Luria–Bertani (LB) broth base or other nutrient-dense broth base
- Long-term storage cryogenic tubes for freezer stocks
- Sterile culture tubes
- Inoculation loops
- Plastic cuvettes
- Sterile pipette tips
- Micropipette

3.2 Protocols for bacterial culturing

Once bacterial strains have been selected and obtained, prepare glycerol freezer stocks as instructed by the bacterial supplier. Store the glycerol stocks at −80 °C. All bacteria should be grown up in media, temperature, and pH conditions best suited for the specific strain. Bacterial growth conditions are usually listed on the purchasing site.

3.2.1 Preparation of iron-deficient media

Siderophore production is stimulated by iron deficiency. To achieve iron-limited conditions, all glassware used for growth should first be rinsed with 4 M hydrochloric acid to remove adsorbed Fe(III). Bacterial media should provide the nutrients necessary for the specific strain to grow but should not include a source of iron. Small batch (~150–500 mL) cultures can be grown to test media conditions before selecting a growth medium that is suitable for growth. For terrestrial bacteria that produce diazeniumdiolate siderophores, iron-deficient M9 minimal medium is a good place to start, and supplementation with sodium pyruvate may accelerate growth. To grow marine bacteria, artificial seawater lacking iron can be used, though some adjustments to specific salt concentrations may be necessary.

1. Rinse an Erlenmeyer or Fernbach flask twice the volume of desired culture (i.e., for a 1 L culture, use a 2 L flask) with 4 M HCl. Rinse out excess acid with ultrapure (e.g., MilliQ, 18 Ω) water.

2. Add all but 10 mL of the intended final volume of ultrapure water. The remaining water will be used to rinse excess salt off of the side of the flask.

3. Add salts and other nutrients.
4. Rinse inside of the flask to ensure all salts are in solution.
5. Stir the media until salts are mostly dissolved.
6. Cover the flask with aluminum foil.
7. Autoclave for 20 min at 121 °C to sterilize.
8. Once the media has cooled to room temperature, add remaining salts and nutrients as needed (e.g., steri-filtered glucose solution) under sterile conditions.
9. Allow media to stand at room temperature for at least 24 h prior to inoculation; an increase in turbidity of the medium may indicate bacterial contamination.

3.2.2 Culture inoculation and monitoring

1. Under sterile conditions, streak bacteria on a nutrient-rich agar plate. Incubate the plate at the temperature suggested for ideal growth conditions (often approximately 30 °C) until single colonies are visible. Store plate at 4 °C until use.
2. To make a starter culture, dispense desired volume of liquid nutrient-dense broth such as Luria-Bertani (LB) broth or tryptic soy broth (TSB) into a sterile culture tube (for a 1 L culture, prepare a 5–7 mL starter culture).
3. Under sterile conditions, use an inoculation loop to transfer a single colony from the plate into the liquid nutrient broth and briefly stir.
4. Incubate the starter culture at suggested temperature while shaking at 180 rpm.
5. Once the starter culture is cloudy, decant the starter culture under sterile conditions into the prepared growth media solution. Cover, and incubate at suggested temperature while shaking at 180 rpm.

3.2.3 Measurement of optical density

Microbial growth can be monitored by measuring the optical density ($OD_{600\,nm}$), or the degree of light scattering caused by the bacteria in the culture, at 600 nm. This wavelength is chosen because it is minimally damaging to the suspended cells. To maximize siderophore production, it is advisable to track the $OD_{600\,nm}$ of the culture throughout the growing period, noting the time since inoculation, to obtain a growth curve. Bacterial growth curves typically progress through a series of consecutive phases including lag phase, log phase, stationary phase, and decline. For maximal siderophore yield, bacterial cultures should be grown until late log/early stationary phase.

1. Swirl bacterial culture to suspend cells.
2. Under sterile conditions, pipette 1 mL of bacterial culture into a cuvette. Cover culture flask and return to incubation conditions.
3. Measure the optical density at 600 nm with a spectrophotometer.
4. Generally, an OD_{600} absorbance of around 1.0 au indicates that a culture has reached late log/early stationary phase.

4. Detection and isolation of the siderophores
4.1 Materials
4.1.1 Equipment
- Top loading balance scale to balance centrifuge bottle masses
- High speed centrifuge that can accommodate at least 1 L total volume
- Refrigerator (4 °C)
- Freezer (−20 °C)
- Orbital shaker
- Rotary evaporator
- Liquid chromatography system for semi-preparative reversed-phase high-performance liquid chromatography (RP-HPLC) equipped with a UV-Visible detector
- C18-AQ, 20 × 250 mm HPLC Column
- Ultra-performance liquid chromatography/mass spectrometry (UPLC-MS, Waters Xevo G2-XS QTOF with positive mode electrospray ionization coupled to an AQUITY UPLC-H-Class system) instrument
- Lyophilizer
- Analytical balance scale to determine yield

4.1.2 Reagents and supplies
- Ultrapure water
- Hexadecyltrimethylammonium bromide (HDTMA)
- Iron chloride hexahydrate ($FeCl_3 \cdot 6H_2O$)
- Hydrochloric acid
- Chrome azurol S (CAS)
- 0.5 M piperazine buffer (pH 5.6)
- 96 well plate
- Micropipette
- Sterile pipette tips
- Ethanol to sterilize surfaces

- Bunsen burner to maintain sterile conditions
- Centrifuge bottles
- Bleach
- Large Erlenmeyer or Fernbach flasks
- Amberlite™ XAD-4 resin
- Borosilicate glass funnel
- Stainless-steel filter
- HPLC grade methanol
- Glass chromatography column (diameter ≈ 2.5 cm) equipped with base fitted with cheesecloth filter
- 1 L round-bottom flask
- 20–60 mL plastic syringe
- 20 μm syringe filter
- 50 mL centrifuge tubes
- Trifluoroacetic acid
- 1 mL Hamilton syringe
- Acetonitrile (CAN)
- Formic acid
- Polypropylene beaker
- Liquid nitrogen

4.2 CAS assay for general siderophore detection

Siderophores in a growing bacterial culture can be detected via the Chrome Azurol S (CAS) assay (Schwyn & Neilands, 1987). This colorimetric method results in a color change when siderophores scavenge Fe (III) from an Fe-CAS-hexadecyltrimethylammonium bromide complex (blue), liberating the CAS dye (red-orange, positive assay response). CAS assays can be performed on agar plates, with the Fe-CAS complex added to the plate medium, or in solution, as detailed here. Unlike agar plate CAS assays, solution CAS assays can provide a quantitative siderophore production measurement. A positive CAS assay response is a simple way to rapidly test for the presence of siderophores. The CAS assay will only detect the presence of metal-free, or apo, siderophores. Perform a CAS assay approximately once per day during bacterial growth to monitor siderophore production. It is convenient to perform the CAS assay at the same time as measuring the optical density, using the separated aliquot of culture solution (see above section, "Optical density").

4.2.1 Preparation of Fe-CAS-hexadecyltrimethylammonium bromide solution

1. Combine the following solutions to achieve final assay conditions of 15 µM $FeCl_3$, 150 µM CAS, 600 µM HDTMA, in 0.5 M piperazine buffer (pH 5.6).
   a. 1.202 mM hexadecyltrimethylammonium bromide (HDTMA) solution
   b. 1 mM $FeCl_3 \cdot 6\,H_2O$ (4.2 mM HCl)
   c. 2 mM Chrome Azurol S in ultrapure water.
   d. Piperazine buffer solution (0.5 M, pH 5.6).
2. Solution will be maroon. Store at room temperature overnight. Solution should be blue within 24 h.

4.2.2 CAS assay procedure

1. Dispense 200 µL bacterial culture into one well of a 96 well plate.
2. Add 100 µL ferric CAS solution, using the micropipette to gently mix the solutions.
3. Wait up to 24 h. Diazeniumdiolate siderophores often exhibit a color change from blue to red within 20 min, while hydroxamate siderophores scavenge Fe(III) from the Fe-CAS complex more slowly.

4.3 Extraction of siderophores

Once the culture has reached late log/early stationary phase and is CAS positive, the supernatant can be separated from the cells by centrifugation. Siderophores are excreted from the bacterial cells, so they are present in the supernatant solution. Note that some siderophores have fatty acid tails, and they may adhere to the cell pellet, requiring further methanolic extraction. To extract low molecular weight organic compounds, including siderophores, from the supernatant solution, Amberlite XAD-4 polymeric resin can be utilized as detailed below.

4.3.1 Solid phase extraction of siderophores from bacterial culture

1. Pour bacterial culture evenly into centrifuge bottles, ensuring that the masses of each bottle are balanced. Centrifuge the culture at 6000 rpm and 4 °C.
2. Decant the supernatant solution into an acid-washed flask. Unless further extraction of the cell pellet is needed, the cell pellets can be bleached and discarded.
3. Prepare fresh XAD-4 resin:
   a. Pour about 300–400 g (300–400 mL volume) dry XAD-4 resin into a funnel fitted with a screen, over a large flask to collect rinse waste.

Discovery, isolation, and characterization of diazeniumdiolate siderophores   201

   b. Rinse resin with 3 L ultrapure water. Dispose of rinse waste, but do not remove resin from the filter funnel.

   c. Repeat with 3 L methanol, then 1 L ultrapure water, and then 1 L methanol.

   d. Store resin in methanol at 4 °C until use.

4. Prepare XAD-4 resin for use:

   a. Pour XAD-4 resin into the glass funnel and flask.

   b. If reusing XAD-4 resin, rinse with 400 mL methanol to remove any residual contaminants. If using freshly prepared resin, skip ahead to the next step.

   c. Rinse resin with 1 L ultrapure water.

5. With ultrapure water, rinse about 100 g (or 100 mL volume) per liter of culture XAD-4 resin into the flask containing the supernatant solution.

6. Place the flask on an orbital shaker at 150 rpm at room temperature and shake for 2–4 h.

7. Remove the flask from the shaker. Swirl contents to create a slurry. Pour supernatant and XAD-4 resin into the borosilicate glass funnel and flask.

8. Rinse with 1 L ultrapure water. Bleach and then dispose of the filtrate.

9. Preparation of column:

   a. Rinse a glass chromatography column (diameter ≈ 2.5 cm) and collection flask (a 1 L round bottom flask works well) with 4 M HCl.

   b. Rinse column and flask with ultrapure water and then with methanol.

   c. Attach column base fitted with a cheesecloth filter.

   d. Replace collection flask with a waste container.

10. Running the column extraction:

   a. Pack the column: use ultrapure water to transfer the rinsed XAD resin into the column. Allow excess water to drip into the waste container.

   b. Once drips have slowed, replace the waste container with the collection flask.

   c. Pour 250 mL 100% methanol into the column to extract adsorbed organic compounds. (Note: The concentration of methanol can be adjusted to the specific elution conditions of the siderophore of interest. Use HPLC elution time, as described below, to estimate an appropriate eluent concentration).

   d. Collect the eluate. Remove the collection flask and replace it with the waste container.

e. To rinse XAD resin, add 50–100 mL 100% methanol to the column. Once all of the solvent has passed through the column, use methanol to transfer the XAD back into its storage container for future use. New XAD resin should be prepared for different bacterial strains, however the existing XAD resin may be reused many times for the same bacterial strain.

11. Add 30 mL of water to the eluate.

12. Concentrate the extracted eluate solution to about 15–30 mL by rotary evaporation, avoiding water bath temperatures above 40 °C. Use a plastic syringe with a 20 μm syringe filter to filter and transfer the concentrated extract to a 50 mL centrifuge tube. Store extract at −20 °C until HPLC separation. Thaw extract before use.

4.4 Purification of siderophores

Once an extract has been prepared, the siderophore can be isolated by semipreparative reverse phase high performance liquid chromatography (HPLC) with a 20 × 250 mm C18-AQ column and UV-Visible detector. For diazeniumdiolate siderophores, it is advisable to monitor the absorbance of HPLC injections at the peptide bond absorption wavelength, 215 nm, as well as the diazeniumdiolate bond absorption wavelength, 250 nm. The eluted compounds can be analyzed by UPLC-MS to determine the peak which corresponds to the siderophore of interest.

4.4.1 Collecting fractions and identifying the siderophore HPLC peak

1. Prepare 3–4–L each of:
   a. Ultrapure water (0.05% trifluoroacetic acid)
   b. HPLC grade methanol (0.05% trifluoroacetic acid)

2. To develop a new HPLC method, start by running a linear gradient of 10% methanol to 80% methanol over 40 min.

3. Inject a small volume (<100 μL) of the extract as a test run.

4. Adjust the method gradient to maximize the peak separation.

5. Inject a larger volume of the extract to achieve high absorbance intensities, without surpassing the detector maximum. Isolate appropriate fractions based on their absorption at 215 and 250 nm.

6. Concentrate fractions to about 10% of their original volume.

7. To test for the presence of general siderophores, CAS assays can be employed following the same procedure as above.

8. Run the concentrated fractions on UPLC-MS. Depending on the system, a linear gradient of 0–100% ACN over ten minutes should be effective for siderophore analysis.
9. See the mass spectrometry section below for siderophore characterization information.
10. Once a potential siderophore mass has been identified, continue to use HPLC separation to collect the peak of interest as it elutes.
11. Concentrate the collected fractions to <10 mL by rotary evaporation.
12. If necessary, ultra-purify the concentrated fraction by repeating steps 10 and 11.

4.4.2 Lyophilization

Upon HPLC ultra-purification of the siderophore, the collected fractions can be combined, concentrated, and dried to yield a solid.

1. Combine ultra-purified fractions in a round bottom flask at least twice the total volume of the combined fractions.
2. Concentrate to a final volume of 5–10 mL by rotary evaporation, avoiding water bath temperatures above 40 °C.
3. Transfer concentrated solution into a pre-weighed 50 mL centrifuge tube. Do not seal the tube.
4. In a polypropylene beaker, flash freeze the contents of the tube in liquid nitrogen.
5. Lyophilize for 24–48 h or until the sample is a fluffy solid (usually white; yellow solids can indicate the presence of ferric siderophores). A 1 L culture will generally yield 3–15 mg of siderophore, though this yield varies based on the strain and growth conditions.

5. Identification and characterization of graminine-containing siderophores

C-Diazeniumdiolate siderophores are readily identifiable by their pH-dependent absorbance band (in gramibactin: λ_{max} = 248 nm, pH 8; λ_{max} = 220 nm), pH 2 and by their characteristic MS ionization pattern (i.e., loss of 30 mu corresponding to the ionization-induced loss of ^{14}NO) (Makris et al., 2022). Upon identifying a diazeniumdiolate siderophore, UPLC-ESI-MS, Marfey's amino acid analysis, and NMR spectroscopy can be employed to characterize the compound's structure.

5.1 Materials

5.1.1 Equipment

- Analytical balance scale
- UPLC-MS (Waters Xevo G2-XS QTOF with positive mode electrospray ionization coupled to an AQUITY UPLC-H-Class system)
- Waters BEH C18 UPLC column
- Liquid chromatography system for analytical reversed-phase high-performance liquid chromatography (RP-HPLC)
- Oven (110 °C)
- Freezer (−20 °C)
- High field (e.g., 500 MHz) nuclear magnetic resonance (NMR) instrument
- UV-Visible spectrophotometer

5.1.2 Reagents and supplies

- $^{15}NH_4Cl$ or other source of ^{15}N
- 99.8% NMR-grade D_2O
- Ultrapure water
- Glass ampoules
- Micropipette
- Pipette tips
- Hydrochloric acid
- Propane tank
- Tweezers
- 1.5 mL centrifuge tubes
- Heated water bath (40 °C)
- Amino acid standards of all relevant stereochemical configurations
- Marfey's reagent
- Acetone
- Sodium bicarbonate ($NaHCO_3$)
- Triethylamine
- Phosphoric acid
- Acetonitrile
- 150 μL Hamilton syringe
- Formic acid
- 5 mM aqueous buffer (MOPS or phosphate, pH 8.0)
- 3 mL quartz cuvette
- 75 mm stir bar
- Mercury (argon) spectral calibration pen lamp

Discovery, isolation, and characterization of diazeniumdiolate siderophores

- Magnetic stirrer
- NMR tube
- Quartz NMR tube
- Deuterated dimethyl sulfoxide (DMSO-d_6)
- 5 mM phosphate buffer in D_2O, pD 8.0

5.2 Mass spectrometry and stable isotope labeling

The mass of the siderophore, in conjunction with the NRPS prediction of the structure, are a starting point for the structural characterization of the compound. In the UPLC-MS chromatogram, search for a peak whose mass is in the range of the mass of the NRPS-predicted structure. Siderophores often will display an m/z $[M - 2H + Fe]^+$ ionized mass, corresponding to the Fe (III)-siderophore complex. For each diazeniumdiolate ligand present, the siderophore mass spectrum will exhibit a characteristic m/z $[M + H - 30]^+$ ion corresponding to a mass loss of 30 Da consistent with ionization of the N–N bond in the Gra residue. Note that the formula mass may include a sodium adduct ion (m/z $[M + Na]^+$).

5.2.1 ^{15}N labeling

The number of nitrogen atoms present in the molecule can be determined by stable isotope labeling. To achieve a fully ^{15}N-labeled siderophore structure, a new culture should be grown, replacing the nitrogen source with ^{15}N (e.g., in M9 media, using $^{15}NH_4Cl$ instead of $^{14}NH_4Cl$ as the sole nitrogen source), so that all nitrogen atoms incorporated into the bacterial metabolites are isotopically labeled. Repeat the culturing and isolation methods, using new XAD-4 resin for the solid phase extraction of organic compounds. Note that ^{15}N-labeled cultures often grow slightly slower. Once the labeled siderophore has been isolated, it should be run on MS under the same conditions as the unlabeled ^{14}N siderophore. The difference in mass between the labeled and unlabeled siderophore ($M_{labeled} - M_{unlabeled}$) is equivalent to the number of nitrogen atoms present in the structure, which provides valuable information about the number and structure of the constituent residues.

5.2.2 Deuterium (^2H) labeling

Similarly to the ^{15}N-labeling detailed above, deuterium exchange can be employed to determine the number of exchangeable protons in the siderophore structure. Exchangeable protons are those which are bound to heteroatoms.

1. Dissolve a small (spatula-tip's worth) aliquot of solid siderophore in 100 µL 99.8% NMR-grade D_2O.
2. Lyophilize to dryness.
3. Add a second aliquot of 100 µL 99.8% NMR-grade D_2O. Lyophilize.
4. Repeat step 3 four more times. Exchangeable protons should now have been replaced by deuterium atoms.
5. Analyze the deuterated siderophore by direct injection mass spectrometry.

5.2.3 Tandem MS/MS

Tandem MS/MS can serve an important role in the structural characterization of linear siderophores (Pluháček et al., 2016). In many cases, however, diazeniumdiolate-containing siderophores are relatively large and often cyclic, convoluting the MS/MS fragmentation pattern. It is therefore recommended to use MS/MS as a useful check for the determined structure, rather than the primary method of characterization. NMR spectroscopy is much more elucidating in these cases. See below for NMR spectroscopic structural characterization details.

5.3 Marfey's amino acid analysis

To unambiguously determine the stereocenter configuration of the amino acid building blocks of the siderophore, Marfey's analysis is performed (Marfey, 1984). HCl is used to hydrolyze the siderophore at elevated temperature, into component L- and D-amino acids. The amino acids in the hydrolysate are derivatized with Marfey's reagent (i.e., 1-fluoro-2-4-dinitrophenyl-5-L-alanine amide), forming diastereomers which can be separated. Analytical HPLC and/or UPLC-MS can then be used to analyze the derivatized hydrolysate in comparison to amino acid standards. The presence of an amino acid with a certain stereo-configuration is indicated by superposition of the peak of the derivatized hydrolysate and the derivatized standard amino acid when co-injected.

5.3.1 Hydrolysis of the siderophore

1. Dissolve 1 mg siderophore in 200 µL ultrapure water.
2. Divide the solution into two glass ampoules.
3. Add 100 µL concentrated HCl to each ampoule to achieve a final concentration of 6 M HCl.
4. Seal ampoules using a propane tank and tweezers.
5. Place the sealed ampoules in an oven at 110 °C for 20–24 h. The solution will be light yellow.

6. Crack open the ampules and combine the contents in a small (1.5 mL) plastic centrifuge tube.
7. Use a hot water bath at 40 °C and gentle air flow to dry the hydrolysate. When nearly dry, add 700 μL ultrapure water. Repeat the evaporation and water addition three times.
8. After final water addition, allow the hydrolysate to evaporate to dryness. Store the dry hydrolysate at −20 °C until use.

5.3.2 Derivatization of the hydrolysate and amino acid standards
1. Dissolve the siderophore hydrolysate and amino acid standards in ultrapure water to yield 1 mg/100 μL solutions of each.
2. To each solution, add 150 μL of 5 mM Marfey's reagent in acetone and 20 μL of 1 M NaHCO$_3$.
3. Heat the solutions in a hot water bath at 40 °C for 1 h.
4. Quench the reaction by adding 15 μL of 4 M HCl to each sample. The solutions should be bright yellow.
5. Dilute each solution with 100 μL ultrapure water.
6. Store solutions at −20 °C until use.

5.3.3 Analytical HPLC
1. Prepare the solvents
   a. Solvent A (50 mM triethylamine phosphate, pH 3.0): To 494.84 mL ultrapure water, add 3.47 mL TEA (triethylamine) and 1.69 mL phosphoric acid. Adjust to pH 3.0 with phosphoric acid.
   b. Solvent B (Acetonitrile)
2. Prepare the solution to dilute samples before injection (dilution solution—v:v 10 acetone: 9 H$_2$O).
3. Analytical HPLC method on a C18 column (250 × 4.6 mm YMC C18-AQ column)—10% to 40% solvent B over 45 min.
4. With a Hamilton syringe, inject derivatized amino acid standards (30 μL standard + 45 μL dilution solution), derivatized siderophore hydrolysate (30 μL hydrolysate + 45 μL dilution solution), respectively, and separately co-inject the two solutions (30 μL standard + 30 μL hydrolysate + 15 μL dilution solution). The absorbance is monitored at 215 nm and 340 nm.
5. In the resulting HPLC traces, look for the unreacted Marfey's reagent peak, the standard amino acid peaks, and peak increase in the co-injection to determine the chirality of amino acids in the siderophore.

5.3.4 UPLC-MS

1. The amino acid standards, siderophore hydrolysate, and a mixture of these two, can also be run on a UPLC-MS (Waters Xevo G2-XS QTOF with positive mode electrospray ionization coupled to an AQUITY UPLC-H-Class system with a Waters BEH C18 column) to confirm the identity of the peaks.
2. Solvent A—ultrapure water + 0.1% formic acid
3. Solvent B—Acetonitrile + 0.1% formic acid
4. LC method—15% to 50% Solvent B over 15 min

5.4 Photoreactivity of the diazeniumdiolate group

Diazeniumdiolate-containing siderophores are photoactive, absorbing at approximately 250 nm at pH 8. The diazeniumdiolate absorbance band is pH dependent, shifting to shorter wavelengths at lower pH ($\lambda_{max, pH 8} \approx 250$ nm shifts to $\lambda_{max, pH 2} \approx 220$ nm) as the ligand is protonated, as is characteristic of the diazeniumdiolate (*N*-nitrosohydroxylamine) group (Gama et al., 2021). The diazeniumdiolate absorbance band of a siderophore can be determined by dissolving the purified siderophore into an aqueous buffer solution (pH 8) and measuring its absorbance by UV-Visible spectrophotometry. Note: Choosing an appropriate aqueous buffer will affect the results of any spectrophotometric analyses because many buffers absorb in the 250 nm range.

UV irradiation destroys the diazeniumdiolate group, causing each Gra residue in the siderophore to lose the equivalent of NO + H. The photolysis reaction of the siderophore can be monitored spectrophotometrically, and its reaction products can be analyzed by UPLC-mass spectrometry and/or NMR spectroscopy. (Note: Direct UV radiation exposure is harmful, so photolysis studies are best performed in a photolysis box or under a closed fume hood with the glass covered.).

5.4.1 Photolysis of diazeniumdiolate siderophores

1. Dissolve siderophore sample in aqueous buffer (e.g., 5 mM MOPS or phosphate buffer, pH 8.0).
2. Transfer solution to a 3 mL quartz cuvette, with a 75 mm stir bar.
3. Measure the absorbance of the "dark" solution before photolysis.
4. In timed intervals, expose the solution to UV light using an Hg(Ar) pen lamp. Keep the cuvette on a stir plate, stirring gently during the photolysis reaction. After each timed interval, measure the absorbance of the solution. The 250 nm band should decrease in absorbance intensity as the photolysis reaction progresses (Fig. 3A).

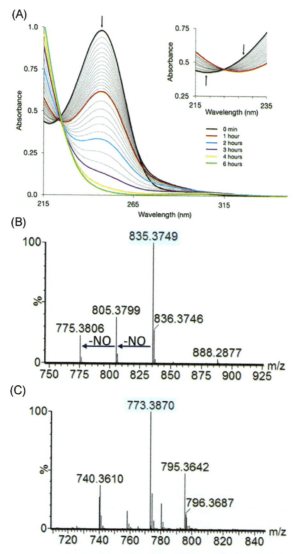

Fig. 3 UV photolysis of apo-Gbt. (A) 254 nm irradiation of apo-Gbt (44 μM) over 6 h in 5 mM MOPS pH 8.0. Inset shows the isosbestic point over the first hour of photolysis; (B) MS of apo-Gbt; (C) MS of the photoproduct of Gbt (Makris et al., 2022).

5. Analyze the photoproduct by UPLC-MS (see Section 5.4.2).
6. *For NMR analysis of the photoproduct, use 99.9% purity D_2O instead of ultrapure H_2O, buffered at pD 8.0 and carry out the photolysis in a quartz NMR tube.

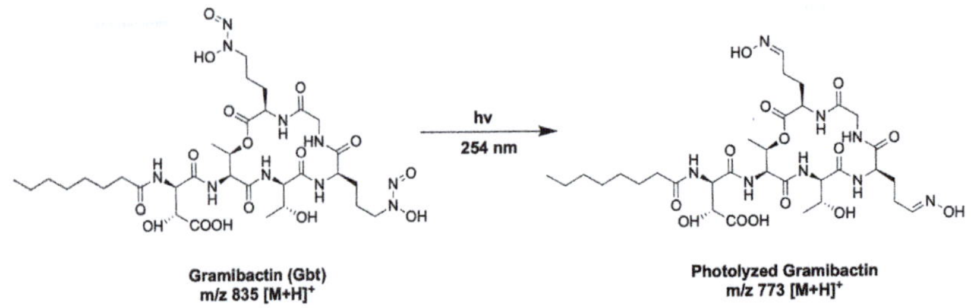

Fig. 4 Photolysis scheme of gramibactin (Gbt). ^{14}N-Gra-enriched-Gbt ehibits a mass loss of 62 in the photoproduct. When isotopically labeled, ^{15}N-Gra-enriched-Gbt observes a mass loss of 64 in the photoproduct.

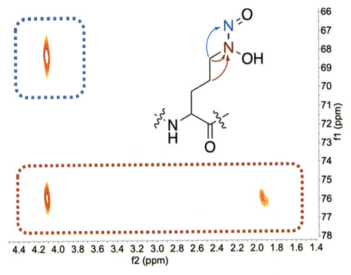

Fig. 5 ^{1}H–^{15}N HMBC NMR spectrum of tistrellabactins A and B showing the presence of C-diazeniumdiolate amino acid Gra. Spectrum collected in DMSO-d_6. Adapted from Makris, C., Leckrone, J. K., & Butler, A. (2023). Tistrellabactins A and B are photoreactive C-diazeniumdiolate siderophores from the marine-derived strain Tistrella mobilis KA081020-065. Journal of Natural Products, 86(7), 1770–1778. https://doi.org/10.1021/acs.jnatprod.3c00230.

5.4.2 MS detection of NO mass loss

The photolysis of diazeniumdiolate siderophores forms E/Z oxime photoproducts (Fig. 4), which differ from the unphotolyzed siderophore by 31 mass units per Gra residue (Hermenau et al., 2018). For a siderophore that contains two Gra residues, like gramibactin, the photoproduct will therefore be 62 mass units lower than the unphotolyzed gramibactin mass (Gbt m/z 835 [M+H]$^+$) (Fig. 3B and C).

Discovery, isolation, and characterization of diazeniumdiolate siderophores 211

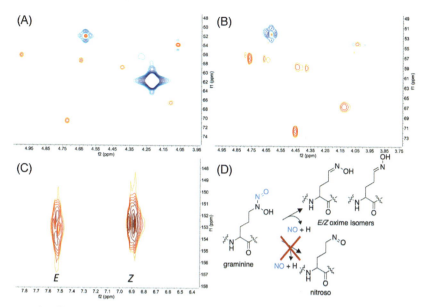

Fig. 6 ^{1}H–^{13}C HSQC NMR spectra show the formation of *E* and *Z* oxime isomers upon photolysis of ^{13}C^{15}N-Gra-enriched-Gbt (pD 8, 6.2 mM, P_i in 99.9% D$_2$O). (A) ^{1}H–^{13}Cδ correlations in Gra residues of apo-Gbt at 4.20 ppm, 61.66 ppm and 4.22 ppm, 61.93 ppm; (B) region of HSQC showing disappearance of the ^{1}H–^{13}Cδ correlations in the photoproduct; (C) new downfield ^{1}H–^{13}C HSQC correlations in photolyzed ^{13}C^{15}N-Gra-enriched Gbt, with chemical shifts consistent with *E/Z* oxime isomers; (D) scheme showing release of the equivalent of NO + H from Gra yields *E* and *Z* oxime isomers but not a nitroso photoproduct. *Adapted from Makris, C., Carmichael, J. R., Zhou, H., & Butler, A. (2022). C-diazeniumdiolate graminine in the siderophore gramibactin is photoreactive and originates from arginine. ACS Chemical Biology, 17(11), 3140–3147. https://doi.org/10.1021/acschembio.2c00593.*

5.5 NMR spectroscopy characterization of the *C*-diazeniumdiolate and the photoproducts

Complete elucidation of *C*-diazeniumdiolate siderophores is achieved through high field NMR spectroscopy. DMSO-d_6 is a suitable solvent for solubilization and data collection. Characterization of *C*-diazeniumdiolate siderophore photoproducts requires buffered D$_2$O (50 mM Na$_2$HPO$_4$, pD 8.0) as the solvent. Spectra are indirectly referenced by the residual solvent peak or ^{2}H lock.

The *C*-diazeniumdiolate in Gra is conclusively identified through ^{15}N-enrichment of the functional group, as described previously. Gra has a distinct ^{1}H–^{15}N HMBC fingerprint, with correlations of Cδ methylene protons to each ^{15}N in the *C*-diazeniumdiolate and Cγ methylene protons to the proximal hydroxylated ^{15}N only (Fig. 5, adapted from Makris et al., 2023).

Chemical shifts of the proximal and distal ^{15}N residues fall in the range of 316–317 ppm and 367–368 ppm, respectively. Without ^{15}N isotopic enrichment, Gra shares a similar NMR fingerprint to ornithine. At the Cδ position in Gra, ^{1}H and ^{13}C resonances are observed at approximately 4.1 ppm and 61 ppm in DMSO-d_6.

The C-diazeniumdiolate group undergoes a photoreaction in the presence of UV light, yielding E/Z oxime isomers. The conversion to E/Z oxime isomers can be followed by NMR, in which the disappearance of Cδ resonances and subsequent appearance of two deshielded protons at 6.9 ppm and 7.5 ppm and ^{13}C resonances in the range of 152–153 ppm are observed (Fig. 6, adapted from Makris et al., 2022). The transformation is most easily observed with an ^{1}H–^{13}C HSQC experiment, which illustrates the disappearance of ^{1}H–^{13}C Cδ Gra residues and the correlation of new ^{1}H and ^{13}C resonances in the photoproduct. TOCSY NMR also may be used to demonstrate the new ^{1}H resonances in the photoproduct are within the same proton spin system as the other Gra ^{1}H resonances.

6. Conclusions and future outlook

The C-diazeniumdiolate group is a newly discovered ligand in siderophores, produced mainly by root-associated bacteria, with one known example discovered in a marine bacterial species. Besides binding Fe(III) with high affinity, the diazeniumdiolate group in Gra is also a potential donor of NO, an important biological signaling molecule. Discovery of novel diazeniumdiolate siderophores is of great interest to study bacterial pathogeny, symbiosis between bacteria and other organisms, and the potential of drug design based on the diazeniumdiolate moiety. In this chapter, we describe methods to discover new diazeniumdiolate siderophores *in silico* using bioinformatics tools together with the known information on characterized siderophore BGCs, isolate the siderophores of interest in the lab, structurally characterize this class of siderophores, and study its photoactivity. We anticipate further exciting developments in the future, including elucidation of the biosynthesis of the diazeniumdiolate group in Gra, as well as the potential incorporation of other diazeniumdiolate-containing amino acids (e.g., alanosine or other new amino acids) into siderophore structures.

Acknowledgements

We are grateful for support from the U.S. National Science Foundation, CHE-2108596.

References

Atkinson, H. J., Morris, J. H., Ferrin, T. E., & Babbitt, P. C. (2009). Using sequence similarity networks for visualization of relationships across diverse protein superfamilies. *PLoS One, 4*(2), e4345. https://doi.org/10.1371/journal.pone.0004345.

Blin, K., Shaw, S., Augustijn, H. E., Reitz, Z. L., Biermann, F., Alanjary, M., ... Weber, T. (2023). antiSMASH 7.0: New and improved predictions for detection, regulation, chemical structures and visualisation. *Nucleic Acids Research, 51*(W1), W46–W50. https://doi.org/10.1093/nar/gkad344.

Dolak, L. A., Castle, T. M., Hannon, B. R., Argoudelis, A. D., & Reusser, F. (1983). Fermentation, isolation, characterization and structure of nitrosofungin. *The Journal of Antibiotics (Tokyo), 36*(11), 1425–1430. https://doi.org/10.7164/antibiotics.36.1425.

Gama, S., Hermenau, R., Frontauria, M., Milea, D., Sammartano, S., Hertweck, C., & Plass, W. (2021). Iron coordination properties of gramibactin as model for the new class of diazeniumdiolate based siderophores. *Chemistry (Weinheim an der Bergstrasse, Germany), 27*(8), 2724–2733. https://doi.org/10.1002/chem.202003842.

Hermenau, R., Ishida, K., Gama, S., Hoffmann, B., Pfeifer-Leeg, M., Plass, W., ... Hertweck, C. (2018). Gramibactin is a bacterial siderophore with a diazeniumdiolate ligand system. *Nature Chemical Biology, 14*(9), 841–843. https://doi.org/10.1038/s41589-018-0101-9.

Hermenau, R., Mehl, J. L., Ishida, K., Dose, B., Pidot, S. J., Stinear, T. P., & Hertweck, C. (2019). Genomics-driven discovery of NO-donating diazeniumdiolate siderophores in diverse plant-associated bacteria. *Angewandte Chemie International Edition, 58*(37), 13024–13029. https://doi.org/10.1002/anie.201906326.

Hrabie, J. A., & Keefer, L. K. (2002). Chemistry of the nitric oxide-releasing diazeniumdiolate ("nitrosohydroxylamine") functional group and its oxygen-substituted derivatives. *Chemical Reviews, 102*(4), 1135–1154. https://doi.org/10.1021/cr000028t.

Iimura, H., Takeuchi, T., Kondo, S., Matsuzaki, M., & Umezawa, H. (1972). Dopastin, an inhibitor of dopamine -hydroxylase. *The Journal of Antibiotics (Tokyo), 25*(8), 497–500. https://doi.org/10.7164/antibiotics.25.497.

Jenul, C., Sieber, S., Daeppen, C., Mathew, A., Lardi, M., Pessi, G., ... Eberl, L. (2018). Biosynthesis of fragin is controlled by a novel quorum sensing signal. *Nature Communications, 9*(1), 1297. https://doi.org/10.1038/s41467-018-03690-2.

Jiao, J., Du, J., Frediansyah, A., Jahanshah, G., & Gross, H. (2020). Structure elucidation and biosynthetic locus of trinickiabactin from the plant pathogenic bacterium *Trinickia caryophylli. The Journal of Antibiotics, 73*(1), 28–34. https://doi.org/10.1038/s41429-019-0246-0.

Li, B., Ming, Y., Liu, Y., Xing, H., Fu, R., Li, Z., ... Chen, J. (2020). Recent developments in pharmacological effect, mechanism and application prospect of diazeniumdiolates. *Frontiers in Pharmacology, 11*, 923. https://doi.org/10.3389/fphar.2020.00923.

Makris, C., Carmichael, J. R., Zhou, H., & Butler, A. (2022). C-diazeniumdiolate graminine in the siderophore gramibactin is photoreactive and originates from arginine. *ACS Chemical Biology, 17*(11), 3140–3147. https://doi.org/10.1021/acschembio.2c00593.

Makris, C., Leckrone, J. K., & Butler, A. (2023). Tistrellabactins A and B are photoreactive C-diazeniumdiolate siderophores from the marine-derived strain *Tistrella mobilis* KA081020-065. *Journal of Natural Products, 86*(7), 1770–1778. https://doi.org/10.1021/acs.jnatprod.3c00230.

Marfey, P. (1984). Determination of D-amino acids. II. Use of a bifunctional reagent, 1,5-difluoro-2,4-dinitrobenzene. *Carlsberg Research Communications, 49*(6), 591. https://doi.org/10.1007/BF02908688.

Morgan, G. L., & Li, B. (2020). In vitro reconstitution reveals a central role for the N-oxygenase PvfB in (dihydro)pyrazine-N-oxide and valdiazen biosynthesis. *Angewandte Chemie International Edition, 59*(48), 21387–21391. https://doi.org/10.1002/anie.202005554.

Murthy, Y. K. S., Thiemann, J. E., Coronelli, C., & Sensi, P. (1966). Alanosine, a new antiviral and antitumour agent isolated from a streptomyces. *Nature, 211*(5054), 1198–1199. https://doi.org/10.1038/2111198a0.

Natori, T., Kataoka, Y., Kato, S., Kawai, H., & Fusetani, N. (1997). Poecillanosine, a new free radical scavenger from the marine sponge Poecillastra spec. aff. tenuilaminaris. *Tetrahedron Letters, 38*(48), 8349–8350. https://doi.org/10.1016/S0040-4039(97)10260-X.

Nishio, M., Hasegawa, M., Suzuki, K., Sawada, Y., Hook, D. J., & Oki, T. (1993). Nitrosoxacins A, B and C, new 5-lipoxygenase inhibitors. *The Journal of Antibiotics (Tokyo), 46*(1), 193–195. https://doi.org/10.7164/antibiotics.46.193.

Plucháček, T., Lemr, K., Ghosh, D., Milde, D., Novák, J., & Havlíček, V. (2016). Characterization of microbial siderophores by mass spectrometry. *Mass Spectrometry Reviews, 35*(1), 35–47. https://doi.org/10.1002/mas.21461.

Sandy, M., & Butler, A. (2009). Microbial iron acquisition: Marine and terrestrial siderophores. *Chemical Reviews, 109*(10), 4580–4595. https://doi.org/10.1021/cr9002787.

Schwyn, B., & Neilands, J. B. (1987). Universal chemical assay for the detection and determination of siderophores. *Analytical Biochemistry, 160*(1), 47–56. https://doi.org/10.1016/0003-2697(87)90612-9.

Sieber, S., Daeppen, C., Jenul, C., Mannancherril, V., Eberl, L., & Gademann, K. (2020). Biosynthesis and structure–activity relationship investigations of the diazeniumdiolate antifungal agent fragin. *Chembiochem: A European Journal of Chemical Biology, 21*(11), 1587–1592. https://doi.org/10.1002/cbic.201900755.

Sieber, S., Mathew, A., Jenul, C., Kohler, T., Bär, M., Carrión, V. J., ... Gademann, K. (2021). Mitigation of *Pseudomonas syringae* virulence by signal inactivation. *Science Advances, 7*(37), eabg2293. https://doi.org/10.1126/sciadv.abg2293.

Tamura, S., Murayama, A., & Hata, K. (1967). Isolation and structural elucidation of fragin, a new plant growth inhibitor produced by a *Pseudomonas*. *Agricultural and Biological Chemistry, 31*(6), 758–759. https://doi.org/10.1080/00021369.1967.10858877.

Tuteja, N., Chandra, M., Tuteja, R., & Misra, M. K. (2004). Nitric oxide as a unique bioactive signaling messenger in physiology and pathophysiology. *Journal of Biomedicine & Biotechnology, 2004*(4), 227–237. https://doi.org/10.1155/s1110724304402034.

Wang, M., Niikura, H., He, H.-Y., Daniel-Ivad, P., & Ryan, K. S. (2020). Biosynthesis of the N–N-bond-containing compound l-alanosine. *Angewandte Chemie International Edition, 59*(10), 3881–3885. https://doi.org/10.1002/anie.201913458.

Wang, M., & Ryan, K. S. (2023). Reductases produce nitric oxide in an alternative pathway to form the diazeniumdiolate group of l-alanosine. *Journal of the American Chemical Society, 145*(30), 16718–16725. https://doi.org/10.1021/jacs.3c04447.

CHAPTER TEN

Linking biosynthetic genes to natural products using inverse stable isotopic labeling (InverSIL)

Tashi C.E. Liebergesell and Aaron W. Puri*
Department of Chemistry and the Henry Eyring Center for Cell and Genome Science, University of Utah, Salt Lake City, UT, United States
*Corresponding author. e-mail address: a.puri@utah.edu

Contents

1. Introduction		216
	1.1 The gene-to-molecule approach for natural product discovery	216
	1.2 Inverse stable isotopic labeling (InverSIL)	216
	1.3 Using InverSIL to link quorum sensing signal synthase genes to their products	217
2. Key resources		218
3. Equipment and reagents		219
	3.1 Inverse labeling of microbial culture	219
	3.2 Natural product extraction	219
	3.3 Analysis of labeled microbial extracts by untargeted mass spectrometry	219
	3.4 Data analysis to identify inverse labeled natural products	220
4. Method		220
	4.1 Inverse labeling of microbial culture	220
	4.2 Natural product extraction	221
	4.3 Analysis of labeled microbial extracts by untargeted mass spectrometry	222
	4.4 Data analysis to identify inverse labeled natural products	222
5. Notes		225
Acknowledgments		226
References		226

Abstract

The sequencing of microbial genomes has far outpaced their functional annotation. Stable isotopic labeling can be used to link biosynthetic genes with their natural products; however, the availability of the required isotopically substituted precursors can limit the accessibility of this approach. Here, we describe a method for using inverse stable isotopic labeling (InverSIL) to link biosynthetic genes with their natural products. With InverSIL, a microbe is grown on an isotopically substituted medium to create a fully substituted culture, and subsequently, the incorporation of precursors of natural isotopic abundance can be tracked by mass spectrometry. This eliminates issues with isotopically substituted precursor availability. We demonstrate the utility of this

Methods in Enzymology, Volume 702
ISSN 0076-6879, https://doi.org/10.1016/bs.mie.2024.06.005
Copyright © 2024 Elsevier Inc. All rights are reserved, including those for text and data mining, AI training, and similar technologies.

approach by linking a *luxI*-type acyl-homoserine lactone synthase gene in a bacterium that grows on methanol with its quorum sensing signal products. In the future, InverSIL can also be used to link biosynthetic gene clusters hypothesized to produce siderophores with their natural products.

1. Introduction

1.1 The gene-to-molecule approach for natural product discovery

The explosion in microbial genome sequencing has revealed many potentially novel biosynthetic gene clusters (BGCs) that may produce natural products with agricultural or biomedical value. However, the availability of these data has also highlighted gaps in researchers' ability to identify the metabolites associated with these BGCs. This has led to the development of gene-to-molecule approaches for linking BGCs with their natural products, many of which combine genomics with differential metabolomics (Caesar, Montaser, Keller, & Kelleher, 2021). One variant of this strategy is the genomisotopic approach, in which researchers predict substrates of the enzymes encoded by biosynthetic genes of interest, and then feed the producing organism isotopically substituted versions of these substrates (Gross et al., 2007). By tracking the incorporation of the isotopically distinct substrates, researchers can identify the natural product(s) associated with the BGC of interest (McCaughey, van Santen, van der Hooft, Medema, & Linington, 2022). Researchers used this approach to identify the lipopeptide orfamide A as the product of a nonribosomal peptide synthetase BGC in the genome of *Pseudomonas fluorescens* (Gross et al., 2007). The researchers predicted the amino acids activated by adenylation domains in the BGC and subsequently fed ^{15}N-substituted versions of these amino acids to *P. fluorescens*. Incorporation of these amino acids into orfamide A was then identified by NMR spectroscopy. In a more recent example, researchers identified the products of cyanobacterial BGCs predicted to incorporate fatty acids by feeding these organisms deuterated fatty acids and tracking their incorporation by mass spectrometry (Figueiredo et al., 2021). These examples illustrate the utility of using isotopically distinct versions of predicted substrates for identifying natural products associated with BGCs of interest.

1.2 Inverse stable isotopic labeling (InverSIL)

While stable isotopic labeling is a proven approach for linking BGCs with their natural products, it may be difficult to acquire the isotopically substituted precursors necessary for this method. One solution is InverSIL, where the producing organism is grown on a medium where at least one element has been substituted for an isotopically distinct version (e.g. ^{13}C or ^{15}N), creating an isotopically substituted culture. This culture can be fed a precursor of interest in its naturally abundant isotopic composition (e.g. predominantly ^{12}C or ^{14}N), and precursor incorporation can be identified by mass spectrometry or NMR spectroscopy. This eliminates issues with the availability of isotopically labeled precursors.

InverSIL has been used to determine the biosynthetic route to natural products including the siderophore gramibactin from the bacterium *Paraburkholderia graminis* (Makris, Carmichael, Zhou, & Butler, 2022). The researchers determined that the non-proteinogenic amino acid graminine within gramibactin is derived from *L*-arginine by growing *P. graminis* on a medium containing $^{15}NH_4Cl$ and then feeding naturally abundant *L*-arginine and detecting its incorporation by mass spectrometry. Separately, InverSIL has also been used to confirm the structure of members of the fimsbactin family of catechol/hydroxamate siderophores produced by *Acinetobacter* spp. (Proschak et al., 2013). The researchers confirmed the presence of a catecholate by growing strains on (^{13}C)glucose and feeding naturally abundant 2,3-dihydroxybenzoic acid. These examples demonstrate the value of InverSIL in natural product studies.

1.3 Using InverSIL to link quorum sensing signal synthase genes to their products

Quorum sensing is a method bacteria use to coordinate group behaviors in a cell density-dependent manner (Papenfort & Bassler, 2016; Schuster, Sexton, Diggle, & Greenberg, 2013). In one well-studied form of quorum sensing used by Gram-negative proteobacteria, the quorum sensing signals are *N*-acyl-homoserine lactones (acyl-HSLs), which are produced by LuxI-type acyl-HSL synthases. The HSL portion of these signals is derived from *L*-methionine via *S*-adenosyl-*L*-methionine (Moré et al., 1996; Schaefer, Val, Hanzelka, Cronan, & Greenberg, 1996). InverSIL has been used to link *luxI*-type acyl-HSL synthase genes with their quorum sensing signal products in ecologically important bacteria that grow on methanol called pink pigmented facultative methylotrophs (Cummings, Snelling, & Puri, 2021; Wallace, Cummings, Roberts, & Puri, 2023). The researchers determined that the signal product of the largest group of acyl-HSL synthases encoded in the genomes of these bacteria was unknown. They subsequently grew strains containing representatives of this synthase gene on (^{13}C)methanol and fed these cultures naturally abundant *L*-methionine. They were then able to identify a new acyl-HSL signal linked to these genes by tracking incorporation of *L*-methionine into metabolites using untargeted mass spectrometry (Wallace et al., 2023). This chapter will describe a detailed method for linking *luxI*-type quorum sensing signal synthase genes with their acyl-HSL products using InverSIL. This method can be modified to link other biosynthetic genes of interest with their natural products (such as siderophores) by substituting the precursor used in the InverSIL experiments.

2. Key resources

Reagent or resources	Source	Identifier
Microbial strain containing biosynthetic gene of interest		
Methylorubrum populi BJ001	Van Aken, Peres, Doty, Yoon, and Schnoor (2004)	Strain BJ001
Chemicals		
Methanol (^{13}C, 99%)	Cambridge Isotope Laboratories, Inc.	CLM-359
L-methionine (natural isotopic abundance)	Millipore Sigma	M9625
Deposited data		
Untargeted mass spectra for *Methylorubrum populi* BJ001 grown on (^{13}C) methanol, ^{12}C-methanol, and (^{13}C)methanol +*L*-^{12}C-methionine	MassIVE https://massive.ucsd.edu/	MSV000094817
Software and algorithms		
ProteoWizard	https://proteowizard.sourceforge.io/download.html Chambers et al. (2012)	Version 3.0.23146
MZmine	https://github.com/mzmine/mzmine/releases/tag/v3.9.0 Schmid et al. (2023)	Version 3.9.0
Metabolite Atlas Python package	https://github.com/biorack/metatlas Yao et al. (2015)	Version 3.0.0
pymzML Python package	https://github.com/pymzml/pymzML Bald et al. (2012)	Version 2.5.10

| Python script for converting .mzML files to .h5 format | https://github.com/purilab/inverse/tree/main/Liebergesell_2024 Liebergesell, Murdock, and Puri (2024) | InverSIL_mz_to_h5.py |
| Python script for InverSIL identification from untargeted metabolomic data | https://github.com/purilab/inverse/tree/main/Liebergesell_2024 Liebergesell, Murdock, and Puri (2024) | Untargeted_InverSIL.py |

3. Equipment and reagents

3.1 Inverse labeling of microbial culture

- Microbial strain containing biosynthetic gene of interest (*see* **Note 1**)
- Sterilized capped test tubes for culture growth
- Isotopically substituted growth medium (*see* **Notes 2 and 3**)
- Precursor of interest in its natural isotopic abundance
- Shaking incubator (e.g. New Brunswick Innova® 44 Stackable Incubator Shaker)

3.2 Natural product extraction

- C_{18} solid phase extraction (SPE) columns (bed weight 50 mg, volume 1 mL)
- Ring stands or vacuum manifold for holding SPE columns
- Sterilized and unsterilized 1.5 mL polypropylene microcentrifuge tubes
- Sterilized and unsterilized 15 mL screw top conical polypropylene tubes
- 1-dram vials with black phenolic screw cap and polyvinyl-faced pulp liner
- MilliQ H_2O
- HPLC-grade methanol
- Tabletop centrifuge for 15 mL conical tubes
- Microcentrifuge for 1.5 mL tubes
- Sonicating water bath
- Nitrogen stream for solvent evaporation (e.g. Organomation 12 Position N-EVAP Nitrogen Evaporator)

3.3 Analysis of labeled microbial extracts by untargeted mass spectrometry

- Liquid chromatography coupled to high resolution mass spectrometry (LC-HRMS) system (e.g. Waters Xevo G2S Q-ToF high-resolution tandem mass spectrometer with attached Acquity I-Class UPLC)

- Autosampler vials for LC-HRMS system
- LC-MS grade methanol
- MilliQ H_2O

3.4 Data analysis to identify inverse labeled natural products
- Computer with the following:
  - o Python 3 distribution (e.g. Anaconda)
  - o ProteoWizard Toolkit
  - o MZmine v 3.9.0
  - o Metabolite Atlas Python package v 3.0.0
  - o pymzML Python package v 2.5.10
  - o Python script "Untargeted_InverSIL.py" for InverSIL identification from untargeted metabolomic data

4. Method

Our example will discuss linking a *luxI*-type acyl-HSL synthase gene in a methanol-utilizing bacterium with its associated quorum sensing signals using InverSIL with (^{13}C)methanol and *L*-methionine. For clarity, in this method we refer to methanol and *L*-methionine in their natural isotopic abundance as ^{12}C-methanol and *L*-^{12}C-methionine, respectively. The genome of *Methylorubrum populi* BJ001 contains one predicted *luxI*-type acyl-HSL synthase gene: gene ID 642653709 in the Integrated Microbial Genomes & Microbiomes (IMG/M) system (Chen et al., 2023).

4.1 Inverse labeling of microbial culture

Timing: ~3 days.

Critical: Sterile technique is required for the steps in this section to avoid contamination.

1. Grow a 6 mL culture of the bacterium of interest to late logarithmic phase in unsubstituted growth medium and record the cell density. *M. populi* BJ001 was grown to an optical density at 600 nm (OD) of ~0.6 on modified AMS medium containing 50 mM ^{12}C-methanol as the sole carbon source (Wallace et al., 2023).
2. Pellet the cells by transferring the entire culture to a sterile 15 mL screw top conical polypropylene tube and centrifuging at 4000 rcf for 10 min. Discard the supernatant.

Gene-to-molecule InverSIL

3. Wash the cells by resuspending the pellet in 1 mL fresh medium without the element to be substituted (e.g. modified AMS with no methanol), transferring the suspension to a sterile 1.5 mL polypropylene tube, and centrifuging the cells at max speed (~15,000 rcf) for 1 min. Discard the supernatant.

4. Using the OD recorded in **Step 1**, resuspend the washed pellet to an OD of 5.0 in fresh medium without the element to be substituted (e.g. modified AMS with no methanol).

5. Add 6 mL of medium to each of three new sterile capped test tubes. Add the following to one tube each: (i) natural isotopic abundance element source (e.g. 50 mM final concentration of ^{12}C-methanol), (ii) isotopically substituted element source [e.g. 50 mM (^{13}C)methanol], (iii) isotopically substituted element source plus natural isotopic abundance precursor [e.g. 50 mM (^{13}C)methanol plus 100 µM final concentration L-^{12}C-methionine] (*see* **Note 4**).

6. Inoculate each of the tubes prepared in **Step 5** with 60 µL of washed cells prepared in **Step 4**, for a starting OD of 0.05.

7. Grow the cells in a shaking incubator for the length of time required for natural product production (usually at least one day after stationary phase is reached). *M. populi* BJ001 was grown for three days at 30 °C.

4.2 Natural product extraction

Timing: 1 h.

1. Pellet the cells by transferring each culture to a 15 mL screw top conical polypropylene tube and centrifuging at 4000 rcf for 10 min. Transfer the supernatant to a new 15 mL tube and reserve (*see* **Note 5**).

2. Resuspend each cell pellet in 600 µL of ice cold 100% methanol and transfer to a new 1.5 mL polypropylene tube. Sonicate tubes for 10 min in a sonicating water bath at room temperature.

3. Centrifuge the sonicated 1.5 mL tubes at max speed (~15,000 rcf) for 1 min.

4. Transfer 500 µL of each clarified methanolic extract (supernatant from **Step 3**) to the reserved supernatant from the same culture reserved in **Step 1**.

5. Equilibrate three SPE columns with 2 mL 100% methanol each. Discard flow through.

6. Wash each SPE column with 1 mL 10% (v/v) methanol in MilliQ H_2O. Discard flow through.

7. Load the supernatants from **Step 1** onto the columns (one supernatant per column) at a rate that produces 1-2 outlet drops per second. Discard flow through.
8. Wash each SPE column with 1 mL 10% (v/v) methanol in MilliQ H_2O. Discard flow through.
9. Elute each SPE column with 2 mL of 100% methanol into a 1-dram glass vial.
10. Dry samples under nitrogen stream and store at -20 °C until analysis.

4.3 Analysis of labeled microbial extracts by untargeted mass spectrometry

Timing: 1 h.
1. Resuspend each sample in 100 µL of 50% (v/v) methanol in MilliQ H_2O. Transfer to an LC-HRMS autosampler vial compatible with your LC-HRMS system.
2. Analyze samples with LC-HRMS. We injected 10 µL onto a Waters Acquity I-class UPLC system coupled to a Waters Xevo G2-S quadrupole time-of-flight mass spectrometer. Please see (Liebergesell, Murdock, & Puri, 2024) for more details about this particular system.
3. Save each LC-HRMS run. For Waters instruments, this is in .raw format.

4.4 Data analysis to identify inverse labeled natural products

Timing: ~1 day.

The LC-HRMS data can be examined manually to identify incorporation of the precursor of interest into natural products detected by the mass spectrometer using a program such as MZmine. Here, we provide an overview of a method to identify inverse labeled natural products in untargeted metabolomic data without any prior knowledge of which natural products are labeled or the number of precursor atoms that are incorporated (Liebergesell, Murdock, & Puri, 2024).
1. Convert each LC-HRMS data file to .mzML format. We converted each .raw file to .mzML using ProteoWizard's MSConvert (Chambers et al., 2012).
2. Use MZmine to detect features in the ^{12}C-medium and (^{13}C)medium conditions, but not the condition with the precursor. We created feature lists using batch mode parameters as described in (Liebergesell, Murdock, & Puri, 2024). Align the feature tables from these two conditions and export the aligned table in .csv format with the row retention time as a common element and feature m/z as the data file element.

3. Convert each of the three LC-MS data files from .mzML format to .h5 format. We used the Metabolite Atlas (Yao et al., 2015) and pymzML (Bald et al., 2012) Python packages with the "InverSIL_mz_to_h5.py" script.
4. Process the aligned feature table from **Step 2** along with the .h5 files from **Step 3** using the Python script "Untargeted_InverSIL.py" (Liebergesell, Murdock, & Puri, 2024).
5. The script output identifies mass spectral features with potential precursor incorporation. Examine the raw mass spectra using a program such as MZmine (Fig. 1) to verify precursor incorporation in features identified by the script (*see* **Note 6**).

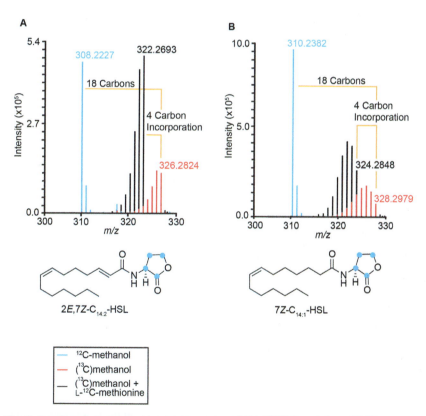

Fig. 1 Overlayed mass spectra and structures of (A) 2E,7Z-C$_{14:2}$-HSL, and (B) 7Z-C$_{14:1}$-HSL from *Methylorubrum populi* BJ001, highlighting the four-carbon incorporation from L-^{12}C-methionine into the HSL portion of these quorum sensing signals.

Using this method, we identified two features that incorporated four carbons when comparing the (^{13}C)methanol condition and the (^{13}C) methanol plus L-^{12}C-methionine condition; which are likely N-(2-*trans*-7-*cis*-tetradecenoyl)-L-homoserine lactone (2E,7Z-C$_{14:2}$-HSL); [M+H]$^+$ calc. 308.2220, obs. 308.2227, 2.27 Δppm, and N-(7-*cis*-tetradecenoyl)-L-homoserine lactone (7Z-C$_{14:1}$-HSL; [M+H]$^+$ calc. 310.2377, obs. 310.2382, 1.61 Δppm) (Fig. 1). These are also the quorum sensing signals produced by the LuxI-type acyl-HSL synthase MlaI in *Methylorubrum extorquens* PA1 (Cummings et al., 2021); which is consistent with the fact that the synthase gene products in BJ001 and PA1 share 95.7% amino acid sequence identity. To confirm that these strains produce the same signals, we compared extracted ion chromatograms of extracts from each strain, revealing that the signals have the same high-resolution mass and retention time (Fig. 2). Therefore, we have used InverSIL to link the *luxI*-type acyl-HSL synthase gene in *M. populi* BJ001 with its quorum sensing signal products: 2E,7Z-C$_{14:2}$-HSL, and 7Z-C$_{14:1}$-HSL.

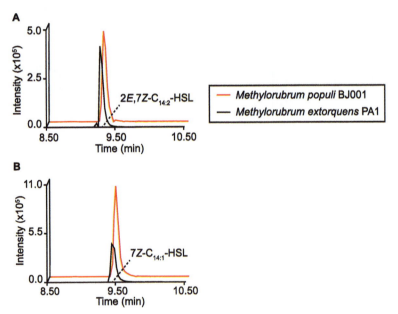

Fig. 2 Extracted ion chromatograms of supernatant extracts of the listed strains for the m/z values (A) 308.2220, corresponding to 2E,7Z-C$_{14:2}$-HSL, and (B) 310.2377, corresponding to 7Z-C$_{14:1}$-HSL. All masses correspond to [M+H]$^+$. Mass tolerance < 5 ppm.

5. Notes

1. Biosynthetic genes of interest can be identified in microbial genomes by a variety of methods. Examples include (1) using the tool antiSMASH (Blin et al., 2023), (2) BLASTing to identify homologs of gene products from characterized BGCs, and (3) searching annotated genomes for Interpro families, Pfam families, KEGG orthology (KO) identifiers, or Enzyme Commission (EC) numbers. These methods will also help predict substrates for the enzymes encoded by these genes.
2. The growth medium is specific to the bacterium of interest and is commonly a chemically defined medium where one element can be provided only in its isotopically substituted form. The strains in this chapter were grown at 30 °C in modified ammonium mineral salts (AMS) medium containing final concentrations of 50 mM methanol and 0.01% (m/v) yeast extract (Wallace et al., 2023). If the strain of interest cannot grow on a chemically defined medium, researchers can also try ISOGRO® ^{13}C medium (Millipore Sigma 606863) or ISOGRO® ^{15}N medium (Millipore Sigma 606871) (Bode et al., 2012).
3. The biosynthetic gene of interest must be transcriptionally active under the chosen growth condition (i.e. not silent). For example, a low-iron medium may activate siderophore production.
4. The cultures containing the isotopically substituted element source with and without the natural abundance precursor of interest will ensure that precursor incorporation can be detected even in cases where there is quantitative incorporation of the precursor into a natural product, as we have observed (Cummings et al., 2021). The culture containing the natural isotopic abundance element source and no precursor is used as a reference sample for the described untargeted metabolomic data analysis workflow.
5. Acyl-HSLs diffuse out of the cell, but we have outlined a protocol for extracting both intra- and extracellular metabolites here to be as untargeted as possible.
6. The size of the ^{13}C-isotopic envelope (i.e. the degree of incomplete ^{13}C-labeling) as well as the relative intensity of the molecular ion will depend on both the purity of the (^{13}C)carbon source as well as the primary metabolism of the bacterium of interest. This will affect the molecular ion identified by the script in each condition, which can influence the predicted number of carbons incorporated.

Acknowledgments

This work was supported by NIH grant R35 GM147018 (to A.W.P.). T.C.E.L. was supported by NIH training grant T32 GM122740.

References

Bald, T., Barth, J., Niehues, A., Specht, M., Hippler, M., & Fufezan, C. (2012). pymzML—Python module for high-throughput bioinformatics on mass spectrometry data. *Bioinformatics, 28*(7), 1052–1053. https://doi.org/10.1093/bioinformatics/bts066.

Blin, K., Shaw, S., Augustijn, H. E., Reitz, Z. L., Biermann, F., Alanjary, M., ... Weber, T. (2023). antiSMASH 7.0: New and improved predictions for detection, regulation, chemical structures and visualisation. *Nucleic Acids Research, 51*(W1), W46–W50. https://doi.org/10.1093/nar/gkad344.

Bode, H. B., Reimer, D., Fuchs, S. W., Kirchner, F., Dauth, C., Kegler, C., ... Grün, P. (2012). Determination of the absolute configuration of peptide natural products by using stable isotope labeling and mass spectrometry. *Chemistry – A European Journal, 18*(8), 2342–2348. https://doi.org/10.1002/chem.201103479.

Caesar, L. K., Montaser, R., Keller, N. P., & Kelleher N, L. (2021). Metabolomics and genomics in natural products research: Complementary tools for targeting new chemical entities. *Natural Product Reports, 38*(11), 2041–2065. https://doi.org/10.1039/D1NP00036E.

Chambers, M. C., Maclean, B., Burke, R., Amodei, D., Ruderman, D. L., Neumann, S., ... Mallick, P. (2012). A cross-platform toolkit for mass spectrometry and proteomics. *Nature Biotechnology, 30*(10), 918–920. https://doi.org/10.1038/nbt.2377.

Chen, I.-M. A., Chu, K., Palaniappan, K., Ratner, A., Huang, J., Huntemann, M., ... Ivanova, N. N. (2023). The IMG/M data management and analysis system v.7: Content updates and new features. *Nucleic Acids Research, 51*(D1), D723–D732. https://doi.org/10.1093/nar/gkac976.

Cummings, D. A., Snelling, A. I., & Puri, A. W. (2021). Methylotroph quorum sensing signal identification by inverse stable isotopic labeling. *ACS Chemical Biology, 16*(8), 1332–1338. https://doi.org/10.1021/acschembio.1c00329.

Figueiredo, S. A. C., Preto, M., Moreira, G., Martins, T. P., Abt, K., Melo, A., ... Leão, P. N. (2021). Discovery of cyanobacterial natural products containing fatty acid residues. *Angewandte Chemie International Edition, 60*(18), 10064–10072. https://doi.org/10.1002/anie.202015105.

Gross, H., Stockwell, V. O., Henkels, M. D., Nowak-Thompson, B., Loper, J. E., & Gerwick, W. H. (2007). The genomisotopic approach: A systematic method to isolate products of orphan biosynthetic gene clusters. *Chemistry & Biology, 14*(1), 53–63. https://doi.org/10.1016/j.chembiol.2006.11.007.

Liebergesell, T., Murdock, E., & Puri, A. (2024). Detection of inverse stable isotopic labeling in untargeted metabolomic data. *ChemRxiv.* https://doi.org/10.26434/chemrxiv-2024-g1ls6-v2.

Makris, C., Carmichael, J. R., Zhou, H., & Butler, A. (2022). C-Diazeniumdiolate graminine in the siderophore gramibactin is photoreactive and originates from arginine. *ACS Chemical Biology, 17*(11), 3140–3147. https://doi.org/10.1021/acschembio.2c00593.

McCaughey, C. S., van Santen, J. A., van der Hooft, J. J. J., Medema, M. H., & Linington, R. G. (2022). An isotopic labeling approach linking natural products with biosynthetic gene clusters. *Nature Chemical Biology, 18*(3), 295–304. https://doi.org/10.1038/s41589-021-00949-6.

Moré, M. I., Finger, L. D., Stryker, J. L., Fuqua, C., Eberhard, A., & Winans, S. C. (1996). Enzymatic synthesis of a quorum-sensing autoinducer through use of defined substrates. *Science, 272*(5268), 1655–1658. https://doi.org/10.1126/science.272.5268.1655.

Papenfort, K., & Bassler, B. L. (2016). Quorum sensing signal–response systems in Gram-negative bacteria. *Nature Reviews Microbiology, 14*(9), 576–588. https://doi.org/10.1038/nrmicro.2016.89.

Proschak, A., Lubuta, P., Grün, P., Löhr, F., Wilharm, G., De Berardinis, V., & Bode, H. B. (2013). Structure and biosynthesis of fimsbactins A–F, siderophores from *Acinetobacter baumannii* and *Acinetobacter baylyi*. *Chembiochem: A European Journal of Chemical Biology, 14*(5), 633–638. https://doi.org/10.1002/cbic.201200764.

Schaefer, A. L., Val, D. L., Hanzelka, B. L., Cronan, J. E., & Greenberg, E. P. (1996). Generation of cell-to-cell signals in quorum sensing: Acyl homoserine lactone synthase activity of a purified *Vibrio fischeri* LuxI protein. *Proceedings of the National Academy of Sciences, 93*(18), 9505–9509. https://doi.org/10.1073/pnas.93.18.9505.

Schmid, R., Heuckeroth, S., Korf, A., Smirnov, A., Myers, O., Dyrlund, T. S., ... Pluskal, T. (2023). Integrative analysis of multimodal mass spectrometry data in MZmine 3. *Nature Biotechnology, 41*(4), 447–449. https://doi.org/10.1038/s41587-023-01690-2.

Schuster, M., Sexton, D. J., Diggle, S. P., & Greenberg, E. P. (2013). Acyl-homoserine lactone quorum sensing: From evolution to application. *Annual Review of Microbiology, 67*(1), 43–63. https://doi.org/10.1146/annurev-micro-092412-155635.

Van Aken, B., Peres, C. M., Doty, S. L., Yoon, J. M., & Schnoor, J. L. (2004). *Methylobacterium populi* sp. Nov., a novel aerobic, pink-pigmented, facultatively methylotrophic, methane-utilizing bacterium isolated from poplar trees (*Populus deltoides×nigra* DN34). *International Journal of Systematic and Evolutionary Microbiology, 54*(4), 1191–1196. https://doi.org/10.1099/ijs.0.02796-0.

Wallace, M., Cummings, D. A., Jr., Roberts, A. G., & Puri, A. W. (2023). A widespread methylotroph acyl-homoserine lactone synthase produces a new quorum sensing signal that regulates swarming in *Methylobacterium fujisawaense*. *mBio, 15*(1), e01999. https://doi.org/10.1128/mbio.01999-23.

Yao, Y., Sun, T., Wang, T., Ruebel, O., Northen, T., & Bowen, B. P. (2015). Analysis of metabolomics datasets with high-performance computing and metabolite atlases. *Metabolites, 5*(3), 431–442. https://doi.org/10.3390/metabo5030431.

CHAPTER ELEVEN

4-Aldrithiol-based photometric assay for detection of methylthioalkylmalate synthase activity

Vivian Kitainda and Joseph Jez*

Department of Biology, Washington University in St. Louis, St. Louis, MO, United States
*Corresponding author. e-mail address: jjez@wustl.edu

Contents

1. Introduction	230
2. Using continuous enzyme assays for MAMS activity	232
2.1 Materials and equipment	232
2.2 Reagents	233
2.3 MAMS spectrophotometric enzyme assay protocol	233
3. Analysis and interpretation of spectrophotometric enzyme assay for *Brassica juncea* MAMS isoforms	236
3.1 pH screen	236
3.2 Buffer screen	236
3.3 Metal cofactor screen	238
3.4 Michaelis-Menten kinetic analysis	238
3.5 Leucine inhibition assays	242
4. Spectrophotometric alternatives to the 4,4'-DTP assay	242
5. Conclusion and outlook	243
References	244

Abstract

In *Brassica* plants, glucosinolates are a diverse class of natural products, of which aliphatic methionine-derived glucosinolates are the most abundant form. Their structural diversity comes from the elongation of some side-chains by up to 9 carbons, which, after the formation of the core glucosinolate structure, can undergo further chemical modifications. Methylthioalkylmalate synthase (MAMS) catalyzes the iterative elongation process for aliphatic methionine-derived glucosinolates. Most biochemical studies on MAMS have been performed using liquid chromatography/ mass spectrometry (LC/MS)-based assays or high-performance liquid chromatography (HPLC)-based assays. The LC/MS- and HPLC-based methods are endpoint assays, which cannot be monitored in real time and require a laborious process for data collection. These analytical methods are inefficient for performing multiple enzymatic

Methods in Enzymology, Volume 702
ISSN 0076-6879, https://doi.org/10.1016/bs.mie.2024.06.016
Copyright © 2024 Elsevier Inc. All rights are reserved, including those for text and data mining, AI training, and similar technologies.

assays needed to determine steady-state kinetic parameters or for mechanistic evaluation of pH-dependence and kinetic isotope effect studies. Although the function of MAMS has long been defined, there is a gap in knowledge as it pertains to biochemical characterization of this plant enzyme. Part of this may be due to the lack of efficient methods that can be used for this type of research. This chapter describes a continuous photometric assay to track MAMS activity in real time using the 4-aldrithiol reagent for reaction detection.

1. Introduction

Plants have evolved specialized metabolic pathways through gene duplication and functional divergence of enzymes involved in primary metabolism (Erb & Kliebenstein, 2020). The result of this process are varied pathways that produce an array of natural products useful to both plants and humans. In plants, glucosinolates are a diverse class of natural products (Blažević et al., 2020; Halkier & Gershenzon, 2006; Kitainda & Jez, 2021). Glucosinolate function stems from their hydrolysis products, which are responsible for the strong flavors of the Brassicaceae, such as Indian mustard (*Brassica juncea*), and serve as plant defense molecules by repelling insects, fighting fungal infections, and discouraging herbivory (Halkier & Gershenzon, 2006; Singh, 2017). Additionally, certain hydrolysis products such as isothiocyanates can potentially serve as cancer prevention agents in humans (Keum, Jeong, & Kong, 2004; Lund, 2003; Thornalley, 2002; Zhang, Talalay, Cho, & Posner, 1992).

The breadth of glucosinolate function is a result of their structural diversity, which comes from the use of aliphatic, aromatic and indole amino acids as precursors and elongation of some side-chains by up to 9 carbons, which, after the formation of the core glucosinolate structure, can undergo further chemical modifications (Halkier & Gershenzon, 2006; Sønderby, Geu-Flores, & Halkier, 2010). Aliphatic methionine-derived glucosinolates are the most abundant form of these compounds (Fig. 1). Methylthioalkylmalate synthase (MAMS) catalyzes the condensation of an amino acid-derived 2-oxo acid, such as 4-methylthio-2-oxobutanoic acid (4-MTOB; also known as α-keto-γ-(methylthio)butyric acid) with acetyl-Coenzyme A (CoA) to form a 2-malate derivative in the iterative elongation process for aliphatic methionine-derived glucosinolates (De Kraker & Gershenzon, 2011; Kliebenstein, Gershenzon, & Mitchell-Olds, 2001; Kumar et al., 2019; Textor, De Kraker, Hause, Gershenzon, & Tokuhisa, 2007). MAMS are evolutionarily derived from α-isopropylmalate synthases

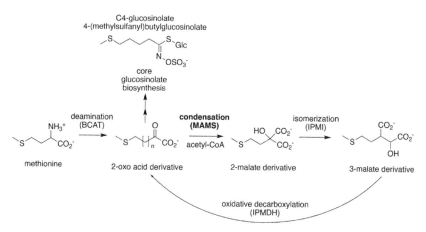

Fig. 1 Methionine-derived glucosinolate biosynthesis pathway. Deamination of methionine by a branched-chain aminotransferase (BCAT) to form a 2-oxo acid derivative is the first step in the pathway. Methylthioalkylmalate synthase (MAMS) catalyzes the condensation reaction of the 2-oxo acid derivative and acetyl-CoA to form a 2-malate derivative. Isomerization to a 3-malate derivate by isopropylmalate isomerase (IPMI) is followed by oxidative decarboxylation catalyzed by isopropylmalate syntase (IPMS) to yield a 2-oxo acid derivative elongated by 2 carbons. The new elongated product can either enter the core glucosinolate pathway, which involves a series of enzymatic steps to elaborate the final C4-glucosinolate (i.e., (4-methylsulfanyl)butylglucosinolate), or reenter the MAMS-IPMI-IPMDH elongation cycle for synthesis of extended aliphatic chain glucosinolates.

(IPMS), which catalyze the first reaction of leucine biosynthesis (de Kraker & Gershenzon, 2011). Even though IPMS is limited to one condensation reaction, IPMS and MAMS both condense 2-oxo acids with acetyl-CoA, placing both enzymes in the Claisen-like condensation subgroup of the DRE-TIM metallolyase superfamily (Kumar, Johnson, & Frantom, 2016). Although MAMS has long been functionally characterized, its X-ray crystal structure has only recently been solved from *B. juncea*, showing that MAMS is dimeric and shares the N-terminal catalytic α/β-barrel domain and the C-terminal α-helical region that forms the CoA binding site with *Mycobacterium tuberculosis* IPMS (Kumar et al., 2019). Furthermore, critical amino acid changes that increase the size of the substrate binding pocket to fit the longer sidechain of the 2-oxo acid substrate in MAMS compared with the smaller substrate of IPMS have been determined to be critical for the evolution of MAMS from IPMS in plants (Kumar et al., 2019).

Most biochemical studies on MAMS have been performed using liquid chromatography/mass spectrometry (LC/MS)-based assays for quantitative analysis or high-performance liquid chromatography (HPLC)-based assays for qualitative studies (Textor et al, 2007; De Kraker & Gershenzon, 2011; Kumar et al, 2019; Kumar, Reichelt, & Bisht, 2022). These analytical methods are inefficient for rapidly performing multiple enzymatic assays needed for evaluating initial velocity studies to determine steady-state kinetic parameters or for mechanistic evaluation of pH-dependence and/or kinetic isotope effect studies. Moreover, the LC/MS- and HPLC-based methods are endpoint assays, which, unlike continuous photometric assays, cannot be monitored in real time and require a laborious process for data collection. Furthermore, since analysis of enzymatic velocity requires data collection from the linear segment of the enzyme reaction progress curves, endpoint assays require verification that changes in reaction conditions do not alter the non-linear part of the progress curve (Bisswanger, 2014). Thus, a photometric assay has been developed to track MAMS activity in real time using the 4-aldrithiol reagent (i.e., 4,4'-dipyridyl disulfide) for reaction detection.

2. Using continuous enzyme assays for MAMS activity

The protocol of the photometric MAMS assay below is based on a similar assay carried out with IPMS from *Mycobacterium tuberculosis* (De Carvalho & Blanchard, 2006). As previously mentioned, MAMS and IPMS carry out a similar chemical reaction that involves the condensation of 2-oxo acids with acetyl-CoA, which releases a free CoA as a byproduct. As the reaction progresses, the buildup of CoA can be detected by an additional reaction of its free thiol group with 4-aldrithiol, which releases a product, 4-pyridinethione (4-PT), that can be detected spectrophotometrically at $A_{324\,nm}$ (Moser et al., 2016) (Fig. 2).

2.1 Materials and equipment

96-well flat bottom transparent plates (e.g., ThermoFisher Scientific, Costar).
Multichannel pipettor.
Spectrophotometer (e.g., Tecan Plate Reader Infinite 200 Pro).
Computer with graphing program that can do curve fitting (e.g., GraphPad Prism, Kaleidagraph, SigmaPlot).

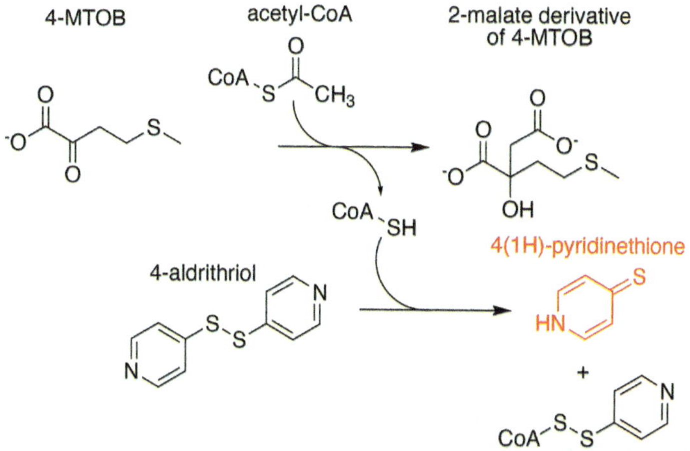

Fig. 2 MAMS reaction and coupling to 4-aldrithiol for spectrophotometric detection. In methionine-derived glucosinolate biosynthesis, MAMS catalyzes the condensation of 4-methylthio-2-oxobutanoic acid (4-MTOB) and acetyl-CoA to form the corresponding 2-malate derivative, which can be used for further extension reactions by MAMS or the core glucosinolate synthesis pathway. Coupling of free CoA produced by MAMS to 4-aldrithiol (4,4'-dipyridyl disulfide; 4,4'-DTP) leads to formation of 4(1H)-pyridinethione (4-PT; red) and $A_{324\,nm}$ signal.

2.2 Reagents

4–Aldrithiol (4,4′–dipyridyl disulfide; 4,4′–DTP; Sigma–Aldrich 143057)—4 to 7 mM stock solution in <0.1 M HCl.

Acetyl–coenzyme A (acetyl–CoA; Sigma–Aldrich A2181)—10 mM stock solution in ddH$_2$O.

4–Methylthio–2–oxobutanoic acid (4–MTOB; Sigma–Aldrich K6000)—30 mM stock solution in ddH$_2$O.

4–(2–Hydroxyethyl)piperazine–1–ethane–sulfonic acid (HEPES)—1 M stock solution in ddH$_2$O; pH 7.5.

Manganese chloride (MnCl$_2$) or magnesium chloride (MgCl$_2$)—1 M stock solution in ddH$_2$O.

Potassium chloride (KCl)—1 M stock solution in ddH$_2$O.

Ultrapure water (ddH$_2$O; MilliQ).

Purified MAMS enzyme (Kumar et al., 2019).

2.3 MAMS spectrophotometric enzyme assay protocol

Step 1. Prepare a homogenous sample of the MAMS enzyme you wish to study. The expression of recombinant MAMS and its purification from

E. coli were previously reported (Kumar et al., 2019). It is advisable that the storage buffer for the enzyme does not contain any reducing reagents such as tris(2-carboxyethyl)-phosphine (TCEP), 1,4-dithiothreitol (DTT), or β-mercaptoethanol (BME). This is because thiol groups of these compounds will increase the background noise and affect photometric data collection, as 4,4′-DTP can react with any exposed free thiol groups, not just CoA. A simple storage buffer such as (25 mM HEPES, 100 mM NaCl) will suffice, unless the stability of the enzyme depends on any extra factors.

Step 2. Initial rates of reaction for the condensation reaction of MAMS are determined using 4,4′-DTP, which reacts with the free thiol group of CoA to release photometrically detectable 4-PT at $A_{324\,nm}$ at 25 °C. A standard 100 μL reaction mix contains 50 mM HEPES at pH 7.5, 20 mM KCl, 20 mM $MgCl_2$, 200 μM 4,4′-DTP, 0.5 mM acetyl-CoA, and 1 mM 4-MTOB. All assays are performed in a 96-well plate format using a Tecan UV–visible plate reader. For each run, 4 wells are used, as follows: well #1 for the control sample (H_2O) and wells #2–4 for three replicates of the 4,4′-DTP assay with protein sample. The control/protein samples are added to wells before the reaction mix; the reaction mix is added right before an assay run. This is where the multichannel pipettor is handy. As a reference, steady-state kinetic assays of *B. juncea* MAMS use a range of 0.02–1 mM acetyl-CoA with 1 mM 4-MTOB held constant and a range of 0.05–1 mM 4-MTOB with 0.5 mM acetyl-CoA held constant. These concentrations are chosen to cover (as much as possible) the range of ~10-fold below K_m to ~10-fold above K_m for each substrate. Furthermore, ~10 μg of protein sample is used per assay to ensure that the enzyme concentration in the 100 μL reaction mix does not exceed the lowest concentration of substrate used.

Based on the above values, data collection by the plate reader can last between 5 and 10 min, as it is around this time that the [product formed] vs. time is no longer linear and substrates depletion begins. The data collected by the plate reader is in the form of ΔA at 324 nm over time ($\Delta A_{324\,nm}$) with the initial rates of activity measured as slope values in the first $30\,s^{-1}$ of the reaction progress curve, which are generally linear (Fig. 3). To get the specific enzymatic rate (nmol min^{-1} mg^{-1}) of the protein sample in each well, the following equation is used:

$$S = (R)\cdot\frac{1}{\varepsilon\cdot b}\cdot V\cdot\frac{1}{m}\cdot 10^9$$

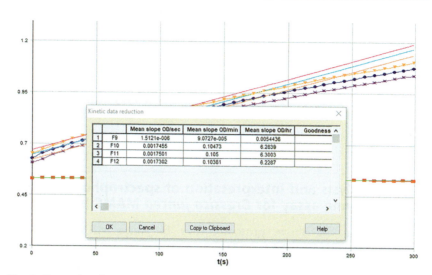

Fig. 3 **Example of 4,4′-DTP continuous assay kinetic data.** Reactions were performed as described in Section 2.3 with resulting change in signal at $A_{324\,nm}$ shown for well 1 (control; red squares) and wells 2–4 (replicates, purple x, blue circle, and orange triangle). Initial rates are shown as lines (green, orange, cyan, red for wells 1–4) with fitted slopes (F9, well 1 control; F10–F12, wells 2–4) in the inset.

where R is rate of $\Delta A_{324\,nm}$ (min^{-1}), ε is the extinction coefficient of the 4-PT product (19,800 M^{-1} cm^{-1} at $A_{324\,nm}$), b (cm) is the path length based on the sample volume and bottom layer of plate well (assuming the plate reader scans from the bottom of the well), V (L) is the assay volume, and m (mg) is the mass of protein sample used in the assay.

Step 3. Using GraphPad Prism (or any other graphing program), the initial enzymatic rates (nmol min^{-1} mg^{-1}) for MAMS measured in Step 2 are plotted vs. initial substrate concentration [S], and the kinetic parameters determined by fitting the data with the Michaelis Menten equation using nonlinear regression. In GraphPad Prism, the equation will take the form:

$$Y = \frac{V_{max} \cdot X}{K_m + X}$$

where Y is the enzyme velocity and X is the substrate concentration [S].

Graphpad Prism can also calculate both apparent K_m and k_{cat} using the equation:

$$Y = \frac{E_t \cdot k_{cat} \cdot x}{K_m + X}$$

where the E_t is the concentration of enzyme sites.

Otherwise, apparent k_{cat} (min^{-1}) can be calculated using the V_{max} and apparent K_m parameters through the equation:

$$k_{cat} = V_{max} \cdot MW \cdot \frac{1}{10^6} = \frac{V_{max}}{E_t}$$

where MW is the molecular weight of MAMS in g/mol. For example, 55,387 g/mol for *B. juncea* MAMS isoform 1A.

3. Analysis and interpretation of spectrophotometric enzyme assay for *Brassica juncea* MAMS isoforms

Based on the above protocol, four *B. juncea* MAMS isoforms (MAMS1A—GenBank/EMBL FM161920, MAMS1B—GenBank/EMBL FM161916, MAMS2A—GenBank/EMBL FM161923, MAMS2B—GenBank/EMBL FM161918) were analyzed for their specific enzyme activities (in parallel to *Brassica rapa* IPMS1.1—GenBank LOC103842698) at different pH values, buffer, and metal cofactors, as well as for steady-state kinetic parameters.

3.1 pH screen

Using HEPES (pH 7.5) as the buffer and Mg^{2+} as the metal cofactor in the reaction mix, a pH range 7–8 was analyzed in 0.5 increments to determine if *B. juncea* MAMS isoforms are affected by pH changes in similar ways in the presence of excess substrates. *B. rapa* IPMS1.1 was also analyzed under the same conditions for comparison as the homolog of MAMS. The measured specific enzymatic rates indicate that, while MAMS1A and MAMS2A seem to function best at pH 7.5, MAMS1B activity is not affected by changes in this pH range (Fig. 4A). On the other hand, IPMS1.1, which carries out a similar condensation reaction between α-ketoisovalerate (α-KIV) and acetyl-CoA, seems to function best at pH 8 (Fig. 4A).

3.2 Buffer screen

Using Mg^{2+} as the metal cofactor and pH 7.5 for the buffer in the reaction mix, HEPES and Tris HCl buffers were used to examine the effect of buffer identity on the *B. juncea* MAMS activities in the presence of excess substrates. As before, *B. rapa* IPMS1.1 was analyzed under the same conditions for comparison. The specific activities for all the MAMS isoforms reveal that the buffer identity plays a significant role in the enzyme function in vitro, as MAMS1A, MAMS1B and MAMS2A all show decreased levels

MAMS Photometric Assay 237

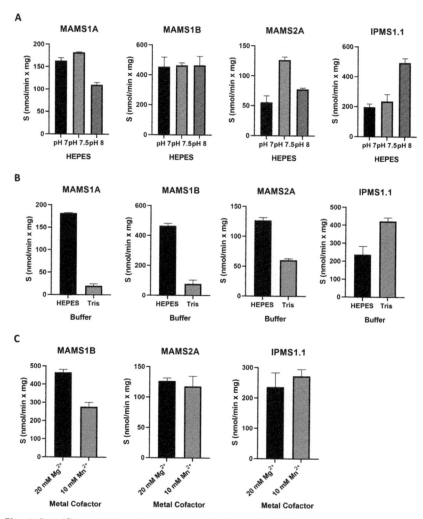

Fig. 4 **Specific activities of *B. juncea* MAMS1A, MAMS1B, and MAMS2A isoforms and *B. rapa* IPMS1.1 under varied conditions.** (A) Varied pH of HEPES with Mg^{2+} as a cofactor. (B) Varying buffer identity at pH 7.5 with Mg^{2+} as a cofactor. (C) Varied metal cofactor with HEPES pH 7.5 buffer. All assays were carried out in the presence of excess substrates (0.5 mM acetyl-CoA and 1 mM 4-MTOB or α-KIV for MAMS and IPMS, respectively). All assay reaction mixes were based on Step 2 of Section 2.3, except for the specified variable reagents. Enzyme activity for the Mn^{2+} cofactor was tested at a lower concentration (10 mM) than that of Mg^{2+} (20 mM) due to noticeably lesser activity of the MAMS isoforms at higher concentrations.

of activity with Tris relative to HEPES (Fig. 4B). The opposite is true, however, for IPMS1.1, which performs significantly better with Tris than with HEPES (Fig. 4B). This information can be used to optimize assay conditions.

3.3 Metal cofactor screen

Using HEPES (pH 7.5) as the buffer in the reaction mix, Mg^{2+} and Mn^{2+} were used to explore metal preferences of *B. juncea* MAMS isoforms, as well as *B. rapa* IPMS1.1. Earlier studies using discontinuous assay techniques indicated that in *Arabidopsis thaliana* MAMS prefers Mn^{2+} as a metal cofactor (De Kraker & Gershenzon, 2011). A 10 mM concentration of Mn^{2+} was used after observing that MAMS activity decreased at 20 mM, which is the effective concentration of Mg^{2+} (data not shown). The results indicate that a 2-fold decrease in metal cofactor concentration for Mn^{2+} still allows for sufficient enzyme activity in MAMS1B, MAMS2A, and IPMS1.1 (Fig. 4C). MAMS2A and IPMS1.1 show similar levels of specific activity with both metals, while MAMS1B has a decreased activity with 10 mM Mn^{2+} compared to 20 mM Mg^{2+}. Since the decrease in Mn^{2+} concentration relative to Mg^{2+} did not have a drastic effect on enzymatic rates overall, all three enzymes appear to prefer Mn^{2+} over Mg^{2+}.

3.4 Michaelis-Menten kinetic analysis

Based on the pH, buffer and metal screen results, apparent steady-state kinetic parameters (k_{cat} and K_m) were determined for the *B. juncea* MAMS isoforms and *B. rapa* IPMS1.1 by fitting the Michaelis Menten equation to initial velocity vs. substrate concentration data obtained using steps 2–3 of the protocol. The reaction mixes for the MAMS isoforms used HEPES (pH 7.5) as the buffer, while the reaction mix for IPMS1.1 used Tris (pH 8.0) as the buffer; Mg^{2+} was used as the metal cofactor in all the assays. Representative velocity versus substrate plots for *B. juncea* MAMS2A and *B. rapa* IPMS1.1 are shown in Fig. 5 with fitted kinetic parameters summarized in Table 1.

Overall, the general trend shared between *Brassica* MAMS and IPMS is that substrate specificity, and therefore enzyme efficiency, is higher for the acetyl-CoA substrate than for the 2-oxo acid substrates (4-MTOB and α-KIV, respectively). Furthermore, the small differences within MAMS isoform groupings (MAMS1A and MAMS1B, MAMS2A and MAMS2B) are enough to affect enzyme efficiency as it applies to the 4-MTOB substrate. *B. juncea* MAMS1B and MAMS2A are more efficient

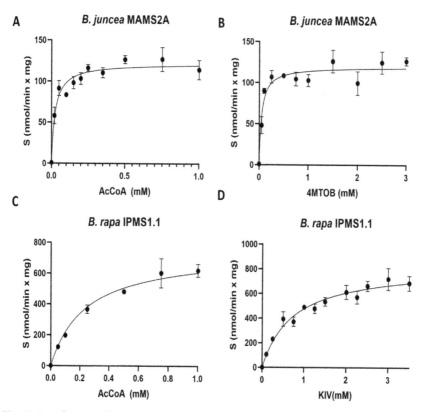

Fig. 5 Steady-state kinetic analysis of *B. juncea* MAMS2A and *B. rapa* IPMS1.1. Initial velocity activity (S) versus substrate concentration data for (A) *B. juncea* MAMS2A with varied acetyl-CoA (AcCoA; fixed 4-MTOB); (B) *B. juncea* MAMS2A with varied 4-MTOB (fixed acetyl-CoA); (C) *B. rapa* IPMS1.1 with varied acetyl-CoA (AcCoA; fixed α-KIV); and (D) *B. rapa* IPMS1.1 with varied α-KIV (fixed acetyl-CoA). Data points are mean ± standard error (n = 3) with fits to the Michaelis-Menten equation.

with 4-MTOB as the 2-oxo acid substrate than MAMS1A and MAMS2B. Further analysis on MAMS substrate preference can be carried out using the 4,4′-DTP assay, especially considering that the *B. juncea* MAMS isoforms have been shown to differ in the length of substrate that they can process. In addition, comparison of the above kinetic values for the *B. juncea* MAMS isoforms determined through the spectrophotometric 4,4′-DTP assay to values determined via a discontinuous LC/MS-based assay (Kumar et al., 2019) further emphasizes the reliability of a continuous enzyme assay for initial rate studies (Table 2). The LC/MS based assay seems to underestimate the catalytic efficiencies of the *B. juncea* MAMS isoforms.

Table 1 Summary of apparent steady-state kinetic parameters of *B. juncea* MAMS isoforms and *B. rapa* IPMS 1.1.

Protein	Substrate	k_{cat} (min^{-1})	K_m (mM)	k_{cat}/K_m (M^{-1} s^{-1})
B. rapa IPMS1.1	acetyl–CoA	73 ± 3	0.24 ± 0.04	5006
B. rapa IPMS1.1	α–KIV	81 ± 4	0.74 ± 0.10	1816
B. juncea MAMS2A	acetyl–CoA	6.7 ± 0.2	0.03 ± 0.01	4553
B. juncea MAMS2A	4–MTOB	6.6 ± 0.2	0.05 ± 0.01	2146
B. juncea MAMS1A	acetyl–CoA	13 ± 1	0.16 ± 0.04	1416
B. juncea MAMS1B	acetyl–CoA	28 ± 2	0.08 ± 0.02	5291
B. juncea MAMS2B	acetyl–CoA	3.5 ± 0.2	0.03 ± 0.01	1964

Average values $\pm$ standard error (n = 3) are shown for apparent steady–state kinetic parameters.

Table 2 Comparison of *B. juncea* MAMS apparent kinetic parameters determined through continuous spectrophotometric and discontinuous LC/MS assays.

Protein	4-Aldrithiol enzyme assay			LC/MS-based assay		
	k_{cat} (min^{-1})	K_m (µM)	k_{cat}/K_m (M^{-1} s^{-1})	k_{cat} (min^{-1})	K_m (µM)	k_{cat}/K_m (M^{-1} s^{-1})
B. juncea MAMS1A	13	157	1416	1.6	126	212
B. juncea MAMS1B	28	88	5291	1.5	198	126
B. juncea MAMS2A	6.7	25	4.553	2.1	69	507
B. juncea MAMS2B	3.5	30	1964	0.8	38	351

Data for the 4-aldrithiol assay are from Table 1 with data for the LC/MS-based assay from Kumar et al. (2019).

3.5 Leucine inhibition assays

Bacterial IPMS are sensitive to feedback inhibition by the amino acid leucine, which binds to an extended C-terminal regulatory domain of the enzyme to down regulate the biosynthetic pathway (De Carvalho, Frantom, Argyrou, & Blanchard, 2009). To determine if the same feedback inhibition happens in *B. rapa* IPMS1.1, the effect of leucine (0.25 mM) on activity were determined using the assay protocol described above (Fig. 6A; Table 3). In comparison to assays of *B. rapa* IPMS1.1 in the absence of leucine, both V_{max} and apparent K_m decrease in the presence of leucine, which indicates that the amino acid is an uncompetitive inhibitor. Uncompetitive inhibitors bind only to the enzyme-substrate complex and not to the free enzyme, which could mean that substrate binding changes the conformation of *B. rapa* IPMS1.1 to allow leucine access the C-terminal regulatory domain.

For comparison, the effect of leucine on *B. juncea* MAMS1B and MAMS2A isoforms, which do not contain the C-terminal regulatory domain, was tested (Fig. 6B). The results show that leucine has no inhibitory effect on either MAMS isoform. The slight increase in the specific activity of MAMS2A in the presence of 0.25 mM leucine, which could be explained by the occurrence of a low IPMS condensation activity, as MAMS has been reported to have condensation activity with leucine when heterologously expressed in *Nicotiana benthamiana* and in *E. coli* (Crocoll, Mirza, Reichelt, Gershenzon, & Halkier, 2016; Mikkelsen, Olsen, & Halkier, 2010; Mirza, Crocoll, Olsen, & Halkier, 2016).

4. Spectrophotometric alternatives to the 4,4′-DTP assay

For the detection of free thiols, Ellman's reagent (5,5′-dithio-bis-(2-nitrobenzoic acid); DTNB) is a well-established and versatile tool for enzyme assays, as well as studies of proteins and proteomes (Schmidt et al., 2023). DTNB-based assays use the stochiometric reaction of DTNB with a thiolate anion to yield a mixed disulfide and the yellow colored dianion of 2-nitro-5-thiobenzoic acid (NTB). The spectrophotometrically detectable NTB product ($A_{409\,nm}$) provides another direct measure of free thiols; however, Ellman's assay is only suitable for alkaline pH values (7.3–10) and comes with disadvantages, such as interferences from certain buffers, the low stability of the Ellman's reagent, variable molar absorption coefficients,

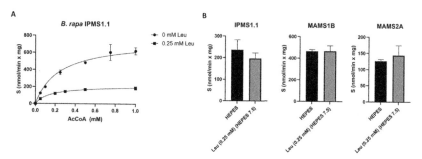

Fig. 6 Effect of leucine on *B. rapa* IPMS1.1, *B. juncea* MAMS1B, and *B. juncea* MAMS2A. (A) Initial velocity activity (S) versus substrate concentration data for *B. rapa* IPMS1.1 with 0 mM (circles) and 0.25 mM (squares) leucine. Data points are mean ± SD (n = 3) with fits to the Michaelis-Menten equation. (B) Comparison of specific activities of *B. rapa* IPMS1.1, *B. juncea* MAMS1B, and *B. juncea* MAMS2A in buffer without leucine (HEPES) and with 0.25 mM leucine. Data is mean ± standard error (n = 3).

Table 3 Summary of apparent steady-state kinetic parameters of *B. rapa* IPMS 1.1 in the absence and presence of leucine.

Protein	Leucine (mM)	V_{max} (nmol min^{-1} mg^{-1})	K_m (mM)
B. rapa IPMS1.1	0	752 ± 27	0.24 ± 0.04
B. rapa IPMS1.1	0.25	214 ± 8	0.15 ± 0.02

Average values ± standard error (n = 3) are shown for apparent steady-state kinetic parameters.

and incomplete reaction with proteins. 4,4′-DTP is an uncharged alternative that is more sensitive due to the higher molar absorption coefficient of its reaction product. It also has an improved stability and can be applied within a broader pH range. It is possible to use Ellman's reagent as a complementary tool to 4,4′-DTP; however, given the pH range within which MAMS is the most active, this is not necessary.

5. Conclusion and outlook

The importance of MAMS in glucosinolate biosynthesis is as a gatekeeper to the extension reactions that lead to varied length of methionine-derived aliphatic glucosinolates, as this protein determines whether products continue elongating the elongation cycle or proceed to the synthesis of the core glucosinolate structure. Although the function of

MAMS has long been defined, there is a gap in knowledge as it pertains to a detailed biochemical characterization of this plant enzyme. Part of this may be due to the lack of efficient methods that can be used for this type of research. Therefore, the more streamlined, continuous enzymatic 4,4′-DTP assay method described in this chapter will hopefully contribute to filling this knowledge gap, especially for future efforts to metabolically engineer glucosinolates in plants and bacteria for plant defense, human health, and nutrition.

References

Bisswanger, H. (2014). Enzyme assays. *Perspectives in Science, 1*, 41–55.

Blažević, I., Montaut, S., Burčul, F., Olsen, C. E., Burow, M., Rollin, P., & Agerbirk, N. (2020). Glucosinolate structural diversity, identification, chemical synthesis, and metabolism in plants. *Phytochemistry, 169*, 112100.

Crocoll, C., Mirza, N., Reichelt, M., Gershenzon, J., & Halkier, B. A. (2016). Optimization of engineered production of the glucoraphanin precursor dihomomethionine in *Nicotiana benthamiana*. *Frontiers in Bioengineering and Biotechnology, 4*, 14.

De Carvalho, L. P. S., & Blanchard, J. S. (2006). Kinetic and chemical mechanism of α-isopropylmalate synthase from *Mycobacterium tuberculosis*. *Biochemistry, 45*, 8988–8999.

De Carvalho, L. P., Frantom, P. A., Argyrou, A., & Blanchard, J. S. (2009). Kinetic evidence for interdomain communication in the allosteric regulation of alpha-isopropylmalate synthase from *Mycobacterium tuberculosis*. *Biochemistry, 48*, 1996–2004.

De Kraker, J. W., & Gershenzon, J. (2011). From amino acid to glucosinolate biosynthesis: Protein sequence changes in the evolution of methylthioalkylmalate synthase in Arabidopsis. *The Plant Cell, 23*, 38–53.

Erb, M., & Kliebenstein, D. J. (2020). Plant secondary metabolites as defenses, regulators, and primary metabolites: The blurred functional trichotomy. *Plant Physiology, 184*, 39–52.

Halkier, B. A., & Gershenzon, J. (2006). Biology and biochemistry of glucosinolates. *Annual Reviews of Plant Biology, 57*, 303–333.

Keum, Y. S., Jeong, W. S., & Kong, A. T. (2004). Chemoprevention by isothiocyanates and their underlying molecular signaling mechanisms. *Mutation Research, 555*, 191–202.

Kitainda, V., & Jez, J. M. (2021). Structural studies of aliphatic glucosinolate chain-elongation enzymes. *Antioxidants, 10*, 1500.

Kliebenstein, D. J., Gershenzon, J., & Mitchell-Olds, T. (2001). Comparative quantitative trait loci mapping of aliphatic, indolic and benzylic glucosinolate production in *Arabidopsis thaliana* leaves and seeds. *Genetics, 159*, 359–370.

Kumar, G., Johnson, J. L., & Frantom, P. A. (2016). Improving functional annotation in the DRE-TIM metallolyase superfamily through identification of active site fingerprints. *Biochemistry, 55*, 1863–1872.

Kumar, R., Lee, S. G., Augustine, R., Reichelt, M., Vassão, D. G., Palavalli, M. H., ... Bisht, N. C. (2019). Molecular basis of the evolution of methylthioalkylmalate synthase and the diversity of methionine-derived glucosinolates. *The Plant Cell, 31*, 1633–1647.

Kumar, R., Reichelt, M., & Bisht, N. C. (2022). An LC-MS/MS assay for enzymatic characterization of methylthioalkylmalate synthase (MAMS) involved in glucosinolate biosynthesis. *Methods in Enzymology, 676*, 49–69.

Lund, E. (2003). Non-nutritive bioactive constituents of plants: Dietary sources and health benefits of glucosinolates. *International Journal for Vitamin and Nutrition Research, 73*, 135–143.

Mikkelsen, M. D., Olsen, C. E., & Halkier, B. A. (2010). Production of the cancer-preventive glucoraphanin in tobacco. *Molecular Plant, 3,* 751–759.

Mirza, N., Crocoll, C., Olsen, C. E., & Halkier, B. A. (2016). Engineering of methionine chain elongation part of glucoraphanin pathway in *E. coli. Metabolic Engineering, 35,* 31–37.

Moser, M., Schneider, R., Behnke, T., Schneider, T., Falkenhagen, J., & Resch-Genger, U. (2016). Ellman's and aldrithiol assay as versatile and complementary tools for the quantification of thiol groups and ligands on nanomaterials. *Analytical Chemistry, 88,* 8624–8631.

Schmidt, R., Logan, M. G., Patty, S., Ferracane, J. L., Pfeifer, C. S., & Kendall, A. J. (2023). Thiol quantification using colorimetric thiol-disulfide exchange in nonaqueous solvents. *ACS Omega, 8,* 9356–9363.

Singh, A. (2017). Glucosinolates and plant defense. In J. Mérillon, & K. Ramawat (Eds.). *Glucosinolates* (pp. 237–246). Cham: Springer.

Sønderby, I. E., Geu-Flores, F., & Halkier, B. A. (2010). Biosynthesis of glucosinolates—Gene discovery and beyond. *Trends in Plant Science, 15,* 283–290.

Textor, S., De Kraker, J. W., Hause, B., Gershenzon, J., & Tokuhisa, J. G. (2007). MAM3 catalyzes the formation of all aliphatic glucosinolate chain lengths in Arabidopsis. *Plant Physiology, 144,* 60–71.

Thornalley, P. J. (2002). Isothiocyanates: Mechanism of cancer chemopreventive action. *Anti-Cancer Drugs, 13,* 331–338.

Zhang, Y., Talalay, P., Cho, C. G., & Posner, G. H. (1992). A major inducer of anticarcinogenic protective enzymes from broccoli: Isolation and elucidation of structure. *Proceedings of the National Academy of Sciences USA, 89,* 2399–2403.

CHAPTER TWELVE

Methods for biochemical characterization of flavin-dependent N-monooxygenases involved in siderophore biosynthesis

Noah S. Lyons[a], Sydney B. Johnson[a], and Pablo Sobrado[a,b,c,*]

[a]Department of Biochemistry, Virginia Tech, Blacksburg, VA, United States
[b]Center for Drug Discovery, Virginia Tech, Blacksburg, VA, United States
[c]Department of Chemistry, Missouri University of Science and Technology, Rolla, MO, United States
*Corresponding author. e-mail address: psobrado@vt.edu

Contents

1. Introduction	248
2. NADPH oxidation	252
2.1 Equipment	253
2.2 Reagents	253
2.3 Procedure	254
2.4 Notes	254
3. Oxygen consumption assays	254
3.1 Equipment	256
3.2 Reagents	256
3.3 Procedure	256
3.4 Notes	256
4. Iodine oxidation assay	257
4.1 Equipment	258
4.2 Reagents	258
4.3 Procedure	259
4.4 Notes	260
5. Determination of hydrogen peroxide formation and reaction uncoupling	260
5.1 Colorimetric measurement of H_2O_2	261
5.2 Continuous measurement of H_2O_2 using catalase	262
5.3 Reagents	262
5.4 Procedure	262
5.5 Notes	262
6. Fmoc-Cl derivatization	263
6.1 Equipment	263
6.2 Reagents	263

6.3 Procedure	264
6.4 Notes	265
7. Determination of reaction stereospecificity	265
7.1 Equipment	267
7.2 Expression and purification of TbADH	267
7.3 Synthesis of (R)-[4-^{2}H]-NADPH	269
7.4 Expression and purification of LmG6PDH	270
7.5 Synthesis of (S)-[4-^{2}H]-NADPH	271
8. Fluorescence anisotropy	272
8.1 Equipment	274
8.2 Reagents	274
8.3 Procedure	274
9. Targeting flavin dynamics for drug discovery	274
10. Summary and conclusions	276
Acknowledgments	276
References	277

Abstract

Siderophores are essential molecules released by some bacteria and fungi in iron-limiting environments to sequester ferric iron, satisfying metabolic needs. Flavin-dependent N-hydroxylating monooxygenases (NMOs) catalyze the hydroxylation of nitrogen atoms to generate important siderophore functional groups such as hydroxamates. It has been demonstrated that the function of NMOs is essential for virulence, implicating these enzymes as potential drug targets. This chapter aims to serve as a resource for the characterization of NMO's enzymatic activities using several biochemical techniques. We describe assays that allow for the determination of steady-state kinetic parameters, detection of hydroxylated amine products, measurement of the rate-limiting step(s), and the application toward drug discovery efforts. While not exhaustive, this chapter will provide a foundation for the characterization of enzymes involved in siderophore biosynthesis, allowing for gaps in knowledge within the field to be addressed.

1. Introduction

Iron is an essential mineral for many biological processes, where its redox state is essential for functionality. Many pathogenic fungi and bacteria, such as *Aspergillus fumigatus* or *Acinetobacter baumannii*, rely on the use of siderophores to transport ferric iron (Fe^{3+}) from an iron-limiting environment due to its low aqueous solubility back to the organism via siderophore-interacting proteins (Moynié et al., 2019; Page, 2019). It has been shown that siderophore biosynthesis is essential for virulence in fungi and bacteria, thus enzymes in this pathway are an attractive drug target.

Methods for Characterizing *N*-hydroxylating Monooxygenases   249

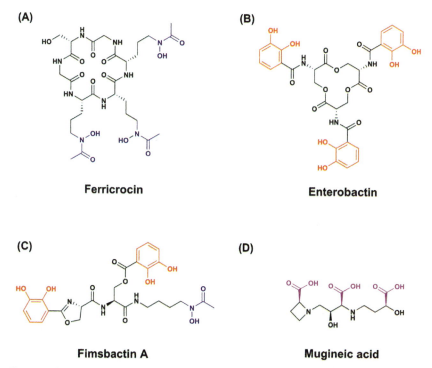

Fig. 1 Siderophores produced by various bacteria, fungi, and plants that contain hydroxamate (blue), catechol (red), or carboxylic acid (magenta) functional groups. These siderophores include (A) ferricrocin from *Aspergillus fumigatus* consisting of hydroxamate groups, (B) enterobactin from Gram-negative bacteria such as *Escherichia coli* that contain catechol groups, (C) fimsbactin A from *Acinetobacter baumannii* that contains both hydroxamate and catechol groups, and (D) mugineic acid from *Hordeum vulgare* that contains carboxylic acid groups.

Siderophores are often characterized by the functional groups involved in the chelation of iron, with the most common moieties being hydroxamates, catecholates, and carboxylates (Bohac, Fang, Giblin, & Wencewicz, 2019; Moynié et al., 2019; Saha et al., 2019; Suzuki et al., 2021). Prototypical examples of such are shown in Fig. 1.

The biosynthesis of siderophores are often encoded within an operon or biosynthetic gene cluster, containing a non-ribosomal peptide synthetase (NRPS) or polyketide synthase and several tailoring enzymes, such as flavin-dependent monooxygenases (FMOs) and transacetylases (Khan, Singh, & Srivastava, 2018). To illustrate the biosynthesis of siderophores in fungi, we will focus on the biosynthesis of ferricrocin in *A. fumigatus* as an example. The first step involves the hydroxylation of L-ornithine into N^5-hydroxy-L-ornithine by the FMO SidA (Chocklett & Sobrado, 2010; Hissen, Wan, Warwas, Pinto, & Moore, 2005). Then, N^5-hydroxy-L-ornithine is acetylated to N^5-acetyl-N^5-hydroxy-L-ornithine by SidL (Blatzer et al., 2011). Last, the NRPS SidC links three N^5-acetyl-N^5-hydroxy-L-ornithine molecules, two glycine molecules, and one serine molecule to form ferricrocin (Happacher et al., 2023). Gene knockout studies have shown that deletion of *sidA* in *A. fumigatus* leads to the production of siderophores that cannot bind iron, and the mutant strain is not virulent (Hissen et al., 2005; Schrettl et al., 2004). Therefore, SidA is essential, rate-determining for siderophore biosynthesis, and remains a vital drug target to fight *A. fumigatus*.

FMOs are a large and catalytically diverse enzyme family that are distinguished into eight classes (A–H) based on their structure and function (Huijbers, Montersino, Westphal, Tischler, & Van Berkel, 2014). N-hydroxylating monooxygenases (NMOs) are a clade within the Class B FMOs, which feature: two Rossmann folds, reduction of the FAD prosthetic group that does not rely on presence of the substrate, and the stabilization of a highly reactive C4a-(hydro)peroxyflavin intermediate by the bound $NADP^+$ (Huijbers et al., 2014; Romero et al., 2012). NMOs can catalyze single or multiple oxidations on primary and secondary amines, most commonly on amino acid substrates (Mügge et al., 2020). Catalysis of NMOs is initiated by the binding of NAD(P)H, followed by a hydride transfer between NAD(P)H and oxidized FAD, reducing the flavin (Paul, Eggerichs, Westphal, Tischler, & van Berkel, 2021). One equivalent of molecular oxygen then reacts with the reduced flavin to generate a reactive C4a-(hydro)peroxyflavin intermediate, which can act as a nucleophile or an electrophile (Palfey & McDonald, 2010). This intermediate then reacts with the amino group of a substrate of interest, oxidizing the molecule to its respective hydroxylamine product. Following flavin dehydration and product release, the enzyme releases the bound $NAD(P)^+$ cofactor and is primed for a subsequent round of catalysis (Robinson, Badieyan, & Sobrado, 2013).

Several NMOs involved in siderophore biosynthesis have been extensively characterized over the last fifteen years (Reis, Li, Johnson, & Sobrado, 2021). Such examples include those acting on amino acids, such as SidA and NbtG, and those acting on aliphatic diamines, such as RhbE, BasC, FbsI, and DesB (Fig. 2) (Binda et al., 2015; Chocklett & Sobrado, 2010; Giddings et al., 2021; Lynch et al., 2001; Lyons, Bogner, Tanner, & Sobrado, 2022; Mihara et al., 2004). A defining characteristic of NMOs is their tight substrate specificity and strict preference for NADPH over

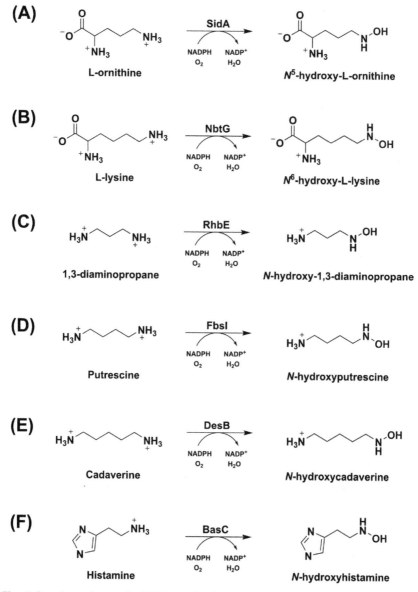

Fig. 2 Reaction schemes for NMOs involved in N-hydroxylation of primary amines. (A) SidA from *Aspergillus fumigatus* and (B) NbtG from *Nocardia farcinica* hydroxylate the amino acids L-ornithine and L-lysine, respectively. NMOs that use aliphatic diamines as substrates include (C) the 1,3-diaminopropane N-hydroxylase RhbE from *Sinorhizobium meliloti*, (D) the putrescine N-hydroxylase FbsI from *Acinetobacter baumannii*, (E) the cadaverine N-hydroxylase DesB from *Streptomyces sviceus*, and (F) the histamine N-hydroxylase BasC from *A. baumannii*.

NADH (Franceschini et al., 2012; Mügge et al., 2020; Reis et al., 2021). This is evident in SidA, as it is selective for L-ornithine over L-lysine (Chocklett & Sobrado, 2010; Franceschini et al., 2012). The same is true for the putrescine N-hydroxylase FbsI, which displays selectivity for putrescine over cadaverine (Lyons et al., 2022).

NMOs involved in siderophore biosynthesis serve as an attractive target for drug discovery efforts. This work lays the foundation for biochemical assays that are useful for high-throughput screening (HTS) and drug design assays. A robust product formation assay is described that allows for the direct measurement of enzymatic activity in the presence or absence of potential inhibitors. Additionally, preliminary studies into the dynamic nature of the FAD prosthetic group will allow for *in silico* approaches to rational drug design, a previous roadblock in computer-aided drug design.

With the growing emergence of multi-drug resistant strains of pathogenic bacteria and fungi, it is vital that siderophore biosynthesis enzymes are characterized such that they can be targeted for drug discovery. While several prototypical enzymes have been characterized previously, there are still gaps in knowledge regarding mechanism, structure, and function. Here, we describe biochemical methods employed in the past and present that are key to characterizing NMOs involved in siderophore biosynthesis. The activity of all NMOs requires molecular oxygen and NADPH, allowing for their enzymatic activity to be determined by monitoring the consumption of oxygen, the oxidation of NADPH, and by measuring the kinetic isotope effect of hydride transfer. The product of the NMO reaction is a hydroxylated amine which can be monitored using a colorimetric product formation assay. These assays in addition to HTS assays developed against these enzymes will be discussed in detail. The goal of this chapter is to allow for the continuous growth of the field while providing succinct methodologies that can be applied to a wide range of flavin-dependent enzymes.

2. NADPH oxidation

NMOs must react with NADPH to form the reduced flavin before the oxidative half-reaction can take place. Because the conversion of NADPH to $NADP^+$ is accompanied with a decrease in the absorbance at 340 nm, a common method used to measure the activity of NMOs is by using UV–visible spectroscopy. Under steady-state conditions, the NMO's concentration is in the low micromolar range (1–2 μM), therefore the absorbance of the flavin would not interfere with the absorbance changes at 340 nm. Although most NMOs display a preference for NADPH, some are also able to use NADH effectively, thus this cofactor can be used in place of NADPH (Chocklett & Sobrado, 2010; Lyons et al., 2022; Shirey, Badieyan, & Sobrado, 2013). An advantage of this method is that it can be utilized immediately following the protein purification process, allowing for activity to be measured across multiple fractions of interest. One set back of this method, however, is that if an enzyme has a high apparent K_D for NADPH (>150 μM), measuring accurate spectral changes becomes an issue due to the absorbance values at 340 nm exceeding 1 AU (at concentrations exceeding 150 μM). Below is a protocol detailing an effective baseline use for this assay using FbsI as an example (Fig. 3).

2.1 Equipment
- UV–visible spectrophotometer (Agilent 8453)
- Quartz cuvette

2.2 Reagents
- Buffer A: 25 mM HEPES, pH 7.5 or 50 mM phosphate, pH 7.5 (Fisher Scientific)
- NADPH (Research Products International)

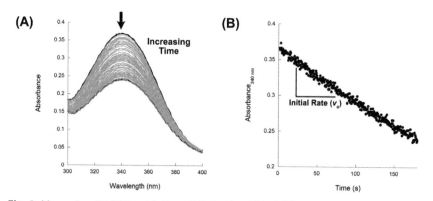

Fig. 3 Measuring NADPH oxidation of FbsI using UV–visible spectroscopy. (A) Spectra of 60 μM NADPH reduced in the presence of 5 μM FbsI over 180 s. The blue trace represents the reduced cofactor at 340 nm while the black trace represents the last trace measured. (B) Sample trace at 340 nm monitoring NADPH consumption over 180 s. The initial rate of the reaction can be determined by measuring the slope of the linear range of the reaction.

2.3 Procedure

1. Thaw out enzyme on ice and turn on UV–visible spectrophotometer. Allow the lamp to warm up for 15–20 min for optimal spectra.
2. Prepare a concentrated stock solution of NADPH (~25 mM) and store on ice.
3. Once the enzyme is thawed, centrifuge at 10,000 × g for 2 min to collect any precipitate that may have formed during the thawing cycle.
4. Set up the UV–visible spectrophotometer in kinetics mode. Set the acquisition time to 180 s and cycle time to 0.5 s, allowing for the maximum amount of data points to be collected.
5. Blank the instrument with buffer, as the enzyme's spectrum should not interfere with that of NADPH.
6. Prepare a 250 μL reaction mix that contains buffer, 5 μM enzyme (based on FAD concentration), and 0–100 μM NADPH (see Note 1).
7. Add NADPH to the cuvette last, cover with parafilm, invert three to five times to mix properly, and start acquiring the spectrum at 340 nm.
8. Once the experiment is complete, record the slope of the reaction in the initial phase of the reaction—usually from 15 to 45 s (Fig. 3).

2.4 Notes

1. The NADPH concentration used in the assay was chosen based on measuring the absorbance at 340 nm (using extinction coefficient: 6.27/(mM cm)) which will be within the linear range of the UV–visible spectrophotometer (0–1 AU).
2. For the purposes of the example here, HEPES is used as the buffering agent, as it has the best activity with FbsI. However, many NMOs are also quite active in phosphate buffer.
3. Reusable cuvettes that are used for this assay should be rinsed thoroughly with DI water in between each measurement to ensure no residual materials are present.

3. Oxygen consumption assays

Measurement of oxygen consumption over time can be an efficient way to measure steady-state kinetic parameters for NMOs, as oxygen consumption is stoichiometric with substrate consumption (Franceschini et al., 2012; Romero, Gómez Castellanos, Gadda, Fraaije, & Mattevi, 2018). Performing oxygen consumption assays provides several advantages over other traditional

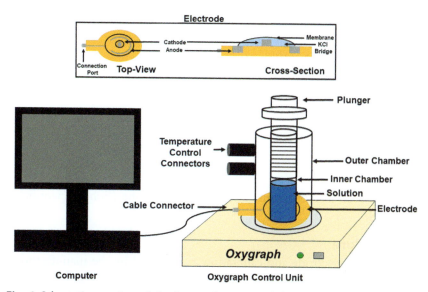

Fig. 4 Schematic overview of the Oxygraph+ system described in this work. The electrode first is coated in KCl and a membrane to maintain a salt bridge for measuring the potential change of oxygen. The electrode is then attached to the chamber and connected to a computer, where the namesake software will allow connection between the control unit and the instrument. The plunger is used to create a closed environment, to which enzyme solution can be introduced through it using a Hamilton syringe.

enzymatic assays. The system described below is a continuous assay, allowing for real-time monitoring of the reaction and modification of assay components if needed (Fig. 4). Measuring and fitting the initial velocities as a function of substrate concentration to the Michaelis–Menten equation (Eq. 1) permits the determination of the apparent steady-state kinetic parameters k_{cat}, K_M, and k_{cat}/K_M. Furthermore, the oxygen consumption assay can be used to probe the enzyme mechanism using pH, viscosity, and solvent kinetic isotope effects (Gadda & Sobrado, 2018; Robinson, Rodriguez, & Sobrado, 2014; Robinson, Klancher, Rodriguez, & Sobrado, 2019). Last, the oxygen consumption assay can be used to measure inhibition by oxidized cofactors (such as NADP$^+$) or potential inhibitor candidates, allowing for determination of apparent K_I, IC$_{50}$, and mode of inhibition (Chocklett & Sobrado, 2010; Martín Del Campo et al., 2016).

$$\frac{v_o}{[E_T]} = \frac{k_{cat}[S]}{K_m + [S]} \tag{1}$$

3.1 Equipment
- Oxygraph+ system (Hansatech Instruments)
- Clark-type polarographic oxygen electrode
- 12.7 mm PTFE magnetic stir bar
- Hamilton syringe
- Buchner flask (1000 mL)
- Vacuum pump

3.2 Reagents
- Buffer A
- Substrate such as L-ornithine, L-lysine, or putrescine (Fisher Scientific)
- NADPH

3.3 Procedure
1. Prepare the electrode and calibration as indicated by the manufacturer.
2. Start by adding buffer to the chamber, followed by substrate and NADPH. Typically, 1 mL reactions are used, with either substrate or NADPH being fixed while varying the other to create pseudo first-order rate conditions.
3. Place the plunger in the chamber to create a closed system and adjust until the bottom of the plunger forms a meniscus with the solution (~1 cm above the reaction mixture). Ensure the system is being consistently stirred using a magnetic stir bar.
4. Wait for the reaction rate to stabilize between $\pm 2\,\mu M/min$. Start the data collection process and allow to equilibrate for 30 s
5. Initiate the reaction by adding enzyme solution with a Hamilton syringe and monitoring the rate of oxygen consumption over a 3-min period (Fig. 5).
6. Record the initial rate of the reaction. Typically, 30 s is sufficient for all components to be mixed after the addition of enzyme.
7. Rinse the chamber several times and store in DI water until it is ready to use again.

3.4 Notes
1. Solutions, except for those containing enzyme, should be at room temperature to minimize changes in oxygen concentration upon mixing (oxygen concentration varies with temperature).
2. Enzyme concentrations used are typically $0.5-2\,\mu M$ (based on FAD concentration). This allows for a measurable rate that can later be normalized to calculate a k_{cat} value.
3. The Buchner flask and vacuum pump are needed to aspirate solutions from the chamber during the rinsing steps.

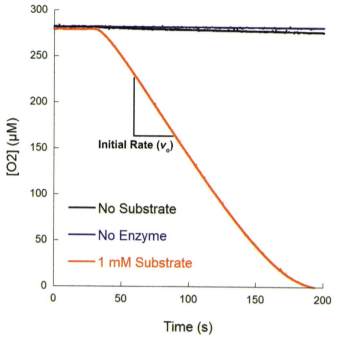

Fig. 5 Representative diagram showing raw traces as measured on the Oxygraph + instrument. When no substrate is present, the observed rate is that of NADPH oxidation in solution. The initial rate of the reaction can be taken by measuring the slope in the linear range of the reaction.

4. Iodine oxidation assay

In conjunction with measuring oxygen concentration as a function of product formation, an assay that directly measures hydroxylated amine products has been utilized in the field for several years. The iodine oxidation assay is a discontinuous, colorimetric assay that allows for steady-state kinetic parameters associated with product formation to be determined. Unlike the oxygen consumption assay, if the reaction catalyzed by the enzyme of interest is uncoupled (hydrogen peroxide is produced instead of product), the observed rate constant is not affected, as hydroxylamine is what is being measured. First developed by Csaky in 1948, this method has been modified over the last couple of decades to decrease assay volume, increase yields of nitrite, and decrease reaction time (Chocklett & Sobrado, 2010; Csáky et al., 1948; Lyons et al., 2022; Meneely & Lamb, 2007; Stehr et al., 1999). Specific elements that have been changed include

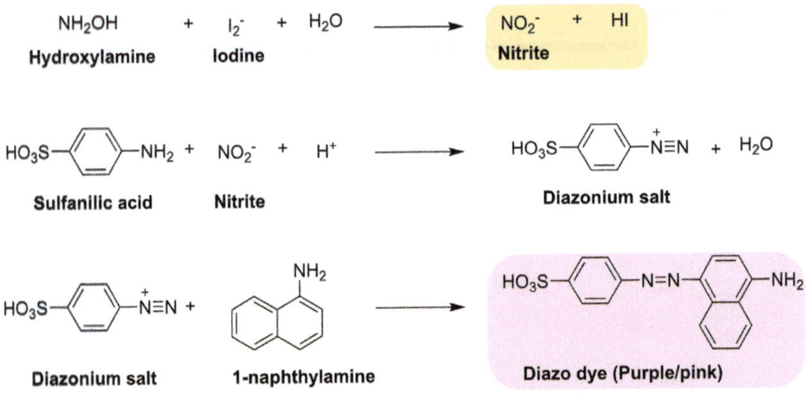

Fig. 6 Proposed reaction schemes for the iodine oxidation reaction. Hydroxylamine present in solution will be oxidized to nitrite following addition of iodine, resulting in an orange color. The resultant nitrite will react with sulfanilic acid to form a diazonium salt, which will conjugate with 1-napthylamine to form the pink diazo dye.

addition of 10% sodium acetate to ensure a proper reaction pH of ~5, increased concentration of perchloric acid to ~0.7 N, changes from sodium arsenate to sodium thiosulfate for quenching of iodine, and longer incubation time for color development. The principle of this assay is that produced hydroxylamine will be oxidized to nitrite by iodine under acidic conditions. Sodium thiosulfate will remove excess iodine, then treatment with 1-napthylamine will produce a pink diazo dye that can be measured at 562 nm (Fig. 6). In a larger scheme, this assay can be applied to HTS of drug libraries. This is because it is a plate-based assay that contains standards along with controls, and it can be modified for use with pipetting robots. In contrast with past methods used for HTS of NMOs, such as fluorescence anisotropy, the iodine oxidation assay can rely strictly on product formation as a means of determining enzyme inhibition.

4.1 Equipment
• Microplate reader (Molecular Devices SpectraMax M5)

4.2 Reagents
• 1 mM hydroxylamine (NH_2OH)
• 2 N Perchloric acid ($HClO_4$)
• 10% w/v sodium acetate
• 1% w/v sulfanilic acid in 25% glacial acetic acid
• 0.5% w/v iodine in 100% glacial acetic acid

- 0.1 N sodium thiosulfate
- 0.6% w/v 1-napthylamine in 30% glacial acetic acid
- Buffer A
- NADPH
- 10 mM putrescine

4.3 Procedure

1. Prepare all solutions in 50 mL conical tubes with DI water unless otherwise noted. The sulfanilic acid, iodine, and 1-napthylamine solutions should be covered in aluminum foil to prevent light exposure and should be incubated at 25 °C overnight to allow the solids to dissolve.
2. In 0.5 mL microcentrifuge tubes, prepare the following NH_2OH standard conditions to a final volume of 120 µL in buffer: 0, 25, 50, 100, 200, and 300 µM.
3. Prepare enzymatic assay reactions to a final volume of 120 µL in separate 0.5 mL microcentrifuge tubes. Typical reaction concentrations include 0.5–2 µM enzyme, 0–1000 µM NAD(P)H, 0–2000 µM substrate, and buffer.
4. Initiate reactions with the addition of enzyme and incubate at room temperature for 5 min with light shaking.
5. Quench both the standards and enzyme reactions with 62.4 µL 2 N $HClO_4$. Centrifuge at $16,000 \times g$ for 2 min and carefully transfer the supernatant to a clean 0.5 mL tube.
6. Transfer 47 µL of the supernatant of each reaction, in triplicate, to a 96-well plate.
   - 47 µL 10% w/v sodium acetate
   - 47 µL 1% w/v sulfanilic acid
   - 19 µL 0.5% w/v I_2
7. Using a multichannel pipette, add the following reagents to each well:
8. Cover the plate with foil and incubate for 15 min in the dark.
9. Add 19 µL 0.1 N sodium thiosulfate and 19 µL 0.6% w/v 1-napthylamine to each well to start the color development process.
10. Cover plate with foil and incubate, with shaking, at room temperature for 45 min.
11. Read plate at 562 nm and record absorbance values. Fit the absorbance values obtained at various NH_2OH concentrations to a linear equation and calculate the amount of NH_2OH produced over time by the enzyme using the standard curve (Fig. 7).

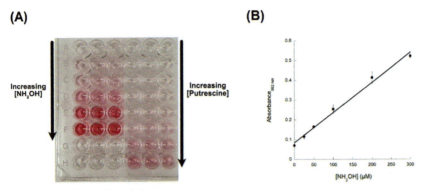

Fig. 7 (A) Development of hydroxylamine standards and FbsI reaction wells into a pink diazo dye following oxidation by iodine. A more intense pink color correlates to a higher concentration of hydroxylamine produced. (B) Sample standard curve generated following color development of hydroxylamine standards.

4.4 Notes

1. Substrates containing primary thiol groups cannot be used in this assay, as it will interfere with the color development steps following addition of iodine. However, molecules containing a secondary thiol (e.g. S-allyl-L-cysteine) will not interfere with the assay.
2. The sulfanilic acid, iodine, and 1-napthylamine solutions can be incubated for 2 h at 37 °C instead of overnight if needed.
3. $HClO_4$ and glacial acetic acid are flammable irritants that should be handled with care inside of a fume hood with proper PPE.

5. Determination of hydrogen peroxide formation and reaction uncoupling

The formation of H_2O_2 is the uncoupling of oxygen activation and product hydroxylation in a NMO's reaction. When NMOs are uncoupled and oxygen consumption or NADPH oxidation assays are used, the kinetic values obtained from these experiments are a combination of H_2O_2 and product formation steps. This can lead to inflated initial rates, causing invalid conclusions. The coupling percentage for an enzyme can be calculated by taking the calculated k_{cat} value obtained from the iodine oxidation assay and dividing it by the k_{cat} value from the oxygen consumption assay. In addition, this can be calculated by the direct measurement of H_2O_2 as described below.

5.1 Colorimetric measurement of H₂O₂

A common method for measuring H₂O₂ is via a colorimetric assay using the Pierce Quantitative Peroxide Assay Kit from ThermoFisher Scientific (See Note 1). In the presence of H₂O₂, addition of the assay solution will yield a purple color that can be measured at 595 nm. The principle of this assay revolves around the reaction of sorbitol with free H₂O₂, forming a peroxyl radical. This radical species can oxidize ferrous sulfate to ferric sulfate. Under the acidic conditions, the ferric iron can coordinate with xylenol orange's carboxylate and oxide groups, forming a purple dye that can be quickly measured using a microplate reader (Fig. 8). This method will allow for a rate of hydrogen peroxide formation to be calculated over a set period. This rate can be divided by the rate obtained by the oxygen consumption assay to determine uncoupling percentage.

This method, which has been used in the characterization of several NMOs, offers rapid screening of reaction conditions to determine what, if any, produces H₂O₂. However, there are a few drawbacks to this method. First, compounds that contain thiol groups cannot be tested using this kit, as it has been observed that this interferes with the color development process. Second, this is a discontinuous assay, which means that conditions cannot be adjusted based on observations until the assay is complete, which

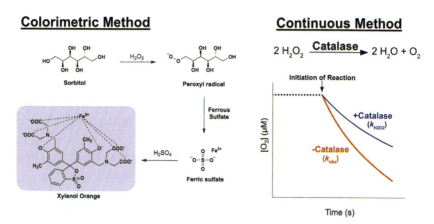

Fig. 8 Overview of the Pierce Quantitative Peroxide assay (left) and catalase method (right) for measuring the amount of H₂O₂ in a given reaction solution. Prior to the initiation of the reaction (dashed line), the reactions in the presence (blue) or absence (red) of catalase should start at the same concentration and should only differ in the observed reaction rates. The rates can then be used to determine uncoupling percentage, as described in Eq. (2).

may present a cost barrier. Last, the system is only linear up to $\sim 200\,\mu M$ H_2O_2. This may present an issue when measuring higher concentrations of H_2O_2.

5.2 Continuous measurement of H_2O_2 using catalase

An alternative method for measuring H_2O_2 in FMOs has been developed to be used in conjunction with the oxygen consumption assay (Valentino et al., 2020). Catalase is a well-known enzyme that turns over two equivalents of H_2O_2 into molecular oxygen and two water molecules. By adding catalase to a reaction that is uncoupled, H_2O_2 that is produced will be consumed by catalase, decreasing the observed rate of oxygen consumption (Fig. 8). Reactions in the presence and absence of catalase can be performed, and the rates of the reactions can be compared to determine coupling efficiency with a given substrate (see Note 2). This method provides the advantage of being a continuous assay that only takes a few minutes to complete. Additionally, there is a much lower cost to performing this assay in comparison to the colorimetric assay.

5.3 Reagents
- 20 mg/mL Catalase (Millipore Sigma)
- Buffer A
- NADPH
- Substrate

5.4 Procedure
1. To the Oxygraph+ reaction chamber, add Buffer A, NADPH (fixed concentration of $250\,\mu M$), substrate (10–$1000\,\mu M$), and catalase (1 mg/mL).
2. Place the plunger in the chamber and allow the reaction rate to stabilize between $\pm\,2\,\mu M/min$. Start the data collection process and allow to equilibrate for 30 s
3. Initiate the reaction by adding enzyme solution with a Hamilton syringe and monitor oxygen consumption over a 3-min period.
4. Calculate the rate of oxygen consumption as described in Section 3.
5. Rinse the chamber several times and store in DI water until it is ready to use again.

5.5 Notes
1. For a detailed procedure using the Pierce colorimetric kit, we direct you to ThermoFisher Scientific's website for more information (https://www.thermofisher.com/order/catalog/product/23280).

2. The uncoupling percentage can be calculated using Eq. (2). In short, the differences of reactions rates in the presence and absence of catalase are divided by two to account for the reaction stoichiometry, resulting in the apparent k_{H2O2}. This in turn can be compared with an observed rate at a specific concentration point to calculate the uncoupling percentage.

$$\Delta O_2 \, (s^{-1}) = \text{Rate}_{\text{No Catalase}} - \text{Rate}_{\text{With Catalase}}$$

$$k_{H2O2} = \frac{\Delta O_2}{2}$$

$$\%\text{Uncoupling} = \frac{k_{H2O2}}{k\text{obs}} \tag{2}$$

6. Fmoc-Cl derivatization

The identity of the products being produced by NMOs of interest can be separated using liquid chromatography for mass spectral or NMR characterization. A major issue, however, is that L–ornithine, L–lysine, and aliphatic diamines do not possess strong spectral properties. Therefore, derivatization with fluorenylmethyloxycarbonyl chloride (Fmoc-Cl) can allow for the measurement of these amine–containing compounds via a photodiode array (PDA) detector coupled to an HPLC. The mechanism of Fmoc-Cl derivatization is a nucleophilic substitution of the acyl chloride with a primary amino group or hydroxylamine, forming a Fmoc-amine conjugate and releasing HCl. The Fmoc-amine products can then be separated via reverse phase HPLC and detected at 254 nm with product identity confirmed by mass spectrometry (Fig. 9). Below is an outlined example of such analysis using FbsI and putrescine as the experimental system.

6.1 Equipment
- HPLC with PDA module (Shimazdu)
- Phenomenex Kinetex C18 column (100×4.60 mm×2.6 μm)

6.2 Reagents
- HPLC–grade acetonitrile (MeCN; Fisher Scientific)
- HPLC–grade water (Fisher Scientific)
- HPLC–grade methanol (MeOH; Fisher Scientific)
- Mobile Phase A: Water with 0.1% formic acid
- Mobile Phase B: Acetonitrile with 0.1% formic acid
- 100 mM borate buffer, pH 8 (Fisher Scientific)

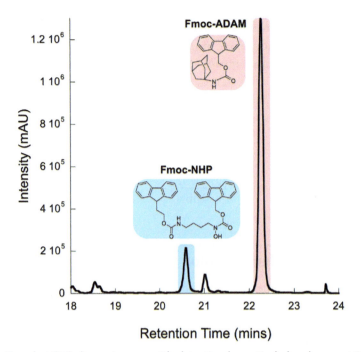

Fig. 9 Sample HPLC chromatogram with detection by optical absorbance at 254 nm showing the expected reaction products following Fmoc-Cl derivatization of FbsI using putrescine as a substrate. The Fmoc-ADAM complex is a byproduct of quenching the reaction following derivatization. The identity of the major peaks was confirmed using standards and/or LC–MS. NHP refers to *N*-hydroxyputrescine.

- 10 mM Fmoc-Cl (Fisher Scientific)
- 100 mM 1-adamantylamine (ADAM) in 1:1 0.2 mM HCl:MeCN (Acros Organics)
- Buffer A
- NADPH
- Putrescine

6.3 Procedure

1. Prepare all solutions and buffers with HPLC-grade reagents to ensure no contamination on the C18 column.
2. Thaw enzyme on ice and centrifuge at 10,000 × g for 2 min to pellet any precipitate. Carefully transfer the supernatant to a clean tube.
3. Set up 300 μL assays consisting of Buffer A, 25 μM putrescine, 500 μM NADPH, and 1 μM FbsI. Negative controls will contain no putrescine, no NADPH, or no enzyme.

4. Incubate at room temperature for 10 min.
5. Take a 50 µL aliquot of the reaction and quench with 100 µL HPLC-grade MeCN. Chill at −20 °C for 10 min
6. Centrifuge at $16,000 \times g$ for 1 min at room temperature and transfer the supernatant to a clean 0.5 mL microcentrifuge tube.
7. Add 50 µL of 100 mM borate buffer, pH 8 followed by 20 µL of 10 mM Fmoc-Cl dissolved in HPLC-grade MeOH.
8. Shake for 5 min at room temperature to allow the derivatization to proceed.
9. Add 20 µL of 100 mM ADAM in 1:1 0.2 mM HCl:MeCN and shake for 15 min at room temperature to remove excess Fmoc-Cl.
10. Inject 10 µL onto the C18 column and run under the following gradient at 0.5 mL/min: 5–100% B, 0–15.00 min; 5% B, 15.01–25.00 min.

6.4 Notes

1. For LC–MS analysis, fractions from a specific peak can be submitted for analysis by collecting a fraction from the HPLC, then either lyophilizing or using a vacuum evaporator to concentrate the reaction product. The sample can be placed in a sealed LC–MS vial before diluting with MeOH prior to loading onto a LC–MS instrument. The method can also be adapted for direct analysis by LC–MS in step 10 of procedure 6.3.
2. If excess Fmoc-Cl is still present in the chromatograms, consider increasing the derivatization time with Fmoc-Cl to 10 min prior to adding ADAM. This will result in a larger Fmoc-ADAM peak but result in less free Fmoc-Cl (Fig. 9).

7. Determination of reaction stereospecificity

The synthesis of deuterated nicotinamide cofactors is attractive to measure kinetic isotope effects, as these cofactors are active participants in many redox reactions and serve as regulators of many metabolic processes (Begley, Kinsland, Mehl, Osterman, & Dorrestein, 2001). The mechanism of hydride transfer from the reduced form of the cofactor to a variety of acceptor molecules such as flavins or cytochromes is widely debated (Birrell & Hirst, 2013; Reis et al., 2021). Because the C–H bond must be cleaved to be transferred to an acceptor, kinetic isotope effect experiments can be a very effective method to study the hydride transfer event in NMOs (Mao & Campbell, 2020).

We present the synthesis of regio- and stereo-specific isotopically labeled (R)-[4-^{2}H]-NADPH and (S)-[4-^{2}H]-NADPH without the restriction of needing commercially available enzymes to carry out the synthesis. A thermostable alcohol dehydrogenase (ADH) from the thermophilic bacterium *Thermoanaerobacter brockii* was previously shown to prefer $NADP^+$ over NAD^+ and was used for the synthesis of (R)-[4-^{2}H]-NADPH (Jeong & Gready, 1994; Korkhin et al., 1998). *T. brockii* ADH (*Tb*ADH) catalyzes the conversion of alcohols to aldehydes coupled to the formation of NADPH from $NADP^+$. Using deuterated propanol, (R)-[4-^{2}H]-NADPH is produced by *Tb*ADH (Fig. 10A). *Tb*ADH was commercially available until recently. Here, we present a developed method for the recombinant expression and purification of *Tb*ADH (Burdette & Zeikus, 1994; Burdette, Vieille, & Zeikus, 1996). In addition, we have conducted the same optimization of the existing purification protocols of *Leuconostoc mesenteroides* glucose 6-phosphate dehydrogenase (*Lm*G6PDH) and the synthesis protocols of (S)-[4-^{2}H]-NADPH (Olavarría, Valdés, & Cabrera, 2012). The typical function of this enzyme is to catalyze the oxidation of glucose-6-phosphate (G6P) to 6-phosphogluconolactone, resulting in the reduction of $NADP^+$ to NADPH (Olive, Geroch, & Levy, 1971). However, isotopically labeled G6P is not commercially available, but D-glucose-1-d is available and can be accepted as a substrate by

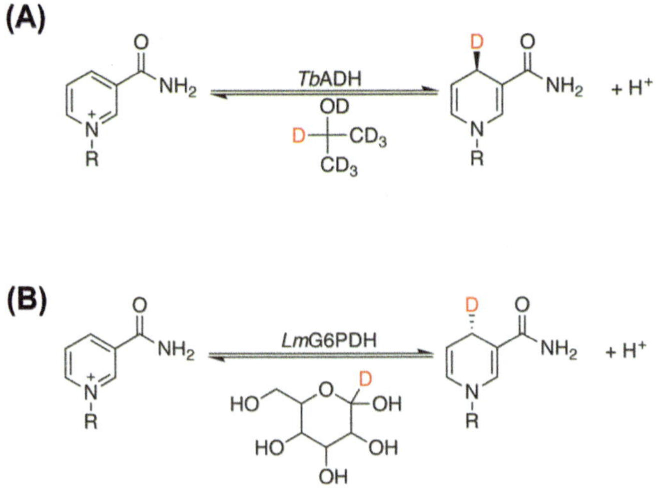

Fig. 10 (A) Reaction of *Tb*ADH with $NADP^+$ and 2-propanol-d$_8$. (B) Reaction of *Lm*G6PDH with $NADP^+$ and D-glucose-1-d.

*Lm*G6PDH (Fig. 10B) (Viola, Cook, & Cleland, 1979). This work presents methods to synthesize regio- and stereo-specific reduced nicotinamide cofactors at a lowered cost with an increased efficiency, aiding in kinetic isotope effect experiments to study hydride transfer.

7.1 Equipment
- Sonicator (Fisher Scientific Sonic Dismembrator Model 500)
- FPLC (AKTA Start, Cytiva)
- HisTrap Fast Flow, 5 mL column
- DEAE Fast Flow, 5 mL column
- 3 kDa centrifugal molecular weight cutoff filter
- 30 kDa centrifugal molecular weight cutoff filter
- Lyophilizer (LabCono FreeZone 2.5 L, −50 °C)

7.2 Expression and purification of TbADH

7.2.1 Reagents
- *Tb*ADH (NCBI Accession Number: CAA46053.1) cloned into pET28a
- Lysogeny broth (LB) Media (10 g tryptone, 10 g NaCl, 5 g yeast extract per 1 L)
- *Escherichia coli* BL21(DE3) cells (ThermoFisher Scientific)
- Kanamycin sulfate (Fisher Scientific)
- Isopropyl β-d-1-thiogalactopyranoside (IPTG) (Fisher Scientific)
- Lysozyme (Fisher Scientific)
- RNAse (Fisher Scientific)
- DNAse (Fisher Scientific)
- Phenylmethylsulfonyl fluoride (PMSF) (Fisher Scientific)
- Buffer A: 25 mM Tris–HCl, pH 7.3, 300 mM NaCl, 5 mM imidazole, 0.1 mM TCEP
- Buffer B: 25 mM Tris–HCl, pH 7.3, 300 mM NaCl, 300 mM imidazole, 0.1 mM TCEP
- Storage Buffer: 25 mM Tris–HCl, pH 7.3, 100 mM NaCl, 0.05 mM $ZnCl_2$

7.2.2 Expression of TbADH protocol
1. Transform the *Tb*ADH-pET28a plasmid into chemically competent *E. coli* BL21(DE3) cells (Day 1).
2. Supplement 50 mL of autoclaved Lysogeny broth media (0.5 g tryptone, 0.5 g NaCl, 0.25 g yeast extract) with 50 μL of a filter sterilized 1000× kanamycin stock and inoculate with one colony from the transformation. Grow overnight at 37 °C, 250 RPM (Day 2).

3. Add 1 mL of a 1000× filter sterilized kanamycin stock and 8 mL of the starter culture generated in step 2 to 1 L of Lysogeny broth media. Grow overnight at 37 °C, 250 RPM until an optical density at 600 nm of 0.8 is reached (Day 3).
4. Induce expression by adding 500 µL of a filter sterilized 1 M IPTG stock to 1 L of culture (Day 3).
5. Lower the temperature to 18 °C. Grow overnight at 18 °C, 250 RPM (Days 3–4).
6. Centrifuge the cultures at 4000 RPM, 4 °C for 20 min to pellet the cells (Day 4).
7. Harvest the pelleted cells and record the weight of the pellet (Day 4). Store the pellet at −70 °C until use (Day 4).

7.2.3 Purification of TbADH protocol
1. Thaw the cell pellet from the expression on ice.
2. Resuspend the pellet in 5 mL/g of cell of Buffer A: 25 mM Tris–HCl, pH 7.3, 300 mM NaCl, 5 mM imidazole, 0.1 mM TCEP.
3. Supplement the cell suspension from step 2 with 25 µg/mL lysozyme, 10 µg/mL RNAse, 10 µg/mL DNAse, and 1 mM PMSF.
4. Stir the solution from step 3 until homogeneous at 4 °C.
5. Sonicate the cell suspension for 15 min at 70% amplitude with the pulse on for 5 s and off for 10 s
6. Centrifuge the lysate at 16,000×g for 45 min at 4 °C to pellet the insoluble material and cell debris.
7. Equilibrate a 5 mL HisTrap Fast Flow column attached to a Cytiva AKTA start FPLC system in Buffer A at 5 mL/min.
8. Load the supernatant onto the equilibrated 5 mL HisTrap Fast Flow column at 2 mL/min and collect the flow-through.
9. Wash the column with 5% of Buffer B: 25 mM Tris–HCl, pH 7.3, 300 mM NaCl, 300 mM imidazole, 0.1 mM TCEP for five column volumes at 5 mL/min.
10. Elute the protein with a linear gradient of 5% Buffer B to 100% Buffer B over 15 column volumes at 5 mL/min. Collect 5 mL fractions.
11. Evaluate the purity of the eluted fractions using SDS–PAGE and pool fractions with pure protein (Fig. 11A).
12. Concentrate the protein sample using a 30 kDa centrifugal molecular weight cutoff filter.
13. Dialyze overnight at 4 °C into 25 mM Tris–HCl, pH 7.3, 100 mM NaCl, 0.05 mM $ZnCl_2$ (storage buffer).

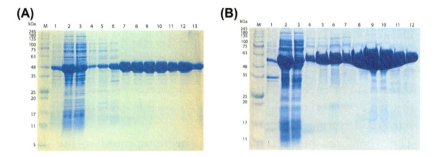

Fig. 11 (A) SDS–PAGE of the purification of TbADH. Lane 1-pellet; Lane 2-supernatant; Lane 3-flow-through; Lane 4-Wash fraction 1 at 5% Buffer B (15 mM imidazole); Lane 5-Wash fraction 2 at 5% Buffer B (15 mM imidazole); Lanes 6–13-Elution fractions (90–300 mM imidazole). Fractions in lanes 8–13 were pooled for further use. The predicted molecular weight of TbADH is ~42 kDa. (B) SDS–PAGE of the purification of LmG6PDH. Lane 1-pellet; Lane 2-supernatant; Lane 3-flow-through; Lanes 4–7-Wash fractions at 5% Buffer B (15 mM imidazole); Lanes 8–12-Elution fractions (60–150 mM imidazole). Fractions in lanes 8–12 were pooled for further use. The predicted molecular weight of LmG6PDH is ~55 kDa.

14. Measure the protein concentration using the Bradford assay.
15. Dilute the protein to 37.5 μM using the storage buffer.
16. Aliquot the protein into 400 μL samples and storage at −70 °C.

7.3 Synthesis of (R)-[4-^{2}H]-NADPH

7.3.1 Reagents
- Buffer: 25 mM Tris–HCl, pH 9.0
- NADP$^+$ (Research Products International)
- 2-propanol (Fisher Scientific)
- 2-propanol-d$_8$ (Sigma-Aldrich)

7.3.2 Protocol
1. Add 2.5 μM TbADH, 5.5 mM NADP$^+$, 1 M 2-propanol-d$_8$ to 6 mL of 25 mM Tris–HCl, pH 9.0. Protect the reaction from the light.
2. Run the reaction at 40 °C with stirring at 125 RPM for 20 min.
3. Monitor the reaction progress by acquiring the UV–visible spectrum and calculating the ratio of the absorbance at 260 to the absorbance at 340 (A_{260}/A_{340}).
4. Remove the enzyme from the reaction using a 3 kDa centrifugal molecular weight cutoff filter after the A_{260}/A_{340} reached ≤2.8 (centrifuge 3× for 15 min at 4100 RPM, 4 °C).
5. Lyophilize the filtrate overnight and store at −70 °C.

7.3.3 Notes

1. Unlabeled NADPH should be synthesized in the same manner as described above with the only modification being the use of 2-propanol in place of the deuterated form. This NADPH should be used in the kinetic isotope effect experiments, as it is a necessary control that ensures that the decreases in the activity are not caused by any contaminates or side products from the synthetic steps.
2. The lyophilized product should only be resuspended in the desired buffer immediately prior to usage to avoid stability issues.

7.4 Expression and purification of LmG6PDH

7.4.1 Reagents

- *Lm*G6PDH (NCBI Accession Number: WP_036092758.1) cloned into pET28a
- Buffer A: 50 mM Tris–HCl, pH 7.8 80 mM $MgCl_2$, 300 mM NaCl, 5 mM imidazole
- Buffer B: 50 mM Tris–HCl, pH 7.8, 80 mM $MgCl_2$, 300 mM NaCl, 300 mM imidazole
- Storage Buffer: 50 mM Tris–HCl, pH 7.8, 10 mM $MgCl_2$

7.4.2 Expression of LmG6PDH protocol

1. Transform the *Lm*G6PDH-pET28a plasmid into chemically competent *E. coli* BL21(DE3) cells (Day 1).
2. Supplement 50 mL of autoclaved Lysogeny broth media (0.5 g tryptone, 0.5 g NaCl, 0.25 g yeast extract) with 50 µL of a filter sterilized 1000× kanamycin stock and inoculate with one colony from the transformation. Grow overnight at 37 °C, 250 RPM (Day 2).
3. Add 1 mL of a 1000× filter sterilized kanamycin stock and 8 mL of the starter culture generated in step 2 to 1 L of Lysogeny broth media. Grow overnight at 37 °C, 250 RPM until an optical density at 600 nm of 0.8 is reached (Day 3).
4. Induce expression by adding 1 mL of a filter sterilized 1 M IPTG stock to 1 L of culture (Day 3).
5. Lower the temperature to 18 °C. Grow overnight at 18 °C, 250 RPM (Days 3–4).
6. Centrifuge the cultures at 4000 RPM, 4 °C for 20 min to pellet the cells (Day 4).
7. Harvest the pelleted cells and record the weight of the pellet (Day 4).
8. Store the pellet at −70 °C until use (Day 4).

7.4.3 Purification of LmG6PDH protocol

1. Thaw the cell pellet from the expression on ice.
2. Resuspend the pellet in 5 mL/g of cell of Buffer A: 50 mM Tris–HCl, pH 7.8 80 mM $MgCl_2$, 300 mM NaCl, 5 mM imidazole.
3. Supplement the cell suspension from step 2 with 25 µg/mL lysozyme, 10 µg/mL RNAse, 10 µg/mL DNAse, and 1 mM PMSF.
4. Stir the solution from step 3 until homogeneous at 4 °C.
5. Sonicate the cell suspension for 15 min at 70% amplitude with the pulse on for 5 s and off for 10 s
6. Centrifuge the lysate at 16,000×g for 45 min at 4 °C to pellet the insoluble material and cell debris.
7. Equilibrate a 5 mL HisTrap Fast Flow column attached to a Cytiva AKTA start FPLC system in Buffer A at 5 mL/min.
8. Load the supernatant onto the equilibrated 5 mL HisTrap Fast Flow column at 2 mL/min and collect the flow-through.
9. Wash the column with 5% of Buffer B: 50 mM Tris–HCl, pH 7.8, 80 mM $MgCl_2$, 300 mM NaCl, 300 mM imidazole for five column volumes at 5 mL/min.
10. Elute the protein with a linear gradient of 5% Buffer B to 100% Buffer B over 15 column volumes at 5 mL/min. Collect 5 mL fractions.
11. Evaluate the purity of the eluted fractions using SDS–PAGE and pool fractions with pure protein (Fig. 11B).
12. Concentrate the protein sample using a 30 kDa centrifugal molecular weight cutoff filter.
13. Dialyze overnight at 4 °C into 50 mM Tris–HCl, pH 7.8, 10 mM $MgCl_2$ (storage buffer).
14. Measure the protein concentration using the Bradford assay.
15. Dilute the protein to 100 µM using the storage buffer.
16. Aliquot the protein into 200 µL samples and storage at −70 °C.

7.5 Synthesis of (S)-[4-^{2}H]-NADPH

7.5.1 Reagents

- Buffer: 100 mM sodium phosphate, pH 8.0
- $NADP^+$
- D-glucose (Fisher Scientific)
- D-glucose-1-d (Sigma-Aldrich)
- Dimethyl sulfoxide (DMSO) (Fisher Scientific)

7.5.2 Protocol

1. Add 10 mM $NADP^+$, 10 mM D-glucose-1-d, 0.5 μM *Lm*G6PDH to 3.2 mL 100% DMSO and 4.8 mL of 100 mM sodium phosphate, pH 8.0.
2. Run the reaction for 18 h at 25 °C with stirring at 125 RPM.
    a. Monitor the reaction progress by measuring the UV–visible spectrum at 340 nm.
3. Once the A_{340} is constant, remove the enzyme using a 3 kDa centrifugal molecular weight cutoff filter (centrifuge 3× for 15 min at 4100 RPM, 4 °C).
4. Load the filtrate onto a 5 mL DEAE FF column equilibrated in Buffer A (100 mM sodium phosphate, pH 8.0) at 2 mL/min. Collect the flow-through.
5. Elute the (S)-[4-^{2}H]-NADPH using a linear gradient of 0% Buffer B-100% Buffer B (1 M ammonium bicarbonate, pH 8.0) over 15 column volumes at 5 mL/min. Collect in 5 mL fractions.
6. Pool fractions with an $A_{260}/A_{340} \leq 2.8$.
7. Lyophilize the sample overnight and store at −70 °C.

7.5.3 Notes

1. As a control, NADPH should be synthesized in the same manner as described above, with the D-glucose substituted for D-glucose-1-d. This NADPH should be used in the kinetic isotope effect experiments to serve as the control as discussed in the previous section.
2. The lyophilized product should only be resuspended in the desired buffer immediately prior to usage to avoid stability issues.

8. Fluorescence anisotropy

NMOs contain a highly conserved structural motif consisting of two Rossmann-fold nucleotide binding domains: one for tightly binding FAD and the other for binding NAD(P)H. Utilizing these features, a fluorescence anisotropy assay has been developed by our group (Qi, Kizjakina, Robinson, Tolani, & Sobrado, 2012). Fluorescence anisotropy is the phenomenon in which a fluorophore emits light of disproportionate intensities along multiple axes of polarization. This technique can be used to measure protein-protein, protein-nucleotide, and protein-ligand interactions (Gijsbers, Nishigaki, & Sánchez-Puig, 2016). For use in NMOs, our group has previously synthesized a fluorophore consisting of ADP linked to the fluorophore TAMRA (Fig. 12A) (Qi et al., 2012). ADP-TAMRA can be used to bind to an enzyme, such as SidA, at the NAD(P)H-binding domain and emit a

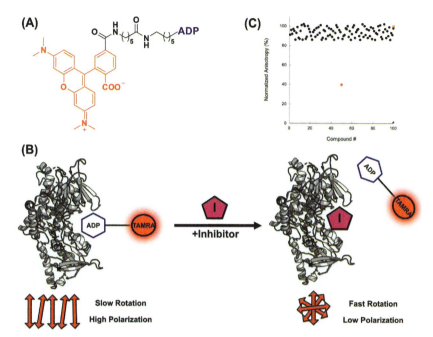

Fig. 12 Overview of a fluorescence anisotropy assay. The fluorophore ADP-TAMRA (A) can bind to the NADPH/NADP+ binding domain of an enzyme, such as SidA (PDB 4B63), and thus exhibits a slow rotation and high anisotropy value (B). If displaced by a molecule, such as an inhibitor, free ADP-TAMRA will be allowed to freely rotate, resulting in a low anisotropy value. (C) Representative plot from a high throughput screening assay showing expected anisotropy values for non-inhibitory compounds (•), an inhibitor (♦), a positive control containing free ADP-TAMRA (▲), and a negative control containing an enzyme:ADP-TAMRA complex (■). More specific examples of this data analysis are referenced here (Martín Del Campo et al., 2016; Qi et al., 2012).

fluorescent signal. ADP-TAMRA that is bound to the enzyme will exhibit higher polarization than the population that is free in solution (Fig. 12B).

The fluorescence anisotropy displacement assay described here can be used as an effective tool for HTS (Martín Del Campo et al., 2016). Scaled up to 96 or 384-well plates, enzyme samples can be incubated with libraries of compounds, and following addition of ADP-TAMRA, will allow for the anisotropy values to be calculated by measuring displacement of the ADP-TAMRA fluorophore (Fig. 12C). This can allow for the K_D for an inhibitor to be calculated, as shown in Eq. (3). While effective at measuring binding events, this method does not directly screen for effects on enzyme activities. This method is most suitable for a preliminary screening to identify candidate molecules for further *in vitro* characterization, such as prospective substrates or competitors for binding within the active site.

8.1 Equipment
- Microplate reader (Molecular Devices SpectraMax M5)
- 96-well black optical polystyrene plate optimized for fluorescence readings (ThermoFisher Scientific)

8.2 Reagents
- ADP-TAMRA (Qi et al., 2012)
- Enzyme solution
- Buffer: 50 mM sodium phosphate, pH 7.0

8.3 Procedure
1. In a 0.5 mL microcentrifuge tube, add 2 µM enzyme, 30 nM ADP-TAMRA, 20 µM inhibitor compound (final DMSO concentration of 2%), and buffer up to a final volume of 100 µL.
2. Transfer 25 µL, in triplicate, to a flat-bottomed 96-well plate optimized for fluorescence readings.
3. In fluorescence polarization mode, perform an endpoint, top read of the assay plate at an excitation wavelength of 544 nm and emission wavelength of 584 nm, using a wavelength cutoff of 570 nm.
4. To find the K_D value, fit data to Eq. (3), where m_1 and m_2 are the minimum and maximum anisotropy values, respectively, m_3 is the slope, and m_4 is the K_D value.

$$y = m_1 + \frac{(m_2 - m_1)x^{m_3}}{m_4^{m_3} + x^{m_3}} \tag{3}$$

9. Targeting flavin dynamics for drug discovery

Recently, contributions investigating the highly dynamic active site of NMOs have revealed new insight into their mechanisms. It has been shown that FAD moves from an *"out"* position following hydride transfer to an *"in"* position during the oxidative half-reaction (Campbell, Robinson, Mena-Aguilar, Sobrado, & Tanner, 2020a; Campbell et al., 2020b). Movement of the flavin in SidA is facilitated by contact between the pyrimidine ring of FAD in the *"in"* conformation with M101, along with movement of a conserved tyrosine loop (Fig. 13A). Y324 has been shown to stabilize the "*out*" conformation via π-stacking interactions with the isoalloxazine ring of FAD. Upon reduction of FAD, the tyrosine loop (N323-Y324-S325) moves ~10 Å to accommodate binding of the reduced nicotinamide cofactor, which is accompanied by a 9.5 Å shift of the flavin to the "*in*" position (Campbell et al., 2020b). Mutation of M101 to alanine shows that the rate of flavin dehydration is decreased 14-fold compared to the wild-type enzyme, promoting the "*out*" conformation *in crystallo* (Campbell et al., 2020a). So, while not mobile like Y324, the M101 residue is vital for stabilizing the "*in*" conformation of the FAD to ensure complete and efficient catalysis. In addition to flavin motion, recent studies with SidA have suggested that the nicotinamide moiety of NADPH may itself rotate to orient the pro-*S* and pro-*R* hydrogens closer to the N5 of the flavin (position of hydride transfer) (Pierdominici-Sottile, Palma, Ferrelli, & Sobrado, 2024). Kinetic isotope experiments have shown that the pro-*R* stereochemistry is required for flavin reduction, though crystal structures show the pro-*S* stereochemistry in closer proximity to the N5 of FAD (Franceschini et al., 2012; Robinson et al.,

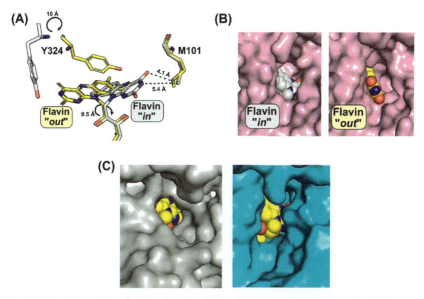

Fig. 13 (A) Active site of SidA showing both the "*in*" (6X0J) and "*out*" (6X0H) conformations of the FAD. Curved arrows indicate the relative distance that either Y324 or FAD rotated between the two different conformations. (B) Differences in space occupancy of the FAD in both the "*in*" (left) and "*out*" (right) conformations of SidA. (C) Differences in space occupancy of the FAD in the "*in*" conformations for NbtG (left, gray, 4D7E) and FbsI (right, cyan, 7US3).

2014; Romero et al., 2012). Through molecular dynamics simulations, it is shown that rotation of the nicotinamide ring around the C1D–N1N bond would allow for the rotation between the two hydrogen stereoisomers (Pierdominici-Sottile et al., 2024).

Previous in silico studies of SidA aimed at developing inhibitors have had largely mixed results, partly because of the dynamic nature of the enzyme's active site (Badieyan & Sobrado, 2013). By understanding the mechanism of flavin motion in this model enzyme, these studies can be applied to other NMOs that are desirable drug targets. The FAD-binding domain undergoes a significant change in occupancy space when transitioning from the *"out"* to the *"in"* conformations (Fig. 13B). The cavity surrounding FAD in the *"in"* conformation is fairly conserved among the NMOs FbsI and NbtG as well (Fig. 13C). This suggests that targeting NMOs in different conformations may allow for the specific screening of candidate inhibitors, as flavin dynamics may be conserved among these enzymes.

10. Summary and conclusions

In summary, this chapter outlines fundamental biochemical assays and techniques that have been employed, and continue to be utilized, for the characterization of NMOs involved in siderophore biosynthesis. These methods mean to serve as a stepping-stone for the development of more advanced assays probing mechanistic questions such as sequential hydroxylation events, flavin dynamics, and inhibition kinetics. Answering key questions regarding the structure and function of NMOs will allow for rational drug design along with the potential for directed evolution to bioengineer these enzymes for biomedical and agricultural purposes. While not discussed here, the measurement of pre-steady-state kinetics is important as well for elucidating the catalytic mechanisms of NMOs. For more detailed procedures on performing anaerobic stopped-flow spectroscopy, readers are directed to previous works referenced here (Moran, 2019; Valentino & Sobrado, 2019).

Acknowledgments

The research presented in this chapter was supported by the National Science Foundation grant CHE-2003658. In addition, we thank the Graduate School at Virginia Tech and Fralin Life Sciences Institute for providing financial support and resources to N.L.

References

Badieyan, S., & Sobrado, P. (2013). *Inhibition of siderophore biosynthesis by targeting, A. fumigatus ornithine hydroxylase: A structure-based virtual screening study. Microbial pathogens and strategies for combating them: Science, technology and education.* Formatex Research Center, 410–413.

Begley, T. P., Kinsland, C., Mehl, R. A., Osterman, A., & Dorrestein, P. (2001). The biosynthesis of nicotinamide adenine dinucleotides in bacteria. *Vitamins and Hormones, 61*, 103–119. https://doi.org/10.1016/s0083-6729(01)61003-3.

Binda, C., Robinson, R. M., del Campo, J. S. M., Keul, N. D., Rodriguez, P. J., Robinson, H. H., et al. (2015). An unprecedented NADPH domain conformation in lysine monooxygenase NbtG provides insights into uncoupling of oxygen consumption from substrate hydroxylation. *Journal of Biological Chemistry, 290*(20), 12676–12688. https://doi.org/10.1074/jbc.M114.629485.

Birrell, J. A., & Hirst, J. (2013). Investigation of NADH binding, hydride transfer, and NAD$^+$ dissociation during NADH oxidation by mitochondrial complex I using modified nicotinamide nucleotides. *Biochemistry, 52*(23), 4048–4055. https://doi.org/10.1021/bi3016873.

Blatzer, M., Schrettl, M., Sarg, B., Lindner, H. H., Pfaller, K., & Haas, H. (2011). SidL, an *Aspergillus fumigatus* transacetylase involved in biosynthesis of the siderophores ferricrocin and hydroxyferricrocin. *Applied and Environmental Microbiology, 77*(14), 4959–4966. https://doi.org/10.1128/AEM.00182-11.

Bohac, T. J., Fang, L., Giblin, D. E., & Wencewicz, T. A. (2019). Fimsbactin and acinetobactin compete for the periplasmic siderophore binding protein BauB in pathogenic *Acinetobacter baumannii. ACS Chemical Biology, 14*(4), 674–687. https://doi.org/10.1021/acschembio.8b01051.

Burdette, D. S., Vieille, C., & Zeikus, J. G. (1996). Cloning and expression of the gene encoding the *Thermoanaerobacter ethanolicus* 39E secondary-alcohol dehydrogenase and biochemical characterization of the enzyme. *Biochemical Journal, 316*(Pt 1), 115–122.

Burdette, D., & Zeikus, J. G. (1994). Purification of acetaldehyde dehydrogenase and alcohol dehydrogenases from *Thermoanaerobacter ethanolicus* 39E and characterization of the secondary-alcohol dehydrogenase (2° Adh) as a bifunctional alcohol dehydrogenase-acetyl-CoA reductive thioesterase. *Biochemical Journal, 302*(1), 163–170. https://doi.org/10.1042/bj3020163.

Campbell, A. C., Robinson, R., Mena-Aguilar, D., Sobrado, P., & Tanner, J. J. (2020a). Structural determinants of flavin dynamics in a class B monooxygenase. *Biochemistry, 59*(48), 4609–4616. https://doi.org/10.1021/acs.biochem.0c00783.

Campbell, A. C., Stiers, K. M., Martin Del Campo, J. S., Mehra-Chaudhary, R., Sobrado, P., & Tanner, J. J. (2020). Trapping conformational states of a flavin-dependent *N*-monooxygenase *in crystallo* reveals protein and flavin dynamics. *Journal of Biological Chemistry, 295*(38), 13239–13249. https://doi.org/10.1074/jbc.RA120.014750.

Chocklett, S. W., & Sobrado, P. (2010). *Aspergillus fumigatus* SidA is a highly specific ornithine hydroxylase with bound flavin cofactor. *Biochemistry, 49*(31), 6777–6783. https://doi.org/10.1021/bi100291n.

Csáky, T. Z., Hassel, O., Rosenberg, T., Lång (Loukamo), S., Turunen, E., & Tuhkanen, A. (1948). On the estimation of bound hydroxylamine in biological materials. *Acta Chemica Scandinavica, 2*, 450–454. https://doi.org/10.3891/acta.chem.scand.02-0450.

Franceschini, S., Fedkenheuer, M., Vogelaar, N. J., Robinson, H. H., Sobrado, P., & Mattevi, A. (2012). Structural insight into the mechanism of oxygen activation and substrate selectivity of flavin-dependent N-hydroxylating monooxygenases. *Biochemistry, 51*(36), 7043–7045. https://doi.org/10.1021/bi301072w.

Gadda, G., & Sobrado, P. (2018). Kinetic solvent viscosity effects as probes for studying the mechanisms of enzyme action. *Biochemistry, 57*(25), 3445–3453. https://doi.org/10.1021/acs.biochem.8b00232.

Giddings, L.-A., Lountos, G. T., Kim, K. W., Brockley, M., Needle, D., Cherry, S., et al. (2021). Characterization of a broadly specific cadaverine N-hydroxylase involved in desferrioxamine B biosynthesis in Streptomyces sviceus. *PLoS One, 16*(3), e0248385. https://doi.org/10.1371/journal.pone.0248385.

Gijsbers, A., Nishigaki, T., & Sánchez-Puig, N. (2016). Fluorescence anisotropy as a tool to study protein-protein interactions. *Journal of Visualized Experiments: JoVE, (116)*, 54640. https://doi.org/10.3791/54640.

Happacher, I., Aguiar, M., Alilou, M., Abt, B., Baltussen, T. J. H., Decristoforo, C., et al. (2023). The siderophore ferricrocin mediates iron acquisition in *Aspergillus fumigatus. Microbiology Spectrum, 11*(3), e0049623. https://doi.org/10.1128/spectrum.00496-23.

Hissen, A. H. T., Wan, A. N. C., Warwas, M. L., Pinto, L. J., & Moore, M. M. (2005). The *Aspergillus fumigatus* siderophore biosynthetic gene sidA, encoding l-ornithine N5-oxygenase, is required for virulence. *Infection and Immunity, 73*(9), 5493–5503. https://doi.org/10.1128/IAI.73.9.5493-5503.2005.

Huijbers, M. M. E., Montersino, S., Westphal, A. H., Tischler, D., & Van Berkel, W. J. H. (2014). Flavin dependent monooxygenases. *Archives of Biochemistry and Biophysics, 544*, 2–17. https://doi.org/10.1016/j.abb.2013.12.005.

Jeong, S. S., & Gready, J. E. (1994). A method of preparation and purification of $(4R)$-deuterated-reduced nicotinamide adenine dinucleotide phosphate. *Analytical Biochemistry, 221*(2), 273–277. https://doi.org/10.1006/abio.1994.1411.

Khan, A., Singh, P., & Srivastava, A. (2018). Synthesis, nature and utility of universal iron chelator – Siderophore: A review. *Microbiological Research, 212–213*, 103–111. https://doi.org/10.1016/j.micres.2017.10.012.

Korkhin, Y., Kalb(Gilboa), A. J., Peretz, M., Bogin, O., Burstein, Y., & Frolow, F. (1998). NADP-dependent bacterial alcohol dehydrogenases: Crystal structure, cofactor-binding and cofactor specificity of the ADHs of Clostridium beijerinckii and *Thermoanaerobacter brockii. Journal of Molecular Biology, 278*(5), 967–981. https://doi.org/10.1006/jmbi.1998.1750.

Lynch, D., O'Brien, J., Welch, T., Clarke, P., Cuív, P. O., Crosa, J. H., et al. (2001). Genetic organization of the region encoding regulation, biosynthesis, and transport of rhizobactin 1021, a siderophore produced by *Sinorhizobium meliloti. Journal of Bacteriology, 183*(8), 2576–2585. https://doi.org/10.1128/JB.183.8.2576-2585.2001.

Lyons, N. S., Bogner, A. N., Tanner, J. J., & Sobrado, P. (2022). Kinetic and structural characterization of a flavin-dependent putrescine N-hydroxylase from *Acinetobacter baumannii. Biochemistry, 61*(22), 2607–2620. https://doi.org/10.1021/acs.biochem.2c00493.

Mao, Z., & Campbell, C. T. (2020). Kinetic isotope effects: Interpretation and prediction using degrees of rate control. *ACS Catalysis, 10*(7), 4181–4192. https://doi.org/10.1021/acscatal.9b05637.

Martín Del Campo, J. S., Vogelaar, N., Tolani, K., Kizjakina, K., Harich, K., & Sobrado, P. (2016). Inhibition of the flavin-dependent monooxygenase siderophore A (SidA) blocks siderophore biosynthesis and *Aspergillus fumigatus* growth. *ACS Chemical Biology, 11*(11), 3035–3042. https://doi.org/10.1021/acschembio.6b00666.

Meneely, K. M., & Lamb, A. L. (2007). Biochemical characterization of an FAD-dependent monooxygenase, the ornithine hydroxylase from *Pseudomonas aeruginosa*, suggests a novel reaction mechanism. *Biochemistry, 46*(42), 11930–11937. https://doi.org/10.1021/bi700932q.

Mihara, K., Tanabe, T., Yamakawa, Y., Funahashi, T., Nakao, H., Narimatsu, S., et al. (2004). Identification and transcriptional organization of a gene cluster involved in bio-synthesis and transport of acinetobactin, a siderophore produced by *Acinetobacter baumannii* ATCC 19606T. *Microbiology (Reading, England), 150*(Pt 8), 2587–2597. https://doi.org/10.1099/mic.0.27141-0.

Moran, G. R. (2019). Anaerobic methods for the transient-state study of flavoproteins: The use of specialized glassware to define the concentration of dioxygen. *Methods in Enzymology, 620*, 27–49. https://doi.org/10.1016/bs.mie.2019.03.005.

Moynié, L., Milenkovic, S., Mislin, G. L. A., Gasser, V., Malloci, G., Baco, E., et al. (2019). The complex of ferric-enterobactin with its transporter from *Pseudomonas aeruginosa* suggests a two-site model. *Nature Communications, 10*(1), 3673. https://doi.org/10.1038/s41467-019-11508-y.

Mügge, C., Heine, T., Baraibar, A. G., van Berkel, W. J. H., Paul, C. E., & Tischler, D. (2020). Flavin-dependent N-hydroxylating enzymes: Distribution and application. *Applied Microbiology and Biotechnology, 104*(15), 6481–6499. https://doi.org/10.1007/s00253-020-10705-w.

Olavarría, K., Valdés, D., & Cabrera, R. (2012). The cofactor preference of glucose-6-phosphate dehydrogenase from *Escherichia coli*—Modeling the physiological production of reduced cofactors. *The FEBS Journal, 279*(13), 2296–2309. https://doi.org/10.1111/j.1742-4658.2012.08610.x.

Olive, C., Geroch, M. E., & Levy, H. R. (1971). Glucose 6-phosphate dehydrogenase from *Leuconostoc mesenteroides*: KINETIC STUDIES. *Journal of Biological Chemistry, 246*(7), 2047–2057. https://doi.org/10.1016/S0021-9258(19)77187-7.

Page, M. G. P. (2019). The role of iron and siderophores in infection, and the development of siderophore antibiotics. *Clinical Infectious Diseases: An Official Publication of the Infectious Diseases Society of America, 69*(Suppl 7), S529–S537. https://doi.org/10.1093/cid/ciz825.

Palfey, B. A., & McDonald, C. A. (2010). Control of catalysis in flavin-dependent monooxygenases. *Archives of Biochemistry and Biophysics, 493*(1), 26–36. https://doi.org/10.1016/j.abb.2009.11.028.

Paul, C. E., Eggerichs, D., Westphal, A. H., Tischler, D., & van Berkel, W. J. H. (2021). Flavoprotein monooxygenases: Versatile biocatalysts. *Biotechnology Advances, 51*, 107712. https://doi.org/10.1016/j.biotechadv.2021.107712.

Pierdominici-Sottile, G., Palma, J., Ferrelli, M. L., & Sobrado, P. (2024). The dynamics of the flavin, NADPH, and active site loops determine the mechanism of activation of class B flavin-dependent monooxygenases. *Protein Science: A Publication of the Protein Society, 33*(4), e4935. https://doi.org/10.1002/pro.4935.

Qi, J., Kizjakina, K., Robinson, R., Tolani, K., & Sobrado, P. (2012). A fluorescence polarization binding assay to identify inhibitors of flavin-dependent monooxygenases. *Analytical Biochemistry, 425*(1), 80–87. https://doi.org/10.1016/j.ab.2012.03.002.

Reis, R. A. G., Li, H., Johnson, M., & Sobrado, P. (2021). New frontiers in flavin-dependent monooxygenases. *Archives of Biochemistry and Biophysics, 699*, 108765. https://doi.org/10.1016/j.abb.2021.108765.

Robinson, R., Badieyan, S., & Sobrado, P. (2013). C4a-hydroperoxyflavin formation in N-hydroxylating flavin monooxygenases is mediated by the 2'-OH of the nicotinamide ribose of NADP+. *Biochemistry, 52*(51), 9089–9091. https://doi.org/10.1021/bi4014903.

Robinson, R., Franceschini, S., Fedkenheuer, M., Rodriguez, P. J., Ellerbrock, J., Romero, E., et al. (2014). Arg279 is the key regulator of coenzyme selectivity in the flavin-dependent ornithine monooxygenase SidA. *Biochimica et Biophysica Acta (BBA) – Proteins and Proteomics, 1844*(4), 778–784. https://doi.org/10.1016/j.bbapap.2014.02.005.

Robinson, R. M., Klancher, C. A., Rodriguez, P. J., & Sobrado, P. (2019). Flavin oxidation in flavin-dependent N-monooxygenases. *Protein Science, 28*(1), 90–99. https://doi.org/10.1002/pro.3487.

Robinson, R. M., Rodriguez, P. J., & Sobrado, P. (2014). Mechanistic studies on the flavin-dependent N^6-lysine monooxygenase MbsG reveal an unusual control for catalysis. *Archives of Biochemistry and Biophysics, 550–551*, 58–66. https://doi.org/10.1016/j.abb.2014.04.006.

Romero, E., Fedkenheuer, M., Chocklett, S. W., Qi, J., Oppenheimer, M., & Sobrado, P. (2012). Dual role of NADP(H) in the reaction of a flavin dependent N-hydroxylating monooxygenase. *Biochimica et Biophysica Acta, 1824*(6), 850–857. https://doi.org/10.1016/j.bbapap.2012.03.004.

Romero, E., Gómez Castellanos, J. R., Gadda, G., Fraaije, M. W., & Mattevi, A. (2018). Same substrate, many reactions: Oxygen activation in flavoenzymes. *Chemical Reviews, 118*(4), 1742–1769. https://doi.org/10.1021/acs.chemrev.7b00650.

Saha, P., Xiao, X., Yeoh, B. S., Chen, Q., Katkere, B., Kirimanjeswara, G. S., et al. (2019). The bacterial siderophore enterobactin confers survival advantage to Salmonella in macrophages. *Gut Microbes, 10*(3), 412. https://doi.org/10.1080/19490976.2018.1546519.

Schrettl, M., Bignell, E., Kragl, C., Joechl, C., Rogers, T., Arst, H. N., et al. (2004). Siderophore biosynthesis but not reductive iron assimilation is essential for *Aspergillus fumigatus* virulence. *The Journal of Experimental Medicine, 200*(9), 1213–1219. https://doi.org/10.1084/jem.20041242.

Shirey, C., Badieyan, S., & Sobrado, P. (2013). Role of Ser-257 in the sliding mechanism of NADP(H) in the reaction catalyzed by the *Aspergillus fumigatus* flavin-dependent ornithine N5-monooxygenase SidA. *Journal of Biological Chemistry, 288*(45), 32440–32448. https://doi.org/10.1074/jbc.M113.487181.

Stehr, M., Smau, L., Singh, M., Seth, O., Macheroux, P., Ghisla, S., et al. (1999). Studies with lysine N6-hydroxylase. Effect of a mutation in the assumed FAD binding site on coenzyme affinities and on lysine hydroxylating activity. *Biological Chemistry, 380*(1), 47–54. https://doi.org/10.1515/BC.1999.006.

Suzuki, M., Urabe, A., Sasaki, S., Tsugawa, R., Nishio, S., Mukaiyama, H., et al. (2021). Development of a mugineic acid family phytosiderophore analog as an iron fertilizer. *Nature Communications, 12*(1), 1558. https://doi.org/10.1038/s41467-021-21837-6.

Valentino, H., Campbell, A. C., Schuermann, J. P., Sultana, N., Nam, H. G., LeBlanc, S., et al. (2020). Structure and function of a flavin-dependent S-monooxygenase from garlic (*Allium sativum*). *The Journal of Biological Chemistry, 295*(32), 11042–11055. https://doi.org/10.1074/jbc.RA120.014484.

Valentino, H., & Sobrado, P. (2019). Performing anaerobic stopped-flow spectrophotometry inside of an anaerobic chamber. *Methods in Enzymology, 620*, 51–88. https://doi.org/10.1016/bs.mie.2019.03.006.

Viola, R. E., Cook, P. F., & Cleland, W. W. (1979). Stereoselective preparation of deuterated reduced nicotinamide adenine nucleotides and substrates by enzymatic synthesis. *Analytical Biochemistry, 96*(2), 334–340. https://doi.org/10.1016/0003-2697(79)90590-6.

CHAPTER THIRTEEN

Siderophore-dependent ferrichelatases

C.E. Merrick, N.M. Gulati, and T.A. Wencewicz[*]

Department of Chemistry, Washington University in St. Louis, St. Louis, MO, United States
[*]Corresponding author. e-mail address: wencewicz@wustl.edu

Contents

1. Introduction		282
2. General methods and statistical analysis		291
3. Overexpression and purification of N-His6-FhuD2		292
	3.1 Equipment	292
	3.2 Buffers, strains, and reagents	292
	3.3 Procedure	293
	3.4 Optional steps	295
	3.5 Notes	296
4. FhuD2 binding assay: intrinsic tryptophan fluorescence quenching		297
	4.1 Equipment	297
	4.2 Buffers and reagents	297
	4.3 Procedure	297
	4.4 Notes	298
5. FhuD2 binding assay: siderophore affinity chromatography		300
	5.1 Equipment	300
	5.2 Buffers and reagents	300
	5.3 Procedure	300
	5.4 Optional steps	301
	5.5 Notes	302
6. Synthesis of DFO-NBD and Fe(III)-siderophores		302
	6.1 Equipment	302
	6.2 Reagents	303
	6.3 Procedure: synthesis of DFO-NBD	303
	6.4 Procedure: synthesis of iron bound siderophores	303
	6.5 Notes	304
7. Ferrichelatase assay		304
	7.1 Equipment	304
	7.2 Buffers and reagents	305
	7.3 Procedure: fluorescence based FhuD2 iron exchange assay	305
	7.4 Procedure: validation by LC–MS	306
	7.5 Notes	307

Methods in Enzymology, Volume 702
ISSN 0076-6879, https://doi.org/10.1016/bs.mie.2024.06.015
Copyright © 2024 Elsevier Inc. All rights are reserved, including those for text and data mining, AI training, and similar technologies.

8. Siderophore EDTA competition assay	307
8.1 Equipment	307
8.2 Buffers and reagents	307
8.3 Procedure	307
8.4 Optional steps	309
8.5 Notes	309
9. *S. aureus* growth studies	309
9.1 Equipment	309
9.2 Strains and reagents	309
9.3 Procedure	310
9.4 Notes	310
10. Summary and conclusions	311
Acknowledgments	312
References	312

Abstract

Iron is a crucial secondary metabolite for bacterial proliferation, but its bioavailability under infection conditions is limited by the low solubility of ferric ion and the host's ability to sequester iron by protein chelation. In these iron limiting conditions, bacteria produce and secrete low molecular weight ferric ion chelators, siderophores, to scavenge host iron. Iron bound siderophores are recognized by surface displayed receptors and internalized by active transport preceding the liberation of the iron payload by reduction or cleavage of the siderophore. The traditional paradigms surrounding the interactions between siderophores and their corresponding receptors have relied on canonical protein-ligand binding models that do not accurately reflect the conditions experienced by siderophore binding proteins (SBPs). Research by the Raymond group suggested that a ligand displacement model does not fully describe the role of SBPs in siderophore transport where the ferric ion can be shuttled between siderophore molecules during the transport process. This work inspired further research by the Wencewicz group, which demonstrated that the *Staphylococcus aureus* SBP FhuD2 can catalyze the transfer of iron from the biological iron source *holo*-transferrin to a SBP bound iron-free siderophore. The discovery of this ferrichelatase activity represents a novel mechanism of receptor mediated active transport which raises the question: is ferrichelatase activity a unique feature of FhuD2 or a previously unappreciated hallmark of SBPs? This chapter highlights a series of protocols for the general functional characterization of SBPs and methodologies to assay ferrichelatase activity with the hopes of providing the tools to answer this question.

1. Introduction

Staphylococcus aureus utilizes a ferric hydroxamate uptake (Fhu) system for the uptake of iron bound xenosiderophores (Fig. 1) (Clancy et al., 2006; Mariotti et al., 2013; Sebulsky, Shilton, Speziali, & Heinrichs, 2003).

Siderophore-dependent ferrichelatases 283

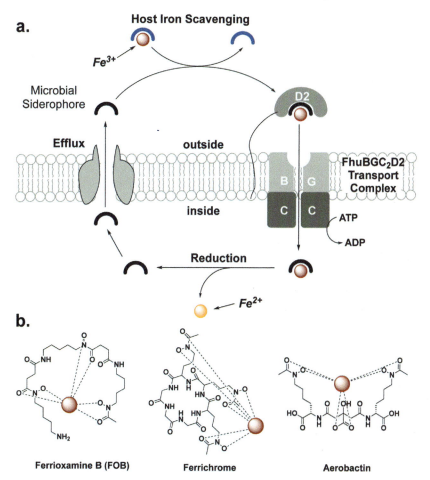

Fig. 1 Overview of the Fhu system in human pathogenic *Staphylococcus aureus*. (A) Scheme of the siderophore transport cycle. (B) Examples of ferric hydroxamate-containing siderophore substrates for Fhu.

The *fhuBGC* operon encodes for the ATPase FhuC and transmembrane proteins FhuB and FhuG that assemble as a functional ABC-type permease, FhuBGC$_2$. This operon is co-transcribed with the gene *fhuD2* encoding for the membrane-anchored siderophore-binding protein (SBP) FhuD2 (Cabrera, Xiong, Uebel, Singh, & Jayaswal, 2001; Clancy et al., 2006; Sebulsky & Heinrichs, 2001; Speziali, Dale, Henderson, Vinés, & Heinrichs, 2006). The lipoprotein FhuD2 is highly conserved, surface-displayed, and required for virulence (Bacconi et al., 2017; Mishra et al., 2012). The N-terminal peptide sequence includes a secretion tag that is cleaved by a signal peptidase during translocation and a lipid anchoring motif. FhuD2 interacts with the FhuBGC$_2$ permease on the extracellular side of the cell membrane to recruit ferric siderophores for internalization. Collectively, the FhuBGC$_2$D2 complex forms a functional siderophore transport system. Expression of the *fhu* operon and *fhuD2* is modified according to iron availability by the ferric uptake regulator (Fur) (Bacconi et al., 2017; Torres et al., 2010). FhuD2 has been shown to be an essential virulence factor linked to establishment of early infection and is the target of vaccine development by Novartis (Mishra et al., 2012).

Many strains of S. *aureus* possess two copies of genes encoding for FhuD-type SBPs known as *fhuD1* and *fhuD2*. Only *fhuD2* is required for virulence; the reason for this is unknown but there are significant sequence differences (41% pairwise protein sequence identity) that suggest there could be important functional and structural differences. Spontaneous resistance to ferrioxamine-based sideromycins such as salmycin produced by *Streptomyces violaceus* occurs via disruption of *fhuD2* transcription and/or single nucleotide polymorphisms located in the encoding DNA sequence for *fhuD2* (Gwinner, 2008). This observation is consistent with the observed conditional essentiality of FhuD2, but not FhuD1, in S. *aureus* human pathogens. This raises the question, why are some SBPs essential for virulence and some are not? The answer to this question is certainly context dependent (e.g., host, iron source). We hypothesize that the essentiality of SBPs depends on the pathogen's requirements for iron and compensatory pathways for obtaining this critical nutrient.

Relative to many SBPs, FhuD2 has several distinguishing features. FhuD2 has a broad substrate specificity and appears to be associated with the scavenging of xenosiderophores including ferrichrome and desferrioxamine B (FDA-approved treatment for iron overload diseases) (FDA, 2022). While S. *aureus* does produce staphyloferrin (siderophore) and staphylopine (metallophore), these metabolites have dedicated uptake pathways (Beasley et al., 2009; Beasley, Marolda, Cheung, Buac, & Heinrichs, 2011; Grim et al., 2017; Song et al., 2018). There is no known endogenous biosynthetic siderophore produced by S. *aureus* that serves as a primary transport substrate for the Fhu pathway. Hence, it is possible that FhuD2 has evolved to capitalize on mixed microbial infections or even utilization of siderophores produced by members of the human microbiome. There is even reason to believe that S. *aureus* can directly utilize desferrioxamine B administered therapeutically given patients receiving this medication for iron overload diseases are statistically more prone to S. *aureus* infection

(Arifin, Hannauer, Welch, & Heinrichs, 2014; Kontoghiorghes, Kolnagou, Skiada, & Petrikkos, 2010; Rahav et al., 2006). Given this apparent circumstantial role, the question arises as to how *fhuD2* and the *fhu* operon can be required for virulence. In 2020, the Wencewicz lab discovered that FhuD2 possesses a unique type of enzymatic activity as a siderophore-dependent ferrichelatase (SDF) (Endicott, Rivera, Yang, & Wencewicz, 2020). While most SBPs are treated as non-catalytic binding proteins that simply pass a transport substrate to an associated ABC permease, there is a distinct need for ferrichelatase activity in the case of SBPs. In the early 2000s, Raymond and coworkers began to imagine an iron shuttling process where ferric ions can somehow be passed between ligands during the transport process (Fukushima et al., 2013; Stintzi, Barnes, Xu, & Raymond, 2000). Since metal exchange between ligands is rate limiting, the discovery of SDF activity in FhuD2, and the related SDF YxeB from *B. cereus*, filled this important knowledge gap with the biocatalyst that accelerates metal exchange between siderophores and biological sources of chelated iron (e.g., *holo*-transferrin) (Endicott et al., 2020).

While there are many metal binding proteins, few have been shown to possess genuine chelatase activity. The best characterized chelatases are involved in heme and chlorophyll biosynthesis where metalation of the porphyrin by ferrous ion or Mg^{2+} is catalyzed by a chelatase. To metalate porphyrins, a mechanism utilizing pyrrole deprotonation and ring distortion has been proposed (Lecerof, Fodje, Hansson, Hansson, & Al-Karadaghi, 2000; Shen & Ryde, 2005). When comparing a siderophore to heme, there are several important differences. First, siderophores selectively form stable complexes with ferric ion over ferrous ion while heme is typically found as the ferrous ion complex inside of cells. Second, the heme cofactor is the final residing place of the metal center which can serve as a redox cofactor and as an oxygen-carrier in aerobic organisms. The siderophore is a transient carrier of ferric ion and is meant to deliver the metal from an external oxidizing environment to an intracellular reducing environment where reduction of the siderophore ferric ion complex by siderophore-interacting proteins, most FAD-dependent, produces a lower affinity siderophore ferrous ion complex that dissociates to recycle the metal siderophore and add ferrous ion to the metabolic iron pool (Cain & Smith, 2021). The specificity of siderophores for ferric ion comes from "hard/soft" acid/base theory where oxygen-rich ligands such as carboxylates, catecholates, and hydroxamates serve as "hard" basic chelating ligands for the "hard" acid ferric ion (Hider & Kong, 2010).

Further, siderophores incorporate the multidentate ligand system into an organic scaffold that present an ideal octahedral coordination sphere for the ferric ion. Hemes present a square planar coordinate mode preferred by ferrous ion and allowing for proximal ligands to bind, which is important to functionality of heme-dependent proteins/enzymes (e.g., hemoglobin, cytochromes). To accommodate the metabolic need for iron and the unique characteristics of the siderophore ligand, SDFs require a different mode of catalysis compared to ferrochelatases.

In the case of FhuD2, the ferrichelatase activity is dependent on a siderophore as the cofactor (Endicott et al., 2020). The siderophore cofactor is "pre-bound" to the SDF in its metal-free form. This presents the FhuD2-siderophore complex to the extracellular space where it is poised to extract metal from biological sources including *holo*-human transferrin ferric ion complex. This requires an initial binding interaction between the FhuD2-siderophore complex and *holo*-transferrin followed by metal transfer from *holo*-transferrin to the bound siderophore. The siderophore is a thermodynamic sink for ferric ion with apparent K_{Fe} ~10 orders of magnitude greater than human transferrin (Holden & Bachman, 2015). FhuD2 provides the kinetic advantage by using a series of conserved Tyr residues to facilitate the metal transfer process and stabilize the transition states for a complicated ligand exchange process. Formation of the ferric siderophore-FhuD2 complex then leads to ungating of the FhuBGC$_2$ permease dictated by the distance of two conserved Glu residues (E97 and E231; all amino acid residue numbers correspond to GenBank sequence ID AAK92086) on the N- and C-terminal globular domains of the SDF. Docking of the SDF in the so-called "closed" conformation with the permease stimulates the cytoplasmic cleavage of ATP to ADP with concomitant influx of the ferric siderophore to the cytoplasmic space. With the siderophore serving as both a cofactor for the chelatase FhuD2 and a transport substrate for FhuBGC$_2$, the replacement of siderophore cofactor is required to complete the turnover process. Cofactor replacement is feasible given the orders-of-magnitude higher concentration of metal-free siderophore (micromolar to millimolar) outside the cell compared to the intracellular environment (nanomolar). FhuD2 and YxeB represent two functionally characterized members of a unique family of SBPs now known as the SDFs (Endicott et al., 2020).

The discovery of SDF activity raises many questions about SBPs. What distinguishes SDFs from ordinary SBPs? Are FhuD2 and YxeB the exception or the standard? Is the SDF model exclusive to Gram-positive bacteria (e.g., extraction of metals from transferrins) or is there a role for SDF activity in Gram-negative bacteria (e.g., shuttling iron from outer membrane receptors to periplasmic SDFs)? Is SDF activity functionally conserved in *S. aureus* and other human pathogens? Is transferrin the only biologically relevant iron source for FhuD2? Or, can FhuD2 extract ferric ion from many different metalloproteins? A structural, mechanistic, and functional understanding of SBPs from diverse bacterial species is needed to address these important questions. Published studies on FhuD2 can serve as a general guide for how to probe for SDF activity in homologous SBPs. The high-resolution structure of FhuD2 has been solved in the solid state (X-ray) and solution state (NMR) including apo- and holo-forms (Mariotti et al., 2013; Podkowa, Briere, Heinrichs, & Shilton, 2014; Sebulsky et al., 2003). The structural studies support the observed broad substrate scope revealing a hydrophobic binding cavity between the N-terminal and C-terminal globular domains. These domains are linked by a central alpha-helix that is flexible enough to allow for a "breathing" motion in this "cradle-type" binding protein. This conformational dynamic presents "closed" and "open" states of FhuD2 that are likely associated with binding and dissociation from the permease FhuBGC$_2$, respectively. Trp197 and Arg199 form stabilizing interactions with the hydroxamate ligands of desferrioxamine B in both the metal free and ferric ion chelated forms. Trp197 undergoes a major conformation change along with several Tyr residues (278, 191, 193) in concert with a transition between the "open" and "closed" forms of FhuD2. Based on these observed conformational dynamics and functional studies using site-directed mutagenesis, a model for the SDF-catalyzed iron shuttle has been proposed (Endicott et al., 2020).

Canonical receptor mediated active transport would involve an Fe(III)-bound siderophore binding FhuD2 causing ATP-dependent internalization. This model, however, fails to account for the affinity of iron-free siderophores for FhuD2 and their excess compared to Fe(III)-siderophores. Stable isotope feeding studies with Ga(III)- and Cr(III)-bound siderophores performed by Raymond et al. were found to be inconsistent with the canonical one-site binding model. Follow up studies with the SBP YxeB from *B. cereus* demonstrated Fe(III) exchange between siderophores promoted by the SBP (Fukushima et al., 2013; Fukushima, Allred, & Raymond, 2014; Stintzi et al., 2000). Together, these results informed a displacement model in which an iron free siderophore is bound to the SBP and is displaced by an Fe(III)-siderophore which is then internalized. Further, an iron shuttle model was proposed in which Fe(III) is transferred from an Fe(III)-siderophore to an iron free siderophore at the SBP which is then internalized. The displacement model does not account for the large excess of iron free siderophores relative to Fe(III)-siderophores meaning that the equilibrium will favor iron free siderophores bound to the SBP. The iron shuttle model better accounts for the equilibrium of the SBP, but relies on an improbable encounter of an Fe(III) bound and iron free siderophore at the SBP. Work by the Wencewicz group supports a revised Raymond iron shuttle model in which the abundant human Fe(III)-transferrin serves as the biological iron source (Endicott et al., 2020). In this model, the SBP catalyzes the otherwise slow transfer of Fe(III) from *holo*-transferrin to an iron free siderophore which is then internalized (Sriyosachati & Cox, 1986). This model accounts for the SBP ligand binding equilibrium favoring the iron free siderophore and utilizes a high probability encounter with Fe(III)-transferrin at the SBP (Fig. 2).

The SBPs from Gram-positive and Gram-negative strains share low pairwise sequence IDs of ~15%. These sequence relationships are captured by phylogenetic analysis (Fig. 3A). It remains to be determined if SDF activity is present in all FhuD homologs. A signature sequence for residues critical to imparting SDF activity needs to be better defined. Site-directed mutagenesis of FhuD2 provides a starting point for defining critical residues. A sequence alignment of FhuD2, YxeB, and related homologs reveals a large degree of sequence diversity in this protein family (Fig. 3B and C). While FhuD1 and FhuD2 are both derived from *S. aureus* these proteins share only 41% pairwise sequence identity. Comparison of SBP homologs across species shows similar pairwise sequence conservations within species type Gram-positive (35%–39% for *S. aureus* and *Bacillus* spp.) or Gram-negative (83% for

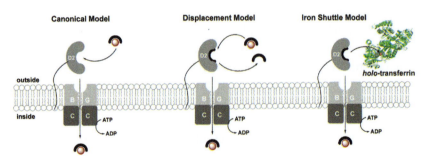

Fig. 2 Three potential membrane transport paradigms for FhuBGC₂D2. The depiction of *holo*-transferrin is derived from PDB 4×1B and visualized using PyMOL v2.3.2 (Schrodinger LLC).

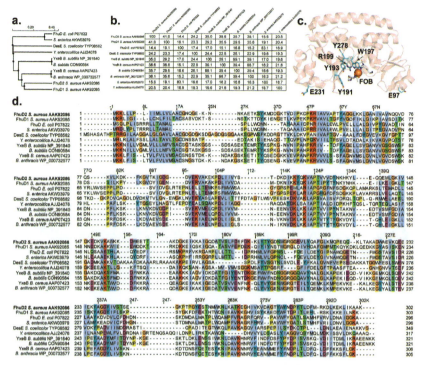

Fig. 3 Comparative sequence analysis of SBPs. (A) Phylogenetic tree of full length SBPs generated using Clustal Omega (https://www.ebi.ac.uk/jdispatcher/msa/clustalo). (B) Percent sequence identity matrix of full length SBPs. (C) X-ray co-crystal structure of truncated S. aureus FhuD2 with FOB (PDB 4FIL) visualized using PyMOL v2.3.2 (Schrodinger LLC). (D) Pairwise sequence alignment of full length SBPs including N-terminal signal sequences generated using Clustal Omega and visualized using Jalview v2.11.3.2 (https://www.jalview.org/).

Escherichia coli and *Salmonella enterica*). The Streptomyces derived DesE SBP shows reduced sequence homology to everything (17%–25% pairwise sequence ID for Gram-positive and Gram-negative strains included in the study). Even with this low degree of pairwise sequence alignment across Gram-positive, Gram-negative, and Actinomyces strains, the analysis of structural alignment shows a high-degree of residue equivalency (TM scores 0.74–0.94) within the SBP protein family (Fig. 4). Hence, analysis of sequence alone appears to be ineffective in predicting substrate specificity and SDF activity within the SBP protein family.

Here, we described a suite of assays that enable the comprehensive exploration of ligand binding, metal affinity, and metal exchange kinetics to determine the substrate specificity and ferrichelatase activity of virtually any

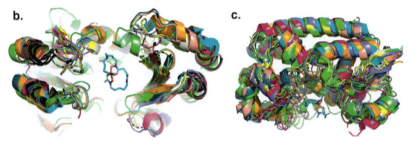

Fig. 4 Structural alignment of SBPs lacking the N-terminal signal sequence peptide. (A) Results from pairwise structural alignment of SBPs using the pairwise structural alignment tool on the RCSB PDB website (https://www.rcsb.org/alignment). The structure alignments were visualized from two angles in panels (B) and (C) using PyMOL v2.3.2 (Schrodinger LLC).

SBP. **Procedure 3** provides a protocol for the heterologous expression of bacterial SBPs including guidance on sequence optimization and placement of affinity tags. **Procedure 4** provides a protocol for measuring the apparent binding affinities of siderophore ligands with SBPs using intrinsic Trp fluorescence quenching. **Procedure 5** provides an alternate means to study reversible ligand binding to SBPs using resin-bound SBP and competing siderophores introduced in flow. Such a protocol can also be scaled for affinity purification of siderophore metabolites from complex mixtures (e.g., bacterial fermentations). **Procedure 6** describes the synthesis of a fluorescent ferrioxamine siderophore and the corresponding ferric ion complex to serve as a chemical probe for iron exchange studies. **Procedure 7** establishes a general assay for ferrichelatase activity using fluorescence quenching monitored under steady-state conditions to determine apparent catalytic efficiency for diverse iron donor substrates. A protocol to validate the fluorescence quenching using LC–MS quantification of metal-free and ferric siderophores is also described. **Procedure 8** provides a protocol for determining the apparent metal-binding specificity constants and metal exchange kinetics for siderophore chelators using a simple optical absorbance-based competition experiment to monitor for ligand exchange between ferric siderophores and EDTA. **Procedure 9** provides a method to evaluate the ability of siderophores to promote or inhibit the growth of *S. aureus* under iron restriction conditions in a chemically defined medium to correlate in vitro findings related to SBPs to whole cell growth phenotypes. The application of these assays toward SBPs will help to define the residues and sequence motifs that underlie SDF activity. An improved understanding of SDF catalysis will help in the study of siderophore-mediated iron acquisition in bacteria and help to inform the design of more effective vaccines and siderophore analogs intended for applications in antibiotic drug delivery, the "Trojan Horse" approach, and siderophore-blocking anti-virulence agents.

2. General methods and statistical analysis

Standard safety protocols for synthetic chemistry and biosafety level-1 bacterial strains should be employed. If **Procedure 9** involves the use of *S. aureus* strains designated as biosafety level-2, then the appropriate facilities and protocols are required. Personal protective equipment including safety goggles, lab coat, and appropriate gloves should be utilized, and all synthesis should be performed in a fume hood. All reactions were performed under ambient air with magnetic stirring unless otherwise stated. Reagents, solvents, buffers, and salts were purchased from Sigma-Aldrich, Thermo Fisher Scientific, or Research Products International unless otherwise stated. An Orion Star A111 pH meter with a PerpHecT Ross micro pH electrode (Thermo Fisher Scientific) was used for all pH measurements. Prep-HPLC and LC–MS utilized variable solvent gradients of water with 0.1% formic acid (A) and acetonitrile with 0.1% formic acid (B). Experimental trials were performed in at least duplicate. Data visualization and analysis was performed with GraphPad Prism v9.3.0b but alternative software packages may be used. Sequence alignments were performed using Clustal Omega and visualized with Jalview v2.11.3.2 (Madeira et al., 2022; Waterhouse, Procter, Martin, Clamp, & Barton, 2009). Structural alignments were performed using resources provided by RCSB PDB (RCSB.org) including structures 4FIL, 6ENK, 1K2V, and 4×1B (Berman et al., 2000; Clarke, Braun, Winkelmann, Tari, & Vogel, 2002; Podkowa et al., 2014; Ronan et al., 2018, Wang et al., 2015). For proteins without a solved structure, AlphaFold predicted structures were utilized (Jumper et al., 2021; Varadi et al., 2022). Protein structures were visualized using The PyMOL Molecular Graphics System Version 2.3.2 (Schrodinger LLC). Experimental procedures, data, and associated content may be viewed in their original format in the corresponding publications (Bar-Ness, Hadar, Chen, Shanzer, & Libman, 1992; Endicott, Lee, & Wencewicz, 2017; Endicott et al. 2020; Rivera, Beamish, & Wencewicz, 2018).

3. Overexpression and purification of N-His6-FhuD2
3.1 Equipment
- 2.8 L baffled flask
- Autoclave (Steris AMSCO 430LS Steam Sterilizer Model LS-1262)
- Cold room (BioCold Environmental)
- Incubator/shaker (Thermo Scientific MaxQ 8000)
- Rotating drum in an incubator (Thermo Scientific Heratherm Incubator)
- Nanodrop Spectrophotometer (NanoDrop 2000c spectrophotometer)
- Centrifuge (Thermo Scientific Sorvall ST 16R Centrifuge) and ultracentrifuge (Beckman Coulter Optima L-90K Ultracentrifuge)
- Fritted glass tube with stopcock and cap
- Falcon tubes (Corning)
- Sonicator (Qsonica) or homogenizer (Emulsiflex C5 cell disruptor (Avestin))
- Gel electrophoresis chamber (Bio-Rad)
- 10,000 MWCO Snakeskin Dialysis Tubing
- 10,000 MWCO Centrifugal Filter (EMD Millipore)
- Cryogenic vials
- FPLC (optional)

3.2 Buffers, strains, and reagents
- Terrific broth (12 g/L casein digest peptone, 24 g/L yeast extract, 9.4 g/L dipotassium phosphate, 2.2 g/L monopotassium phosphate, 4 mL/L glycerol)
- Luria broth (10 g/L casein digest peptone, 10 g/L sodium chloride, 5 g/L yeast extract)
- Kanamycin sulfate (Chem-Impex)
- Glycerol stock of *E. coli* BL21-Gold(DE3) with pET28a-*fhuD2*
- Isopropyl β-d-1-thiogalactopyranoside (IPTG) (GoldBio)
- Lysis buffer (50 mM K_2HPO_4, 500 mM NaCl, 20 mM imidazole, 10% glycerol, 5 mM β-mercaptoethanol (BME), pH 8)

Siderophore-dependent ferrichelatases 293

- Elution buffer (50 mM K_2HPO_4, 500 mM NaCl, 300 mM imidazole, 10% glycerol, 5 mM BME, pH 8)
- Dialysis buffer (50 mM K_2HPO_4, 150 mM NaCl, 1 mM dithiothreitol (DTT), pH 8)
- Ni-NTA resin (Invitrogen)
- Any *k*D SDS–PAGE gels (Bio-Rad)
- Coomassie blue stain
- Liquid nitrogen
- Thrombin cleavage kit (Invitrogen; optional)
- Thrombin cleavage buffer (20 mM Tris–HCl, 150 mM NaCl, 1 mM $CaCl_2 \cdot 2H_2O$, 1 mM DTT, pH 8.4; optional)
- SIGMAFAST Protease Inhibitor Cocktail Tablets, EDTA-Free (optional)

3.3 Procedure

Day 1.

1. In a foil topped 2.8 L baffled flask, autoclave 1 L of terrific broth at 121 °C for 15 min and allow the solution to cool to room temperature.
2. Prepare an overnight culture by adding a sterile loop of *E. coli* BL21-Gold(DE3) with pET28a-*fhuD2* to 5 mL of Luria broth with 50 μg/mL of kanamycin sulfate. Allow the bacteria to grow overnight at 37 °C with gentle shaking.

Day 2.

1. Preheat an incubator/shaker to 37 °C.
2. Add 1 mL of overnight culture to 1 L of sterilized terrific broth followed by 1 mL of 50 mg/mL kanamycin.
3. Incubate the culture at 37 °C with shaking at 225 rpm while monitoring the optical density (OD) at 600 nm. Once the OD reaches 0.4–0.8, cool the cell culture at 4 °C for 20 min. Equilibrate a refrigerated incubator shaker to 15 °C.
4. Add 1 mL of 0.5 M IPTG solution to the culture.
5. Incubate the culture at 15 °C with shaking at 225 rpm for 16–18 h.
6. Prepare 1 L of lysis and elution buffers and store at 4 °C.

Day 3.

1. Pack a fritted glass tube (with a stopcock and cap) with Ni-NTA resin for a working resin volume of approximately 2.5 cm × 3 cm. Keep the resin at 4 °C for the remainder of the protocol.
2. Wash the resin with three column volumes of water followed by lysis buffer.

3. Distribute the contents of the culture into centrifuge bottles (masses within 0.5 g) and centrifuge at 5000 rpm for 15 min. Discard the supernatant and resuspend the pellets in 15 mL lysis buffer. Transfer the suspension to a 50 mL falcon tube over ice.

4. Sonicate the ice cooled cells for 2 min (4 cycles of 30 s on 30 s off) at 40% amplitude.

   Or

   Flash freeze the cell suspension in liquid nitrogen and allow to thaw. Lyse the cells by passing the suspension through a homogenizer using additional lysis buffer as needed to reach an appropriate working volume. The sample may need to be passed through more than once to achieve complete lysis.

5. Transfer the cell lysates to ultracentrifuge tubes (masses within 0.01 g) and centrifuge at 48,380 × g for 30 min at 4 °C. Collect the supernatants. Discard the cell debris.

6. Load the solution onto the pre-equilibrated nickel-NTA column and incubate at 4 °C on a nutating mixer for 30 min.

7. Collect the initial flow through in a 50 mL falcon tube. Remove unbound protein by washing the column with 40 mL lysis buffer twice and collecting the eluent in 50 mL falcon tubes. Elute the protein of interest by addition of 45 mL elution buffer collecting in 15 mL falcon tubes.

8. Analyze fractions by SDS–PAGE for correct molecular weight and purity with Any kD gels by running the gel at 150 V for 10 min then 185 V for 35 min followed by Coomassie blue staining (Fig. 5).

9. Prepare 1.8 L of dialysis buffer and store at 4 °C.

Day 4/5.

1. Transfer eluted protein to 10,000MWCO Snakeskin dialysis tubing and place into a beaker with 1.8 L of dialysis buffer. Dialyze the protein solution at 4 °C for ~18 h.

2. Transfer the protein solution to the top of a 10,000MWCO centrifuge filter and centrifuge at 5000 rpm for 15 min. Refill the top with additional protein solution and repeat until all protein has been loaded and the final volume reaches ~2 mL.

3. Measure the final concentration by optical absorbance at 280 nm to determine protein concentration and prepare protein beads by dropwise addition (~50 μL) to liquid nitrogen in a vacuum insulated glass Dewar flask. Collect and store beads in a cryogenic vial at −80 °C.

Siderophore-dependent ferrichelatases 295

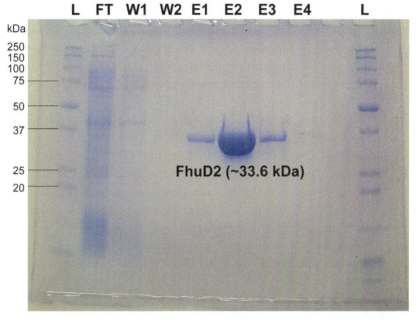

Fig. 5 SDS–PAGE analysis of N-His$_6$-FhuD2 purified using Ni-NTA affinity chromatography after heterologous expression in *Escherichia coli* BL21. The gel (Bio-Rad any kD) was visualized with Coomassie blue staining against a molecular weight standard (ThermoFisher PageRuler Prestained Protein Ladder). This is a fresh protein prep made to validate this protocol.

3.4 Optional steps

For FhuD2, these steps have been investigated and are not essential for the protocols described in this chapter. For other SBPs, these steps may be required or may improve results of assays.
1. Addition of protease inhibitor.
    a. Add 1 tablet of SIGMAFAST protease inhibitor cocktail to 100 mL of lysis buffer.
    b. Utilize lysis buffer with protease inhibitor during initial resuspension of cell pellets in lysis buffer to prevent proteolytic degradation of the target protein.
2. The protocol above provides N-His$_6$-FhuD2, referred to simply as FhuD2 in this chapter. Removal of the His$_6$-tag can be achieved if necessary, using the following steps.
    a. Prepare 1.8 L thrombin cleavage buffer and store at 4 °C
    b. Load purified N-His$_6$-FhuD2 into 10,000MWCO snakeskin dialysis tubing and place into a beaker containing 1.8 L thrombin cleavage buffer. Dialyze for ~18 h at 4 °C.

c. Mix 1 mg of protein with 100 µL thrombin agarose suspension, 100 µL 10× cleavage buffer (provided in kit), and distilled deionized water for a working volume of 1 mL. Incubate the reaction at room temperature for 2 h with gentle agitation.

d. Remove the resin by gentle centrifugation (500 × g, 5 min) and load the supernatant onto ~1 mL of Ni–NTA resin. Discard the pellet.

e. Collect the flow through and wash the resin with thrombin cleavage buffer (2 × 5 mL).

f. Analyze fractions by SDS–PAGE for correct molecular weight and purity with Any kD gels by running the gel at 150 V for 10 min then 185 V for 35 min followed by Coomassie blue staining. Include a mix of "tagged" and "untagged" FhuD2 in one gel lane to visualize and confirm the molecular weight separation.

g. Combine fractions containing pure untagged FhuD2. Transfer the protein solution to the top of a 10,000MWCO centrifuge filter and centrifuge at 5000 rpm for 15 min. Refill the top with additional protein solution and repeat until all protein has been loaded and the final volume reaches ~2 mL.

h. Measure the final concentration by optical absorbance at 280 nm to determine protein concentration and prepare protein beads by drop-wise addition (~50 µL) to liquid nitrogen in a vacuum insulated glass Dewar flask. Collect and store beads in a cryogenic vial at −80 °C.

3. If desired, FhuD2 or other SBPs of interest can be purified further by standard SEC protocols. The proteins are highly soluble as truncated constructs lacking the N-terminal signal peptide sequence and including an N-terminal hexahistidine-containing sequence.

3.5 Notes

1. Prior to heterologous expression of SBPs, it is recommended to analyze the sequence of interest using the SignalP 6.0 software (https://services. healthtech.dtu.dk/services/SignalP-6.0/) to determine the N-terminal site of signal peptidase cleavage and potential sites for lipid anchoring (Teufel et al., 2022). For optimal expression and protein solubility, it is recommended to express the SBPs without the signal peptide and replace this portion of the sequence with the N-terminal His_6-tag to facilitate purification on Ni–NTA resin.

2. When necessary, codon optimize the nucleotide sequence for expression in an E. coli heterologous host (Koblan et al., 2018).

3. Site-directed mutagenesis can be utilized to generate point mutants using standard QuickChange protocols (Bachman, 2013; Endicott et al., 2020). These mutant proteins can be purified as described above and used in all protocols described in this chapter.
4. If sonication is utilized for cell lysis, take care to properly align the sonicator probe such that it is submerged in the suspension without touching the walls or bottom of the falcon tube.
5. Care should be taken to avoid overconcentrating the protein which can result in protein precipitation. Shorter centrifuging intervals can be utilized once all protein has been loaded onto the centrifuge filter.
6. Pre-cooling cryogenic vials by submerging in liquid nitrogen can help prevent beads from sticking together once collected.
7. Carefully vent cryogenic vials to remove excess nitrogen gas prior to capping and storing in freezer.

4. FhuD2 binding assay: intrinsic tryptophan fluorescence quenching

4.1 Equipment
- Nanodrop spectrophotometer (NanoDrop 2000c spectrophotometer)
- Cuvette fluorimeter (PerkinElmer LS 55 Luminescence Spectrometer) or plate reader (Tecan Infinite M Plex)
- Blacked out quartz cuvette or well plate (Hellma; flat bottom)
- Calibrated micropipettes
- Analysis software

4.2 Buffers and reagents
- TBS buffer (25 mM Tris–HCl, 8 g/L NaCl, 0.2 g/L KCl, pH 7.4)
- Purified recombinant FhuD2
- Siderophore/substrate of interest

4.3 Procedure
1. Prepare 500 mL TBS buffer and store at room temperature.
2. Thaw a protein bead at 4 °C. Perform an initial dilution with TBS buffer to reach an $A_{280\,nm}$ of 1.0–1.5. Calculate the concentration using the molar extinction coefficient of FhuD2 ($\varepsilon_{280\,nm}$ = 55,350/(M cm); calculated using Expasy ProtParam (Duvaud et al., 2021)) and dilute to 100 nM in TBS buffer accordingly.

3. To minimize changes in protein concentration when adding substrate, prepare substrate solutions using 100 nM FhuD2 rather than water. Prepare stock solutions of substrate at 4 μM in 100 nM FhuD2 in TBS buffer.

4. Perform fluorescence emission scans utilizing a cuvette fluorimeter or a plate reader. Scan emission wavelengths of 300–400 nm with an excitation wavelength of 280 nm, slits for both excitation and emission at 10 nm, and a scan speed of 400 nm/min. Blacked out quartz is the ideal material for cuvettes and plates. Utilize a sample volume based on the recommended working volume for a given cuvette/plate; volumes in this methodology assume a 300 μL working volume.

5. Ensure that the z-plane is properly adjusted to coincide with the desired height on the cuvette or provide optimal signal for the plate.

6. Blank the instrument by scanning with TBS buffer. If the instrument cannot be blanked, scan TBS buffer and subtract this from all sample signals.

7. Perform a scan of 100 nM FhuD2; if using a cuvette, first thoroughly rinse with 100 nM FhuD2 to ensure an accurate reading.

8. Remove 2 μL of sample and add 2 μL of 4 μM substrate in 100 nM FhuD2 to achieve 26.7 nM substrate in 100 nM FhuD2 while maintaining the sample volume. Thoroughly mix taking care to avoid bubbles prior to scanning.

9. Repeat step 8 (can vary the volume removed/added and the substrate stock solution concentration to achieve different final ligand concentrations) until increasing concentration of substrate no longer decreases the maximum fluorescence intensity (typically at ~340 nm).

10. Repeat steps 5–9 for independent replicates in a cuvette. Replicate samples can be run in parallel in a well plate with a plate reader.

11. Normalize the background corrected maximum fluorescence values. Set the highest fluorescence reading to 1 and the lowest to 0.

12. Calculate the apparent K_d of FhuD2 for the substrate using a nonlinear fit to a one binding site model (Fig. 6).

4.4 Notes

1. Black cuvettes/plates minimize light contamination and provide more accurate fluorescence measurements. Care should be taken with non-quartz materials to avoid unwanted protein and/or ligand interactions with the cuvette/plate and fluorescence from the material (i.e., polystyrene).

2. Fluorescence quenching studies can be performed on both metal-free and metal-chelated forms of siderophores.

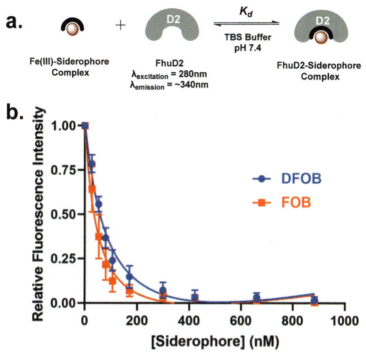

Fig. 6 FhuD2 binding assay using intrinsic tryptophan fluorescence quenching. (A) General schematic of the binding equilibrium. (B) Example binding curves for DFOB and FOB. *Data reproduced from Endicott, N. P., Lee, E., & Wencewicz, T. A. (2017). Structural basis forxenosiderophore utilization by the human pathogen Staphylococcus aureus. ACS Infectious Diseases, 3(7), 542–553. https://doi.org/10.1021/acsinfecdis.7b00036.*

3. FhuD2 has a non-negligible background rate of fluorescence quenching in TBS buffer at room temperature. As such, time between scans should be kept to approximately 2 min to minimize the impact of the background quenching. A correction for background quenching rate can be included as needed for prolonged experiments.
4. Total fluorescence quenching may vary greatly depending on the substrate assayed.
5. FhuD2 can bind a variety of hydrophobic ligands. The apparent K_d values obtained from intrinsic tryptophan fluorescence quenching reflect ligand binding but does not inform on the nature of the ligand binding nor does it provide a direct assessment of k_{on} and k_{off}. To confirm that ligand binding results in the formation of a stable FhuD2-ligand complex we recommend validation using **Procedure 5**.

5. FhuD2 binding assay: siderophore affinity chromatography

5.1 Equipment
- Fritted glass column
- Cold room (BioCold Environmental)
- Nanodrop spectrophotometer (NanoDrop 2000c spectrophotometer)
- Gel electrophoresis chamber
- LC–MS (Agilent 6130 quadrupole with G1329A autosampler, G1311A quaternary pump, G1322A degasser, G1316A thermostatted column compartment, G1315 diode array detector, Gemini 5 μm NX-C18 110 Å LC column 50 × 4.6 mm with guard column)
- Prep-HPLC (Hewlett Packard series 1050, Kinetex 5 μm C18 100 Å LC column 150 × 21.2 mm with guard column)
- C18 silica or Lyopholizer (Labconco)

5.2 Buffers and reagents
- Ni-NTA agarose resin (Invitrogen)
- EtOH, MeOH, deionized distilled H_2O
- SBP Buffer (50 mM K_2HPO_4, 150 mM NaCl, 1 mM dithiothreitol (DTT) or BME, pH 8)
- Purified recombinant FhuD2
- Siderophore of interest and sacrificial siderophore (ferrioxamine B, N-succinyl-ferrioxamine B, or N-acetyl-ferrioxamine B)
- DEAE or Cellex P (Bio-Rad) resin
- Ammonium acetate and ammonium hydroxide

5.3 Procedure
1. Suspend fresh Ni-NTA agarose resin in a 1:1 EtOH:H_2O mixture. Pack a fritted glass column with the agarose suspension for a working resin volume of 2.3 cm × 1 cm. Keep the resin at 4 °C for the remainder of the protocol.
2. Wash the resin with excess water and pre-equilibrate with SBP buffer.
3. Thaw a sufficient number of FhuD2 protein beads at 4 °C to reach 500 μL of 1.5 mM protein solution. Dilute to 3 mL in SBP buffer and add to the Ni-NTA agarose resin.
4. Gently mix the column contents on a nutating mixer for 30 min and drain excess SBP buffer. Wash the column with excess SBP buffer and monitor fractions by SDS–PAGE. Continue washing the SBP-resin until FhuD2 is no longer detected in the eluent.

5. Load the siderophore of interest onto the resin by adding 5 mL of 0.1 mg/mL siderophore in SBP buffer with gentle mixing on a nutating mixer for 20 min (Fig. 7).
6. Drain excess SBP buffer and wash the resin five times with 15 mL of SBF buffer. Confirm via LC–MS analysis that the siderophore of interest is not detectable by the end of the washes.
7. Elute the siderophore of interest by adding 5 mL of 0.1 mg/mL competing siderophore in SBP buffer with gentle mixing on a nutating mixer for 20 min. Analyze the eluent by LC–MS to confirm the displacement of the siderophore of interest from the SBP resin (0%–100% B over 10 min, 10 min at 100% B, and re-equilibration to 100% A over 5 min).

5.4 Optional steps

The siderophore affinity chromatography protocol can be used on a preparative scale to purify siderophores from complex mixtures (e.g., bacterial fermentations) according to these optional steps. This process has been applied to the purification of salmycin from *S. violaceus* (Rivera et al., 2018). The volumes in the procedure below can be adjusted for preparative scale. Current volumes represent addition of this step to procedure 5.3.

1. The siderophore of interest can be purified by reverse phase (RP)-C18 preparative high-performance liquid chromatography (HPLC).

   Or

   If the competing siderophore is chosen such that it has a different net charge than the siderophore of interest, then ion exchange chromatography can provide a facile purification.

   a. Pack a fritted glass column with DEAE or Cellex P resin for a working volume of $2 \, cm \times 7 \, cm$. Anionic siderophores will bind to DEAE while cationic siderophores will bind to Cellex P. Neutral siderophores should not bind either resin. Wash the resin with 50 mL of 10 mM aqueous ammonium acetate (pH 7) followed by 50 mL water.

   b. Desalt the SBP resin elution by C18 chromatography or lyopholization followed by trituration with methanol. Reconstitute the sample in pure water.

   c. Load a 5 mL aliquot of siderophore solution onto the resin and collect the flow through. Analyze the flow through by LC–MS to confirm siderophore elution (cationic or neutral siderophores on DEAE and anionic or neutral siderophores on Cellex P).

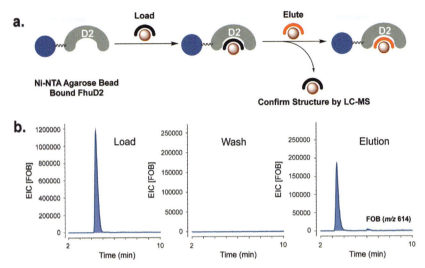

Fig. 7 FhuD2 binding assay using siderophore-affinity chromatography. The scheme (A) depicts resin-immobilized FhuD2 with competing ligands and detection of displaced siderophores by LC–MS. (B) Sample LC–MS extracted ion traces. *Data reproduced from Endicott, N. P., Rivera, G. S. M., Yang, J., & Wencewicz, T. A. (2020). Emergence of ferrichelatase activity in a siderophore-binding protein supports an iron shuttle in bacteria. ACS Central Science, 6(4), 493–506. https://doi.org/10.1021/acscentsci.9b01257.*

    d. Wash the resin with 5 mL of water twice. Elute the bound siderophores with 5 mL of 4% aqueous ammonium hydroxide. Confirm by LC–MS that the elution contains the bound siderophore while the water washes do not.

5.5 Notes
1. The use of DTT in the SBP buffer can reduce the Ni-NTA resin (seen as a change from blue to orange). BME can be substituted for DTT to avoid this.
2. Prep-HPLC may be limited by similar retention times of the siderophore of interest and competing siderophore and the large excess of competing siderophore limits injection volumes and peak resolution.
3. Siderophore-affinity chromatography can be performed on both metal-free and metal-chelated forms of siderophores.

6. Synthesis of DFO-NBD and Fe(III)-siderophores
6.1 Equipment
- Magnetic stir/hot plate
- Oil bath

Siderophore-dependent ferrichelatases 303

- Round bottom flask
- Rotary evaporator with vacuum pump (BUCHI)
- Prep-HPLC (Hewlett Packard series 1050, Kinetex 5 μm C18 100 Å LC column 150 × 21.2 mm with guard column)

6.2 Reagents
- Desferrioxamine B (DFOB) mesylate
- Solvents: MeOH, dichloromethane, acetonitrile, diethyl ether, acetone
- $NaHCO_3$
- 4-Chloro-7-nitro-1,2,3-benzoxadiazole (NBD-Cl)
- Dry ice
- Silica gel
- Iron free siderophore
- $Fe(acac)_3$

6.3 Procedure: synthesis of DFO-NBD
1. Dissolve DFOB mesylate (656 mg, 1 mmol) in MeOH (10 mL) in a round bottom flask.
2. Add 0.1 M $NaHCO_3$ (20 mL) and NBD-Cl (200 mg, 1 mmol) in MeOH (7.5 mL) (Fig. 8A).
3. Heat the reaction at 65 °C for 1 h.
4. Concentrate via rotary evaporation under reduced pressure.
5. Purify the product (DFOB-NBD) by RP-C18 prep-HPLC (5% B to 95% B over 20 min).

   Or

   Purify by silica gel column chromatography (10% methanol in dichloromethane) followed by recrystallization in methanol/acetonitrile.
6. Confirm product purity by LC–MS and NMR according to published data (Endicott et al., 2020).

6.4 Procedure: synthesis of iron bound siderophores
1. Dissolve the iron free siderophore (DFOB or DFOB-NBD) in MeOH at 40 °C (2 mM) in a round bottom flask.
2. Add 1.1eq of $Fe(acac)_3$ to the siderophore and stir for 2 h. The solution should be a clear orange (Fig. 8B).
3. Remove the MeOH by rotary evaporation under reduced pressure to yield an orange film.
4. Dissolve the crude product in a minimal volume of MeOH and precipitate by addition of Et_2O. Triturate with Et_2O.

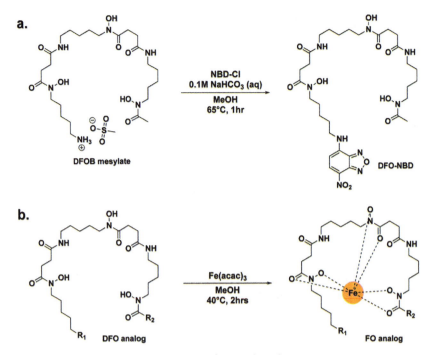

Fig. 8 Synthesis of (A) DFO-NBD and (B) ferric siderophores.

5. Dry under vacuum to yield the siderophore-Fe(III) complex as an orange powder.
6. Confirm compound purity and structure of products (FOB or FOB-NBD) by LC–MS and optical absorbance (λ_{max} ~430 nm).

6.5 Notes
1. In the synthesis and handling of iron free siderophores, washing all glassware with acid or EDTA can minimize iron contamination. Recommended to soak glassware in 1 M HCl for several hours followed by thorough washing with distilled/deionized H_2O.

7. Ferrichelatase assay
7.1 Equipment
- Nanodrop spectrophotometer (NanoDrop 2000c spectrophotometer)
- Analysis software
- Fluorimeter (PerkinElmer LS 55 Luminescence Spectrometer) or plate reader (Tecan Infinite M Plex)

Siderophore-dependent ferrichelatases

- Blacked out quartz cuvette or well plate (Hellma; flat bottom)
- LC–MS (Agilent 6130 quadrupole with G1329A autosampler, G1311A quaternary pump, G1322A degasser, G1316A thermostatted column compartment, G1315 diode array detector, Gemini 5 µm NX-C18 110 Å LC column 50 × 4.6 mm with guard column)

7.2 Buffers and reagents

- HEPES buffer (10 mM HEPES, 100 mM KCl, 600 mM NaCl, pH 7.4)
- Purified recombinant FhuD2
- Iron free siderophore (DFO-NBD)
- Fe(III) source (FOB, FOE, Fe(III)-Dan, or ferric *holo*-transferrin)
- Quinoline

7.3 Procedure: fluorescence based FhuD2 iron exchange assay

1. Thaw a FhuD2 protein bead at 4 °C. Perform an initial dilution with HEPES buffer to reach an $A_{280\,nm}$ of 1.0–1.5. Calculate the concentration using the molar extinction coefficient of FhuD2 ($\varepsilon_{280\,nm} = 55{,}350/(M\,cm)$; calculated using Expasy ProtParam (Duvaud et al., 2021)).
2. Prepare 2 mM stock solutions of DFO-NBD and Fe(III) sources (FOB, FOE, Fe(III)-Dan, or ferric *holo*-transferrin) in HEPES buffer.
3. Add DFO-NBD (2 µM final concentration) to a solution of Fe(III) source (0.5–4 µM final concentration) with or without FhuD2 (100 nM final concentration) in HEPES buffer for a working volume of 200 µL.
   a. For samples containing FhuD2, pre-equilibrate with the Fe(III) source for 15 min before adding DFO-NBD.
4. Scan fluorescence emission from 500 to 600 nm (10 nm emission slit) with a 475 nm excitation using a 100 nm/min scan speed. Take scans once per minute for 10 min then at the 15, 30, and 60 min time points.
5. Plot relative fluorescence intensity normalized to the decay of 2 µM DFO-NBD in HEPES and generate Michaelis–Menten plots using the linear portion of the data to determine velocity assuming steady-state (Fig. 9).
6. Create fluorescence difference plots with the percent change in fluorescence at each time point in the presence of FhuD2. The apparent inhibitory effect of iron free siderophores on the Fe(III) exchange from *holo*-transferrin to DFO-NBD can also be investigated in these conditions.

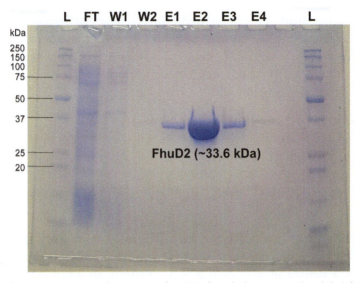

Fig. 9 Fluorescence quenching assay for SBP ferrichelatase activity. (A) Schematic showing FhuD2-catalyzed exchange of ferric ion from *holo*-transferrin to DFO-NBD. (B) Sample plot of relative fluorescence quenching at 560 nm with and without SBP added. (C) Sample Michaelis–Menten plot of steady-state kinetic data. *Data reproduced from Endicott, N. P., Rivera, G. S. M., Yang, J., & Wencewicz, T. A. (2020). Emergence of ferrichelatase activity in a siderophore-binding protein supports an iron shuttle in bacteria. ACS Central Science, 6(4), 493–506. https://doi.org/10.1021/acscentsci.9b01257.*

7.4 Procedure: validation by LC–MS

1. Thaw a FhuD2 protein bead at 4 °C. Perform an initial dilution with HEPES buffer to reach an $A_{280\,nm}$ of 1.0–1.5. Calculate the concentration using the molar extinction coefficient of FhuD2 ($\varepsilon_{280\,nm}$ = 55,350/(M cm); calculated using Expasy ProtParam (Duvaud et al., 2021)).
2. Add a solution of ferric holo-transferrin (50 μM final concentration) to iron free siderophore (50 μM final concentration) with or without FhuD2 (10 μM final concentration) in HEPES buffer for a working volume of 200 μL.
   a. For samples containing FhuD2, pre-equilibrate with ferric holo-transferrin for 15 min.
3. Analyze the solution by LC–MS (5%–95% B over 20 min, 3 min hold at 100% B, and re-equilibration to 5% B over 2 min) with a quinoline internal standard (25 μM final concentration).

4. Quantify [M+H]$^+$ ion counts corresponding to Fe(III) siderophores relative to the quinoline internal standard.

7.5 Notes
1. LC–MS quantification may need to be altered if the siderophore of interest does not ionize well in positive mode.
2. While DFOB and *holo*-transferrin are commercially available, FOE and danoxamine (Fe(III)-DanFe) were obtained through chemical and/or enzymatic synthesis (Endicott et al., 2017, 2020). The procedure is not limited to these iron sources. The experimentalist can choose virtually any iron source of interest for this assay.

8. Siderophore EDTA competition assay
8.1 Equipment
- Methacrylate cuvette
- UV–Vis spectrophotometer (Agilent Cary 60 UV–Vis)
- Analysis software

8.2 Buffers and reagents
- Fe(III)-siderophore (ferrioxamine type)
- HEPES buffer (10 mM HEPES, 600 mM NaCl, 100 mM KCl, pH 7.4)
- EDTA

8.3 Procedure
1. Dissolve the ferrioxamine siderophore-Fe(III) complex in HEPES buffer and transfer to a methacrylate cuvette to give an $A_{430\,nm}$ of ~0.1 to 0.2.
2. Add 1.2eq EDTA for a final volume of 1 mL. Scan optical absorbance at 430 nm continuously (1 scan/s) for 90 min or until equilibrium is reached. The $A_{430\,nm}$ should decrease over time with pseudo first-order kinetics.
3. Calculate Fe(III) exchange $t_{1/2}$ by fitting the exponential decay plot (Fig. 10) to a pseudo first-order rate equation.
4. Calculate apparent K_{Fe} values using the following equations:
    a. $K_L = \frac{[FeL]}{[Fe^{3+}][L]}$ for the equilibrium $[Fe^{3+}] + [L] \rightleftharpoons [FeL]$
    b. $K_{FeEDTA} = \frac{[FeEDTA]}{[Fe^{3+}][EDTA]}$ for the equilibrium $[Fe^{3+}] + [EDTA] \rightleftharpoons [FeEDTA]$
       Where $K_{FeEDTA} = 10^{25.1}$ (Martell & Smith, 1974)

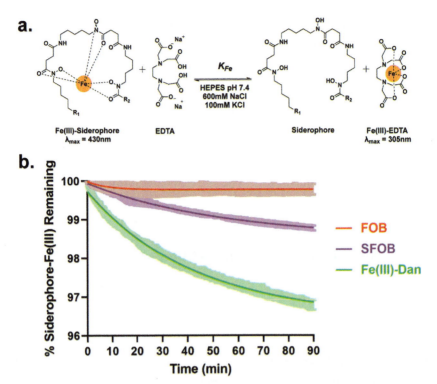

Fig. 10 Determination of apparent siderophore ferric ion stability constants. (A) General equilibrium of ferric ion exchange between a ferrioxamine siderophore analogue and EDTA. The error (standard deviation) for three independent trials is represented by the shaded regions. (B) Sample decay plots of siderophore-Fe(III) complexes FOB, N-succinyl-FOB (SFOB), and Fe(III)-danoxamine (Dan). *Data reproduced from Endicott, N. P., Lee, E., & Wencewicz, T. A. (2017). Structural basis for xenosiderophore utilization by the human pathogen Staphylococcus aureus. ACS Infectious Diseases, 3(7), 542–553. https://doi.org/10.1021/acsinfecdis.7b00036.*

c. $K_{Exchange} = \dfrac{K_L}{K_{FeEDTA}}$ for the equilibrium $[FeEDTA] + [L] \rightleftharpoons [FeL] + [EDTA]$

d. $K_{Exchange} = \dfrac{[FeL][EDTA]}{[FeEDTA][L]}$

e. $\Delta = \dfrac{Abs_{FeL} - Abs_{FeL+EDTA}}{\varepsilon_L}$ where $\varepsilon_L = 3000 M^{-1} cm^{-1}$ for siderophore-Fe(III) complexes (Hider & Kong, 2010)

f. $K_L = K_{FeEDTA} \times \dfrac{[FeL][EDTA]}{[FeEDTA][L]}$

g. $[FeL] = \dfrac{Abs_{FeL}}{\varepsilon_L}$

h. $[EDTA] = [EDTA]_T - \Delta$ where $[EDTA]_T$ = total EDTA added

i. $[FeEDTA] = \Delta$
   j. $[L] = \Delta$
   k. $K_{Fe} = $ apparent K_L

8.4 Optional steps
1. A stoichiometric concentration of SBP can be included in these assays to determine the effect of SBP binding on the thermodynamic stability of siderophore ferric ion complexes.

8.5 Notes
1. Siderophores with a high apparent K_{Fe} may not display significant changes in absorbance at 430 nm during the 90 min window. In these instances, the assay window and/or EDTA concentration can be increased as needed. The cuvette should be sealed to minimize evaporation if using an extended observation time. Conventional measurements are made a room temperature and physiological pH, but variable temperatures and pH may be employed.

9. S. aureus growth studies
9.1 Equipment
- Autoclave (Steris AMSCO 430LS Steam Sterilizer Model LS-1262)
- Rotating drum in an incubator (Thermo Scientific Heratherm Incubator)
- Sterile filter
- Nanodrop spectrophotometer (NanoDrop 2000c spectrophotometer)
- Plate reader (SpectraMax Plus 384 Microplate Reader or Tecan Infinite M Plex)
- 96-well plate (round bottom, polystyrene) and lid
- Calibrated Micropipettes
- Biosafety Cabinet (LabGard Class II, Type A2)

9.2 Strains and reagents
- Luria broth (10 g/L casein digest peptone, 10 g/L sodium chloride, 5 g/L yeast extract)
- Glycerol stock of S. aureus ATCC 11632 (biosafety level 2)
- Fe-deficient TMS media (12.1 g/L Tris-base, 16.6 g/L succinic acid, 10 g/L casamino acids, 40 mL salt solution [145 g NaCl, 92.5 g KCl, 27.5 g NH_4Cl, 3.55 g cysteine, 4.23 g thiamine, 0.31 g nicotinic acid,

2.5 mg biotin, 23.9 g MgCl$_2$, and 2.78 g of CaCl$_2$ in 100 mL H$_2$O], 200 µM 2,2′-dipyridyl, pH 7.4) (Madsen, Johnstone, & Nolan, 2015)

9.3 Procedure

1. Autoclave Luria broth at 121 °C for 15 min and allow the solution to cool to room temperature.
2. Prepare an overnight culture by adding a sterile loop of *S. aureus* ATCC 11632 from a frozen glycerol stock to 5 mL of sterile Luria broth. Allow the bacteria to grow overnight at 37 °C on a rotating drum.
3. Prepare Fe-deficient TMS media and filter sterilize.
4. Prepare a 0.5 McFarland standard (OD of ~0.8 to 1 at 600 nm) in TMS media as the plate inoculum (Murray, Baron, Pfaller, Tenover, & Yolken, 1999).
5. Fill each well of a sterile 96-well plate with 50 µL TMS media. Add 50 µL of siderophore in TMS media to the first column of the plate for a final concentration of 2560 µM.
6. Perform a two-fold serial dilution down the length of the plate.
7. Treat each well with 50 µL of inoculum for a final siderophore concentrations of 1280–1.25 µM and a working volume of 100 µL. Cover the plate with a sterile lid and continuously measure the OD at 600 nm for 90 h at 37 °C with agitation before measurement (Fig. 11). Utilize TMS media as a blank.

9.4 Notes

1. This protocol assays the ability of a given siderophore to be utilized by *S. aureus* and rescue growth in iron restricted medium.
2. If using a biosafety level 2 strain of bacteria, then all strain transfers and inoculations should be carried out in a certified biosafety cabinet. General lab protocols for the handling of biosafety level 2 pathogens should be followed. Labs working with biosafety level 2 pathogens should obtain proper approval from the institution's biosafety oversight committee.
3. The defined TMS growth medium is suitable only for *S. aureus*. To adapt this protocol to other strains of interest an appropriate growth medium will need to be utilized.
4. If insufficient bacterial growth is observed, the concentration of 2,2′-dipyridyl should be adjusted (decrease the concentration if more growth is desired). The amount of 2,2′-dipyridyl needed to induce iron deficiency depends on the iron content of the water source and trace metals

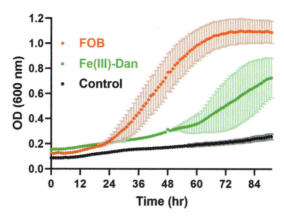

Fig. 11 Sample growth curves demonstrating the recovery of *Staphylococcus aureus* ATCC 11632 growth in iron deficient TMS minimal media containing 200 μM 2,2′-dipyridyl by ferric siderophore complex supplementation. The error (standard deviation) for three independent trials is represented by the shaded regions. *Data reproduced from Endicott, N. P., Lee, E., & Wencewicz, T. A. (2017). Structural basis for xenosiderophore utilization by the human pathogen Staphylococcus aureus. ACS Infectious Diseases, 3(7), 542–553. https://doi.org/10.1021/acsinfecdis.7b00036.*

in the ingredients used to prepare the growth medium. If necessary, a titration experiment with 2,2′-dipyridyl can be performed to identify optimal concentration for desired growth phenotype. ICP-MS can also be used to standardize the iron content of growth media, but it is recommended to optimize 2,2′-dipyridyl based on observed bacterial growth rates rather than a measurement of iron concentrations.

10. Summary and conclusions

This chapter outlines the methodologies utilized to isolate recombinant FhuD2, characterize its ligand binding and ferrichelatase properties, and apply this knowledge to design SBP affinity chromatography. N-His$_6$-FhuD2 was overexpressed in *E. coli* using a pET28a-*fhuD2* vector and purified by Ni-NTA affinity chromatography. The apparent K_d of FhuD2 for various substrates has been determined by intrinsic tryptophan fluorescence quenching. FhuD2's ferrichelatase activity was demonstrated by DFO-NBD fluorescence quenching and LC–MS validation. Immobilization of FhuD2 on an agarose resin provided a unique, complementary methodology to assess ligand affinity which can also be utilized as a purification technique for SBP substrates. General methods for the preparation of siderophore substrates, the assaying of their iron binding affinity, and their ability to rescue *S. aureus* growth in iron restricted media are also discussed. These studies are important to validate siderophore activity determined in vitro given that cellular features introduce variables that determine siderophore selectivity. For example, *S. aureus* utilizes ferrioxamine siderophores that carry a net positive or neutral charge preferentially over anionic siderophores presumably due to electrostatic repulsion of anionic molecules by the anionic surface of the *S. aureus* cell envelope (Endicott et al., 2017). Combined, FhuD2 possesses ferrichelatase activity and binds to a broad scope of hydroxamate siderophores. The methodologies described here may be facilely adapted to the study of other SBPs to assess substrate pools, apparent K_d's, and ferrichelatase activity.

Acknowledgments

We would like to thank the National Science Foundation (NSF CAREER-1654611 Award to TAW), Alfred P. Sloan Foundation (Sloan Research Fellowship to TAW), Research Corporation for Science Advancement (Cottrell Scholars Award to TAW), The Camille and Henry Dreyfus Foundation (Camille Dreyfus Teacher-Scholar Award to TAW), Oak Ridge Associated Universities (Ralph E. Powe Junior Faculty Enhancement Award to TAW), and the Incubator for Transdisciplinary Futures (WUSTL College of Arts & Sciences) for funding and support. This material is based upon work supported by the National Science Foundation Graduate Research Fellowship Program under Grant No. DGE 2139839 to CEM. Any opinions, findings, and conclusions or recommendations expressed in this material are those of the author(s) and do not necessarily reflect the views of the National Science Foundation.

References

Arifin, A. J., Hannauer, M., Welch, I., & Heinrichs, D. E. (2014). Deferoxamine mesylate enhances virulence of community-associated methicillin resistant *Staphylococcus aureus*. *Microbes and Infection, 16*(11), 967–972. https://doi.org/10.1016/j.micinf.2014.09.003.

Bacconi, M., Haag, A. F., Chiarot, E., Donato, P., Bagnoli, F., Delany, I., et al. (2017). In vivo analysis of *Staphylococcus aureus*-infected mice reveals differential temporal and spatial expression patterns of fhuD2. *Infection and Immunity, 85*(10), https://doi.org/10.1128/IAI.

Bachman, J. (2013). *Site-directed mutagenesis. Methods in enzymology, Vol. 529*, Academic Press Inc, 241–248. https://doi.org/10.1016/B978-0-12-418687-3.00019-7.

Bar-Ness, E., Hadar, Y., Chen, Y., Shanzer, A., & Libman, J. (1992). Iron uptake by plants from microbial siderophores' a study with 7-nitrobenz-2 oxa-1,3-diazole-desferrioxamine as fluorescent ferrioxamine B analog. *Plant Physiology, 99*, 1329–1335. https://academic.oup.com/plphys/article/99/4/1329/6088356.

Beasley, F. C., Marolda, C. L., Cheung, J., Buac, S., & Heinrichs, D. E. (2011). *Staphylococcus aureus* transporters Hts, Sir, and Sst capture iron liberated from human transferrin by staphyloferrin A, staphyloferrin B, and catecholamine stress hormones, respectively, and contribute to virulence. *Infection and Immunity, 79*(6), 2345–2355. https://doi.org/10.1128/IAI.00117-11.

Beasley, F. C., Vinés, E. D., Grigg, J. C., Zheng, Q., Liu, S., Lajoie, G. A., et al. (2009). Characterization of staphyloferrin A biosynthetic and transport mutants in *Staphylococcus aureus*. *Molecular Microbiology, 72*(4), 947–963. https://doi.org/10.1111/j.1365-2958.2009.06698.x.

Berman, H. M., Westbrook, J., Feng, Z., Gilliland, G., Bhat, T. N., Weissig, H., et al. (2000). The protein data bank. *Nucleic Acids Research, 28*(1), http://www.rcsb.org/pdb/status.html.

Cabrera, G., Xiong, A., Uebel, M., Singh, V. K., & Jayaswal, R. K. (2001). Molecular characterization of the iron-hydroxamate uptake system in *Staphylococcus aureus*. *Applied and Environmental Microbiology, 67*(2), 1001–1003. https://doi.org/10.1128/AEM.67.2.1001-1003.2001.

Cain, T. J., & Smith, A. T. (2021). Ferric ion reductases and their contribution to unicellular ferrous iron uptake. *Journal of Inorganic Biochemistry, 218*, 111407. https://doi.10.1016/j.jinorgbio.2021.111407.

Clancy, A., Loar, J. W., Speziali, C. D., Oberg, M., Heinrichs, D. E., & Rubens, C. E. (2006). Evidence for siderophore-dependent iron acquisition in group B streptococcus. *Molecular Microbiology, 59*(2), 707–721. https://doi.org/10.1111/j.1365-2958.2005.04974.x.

Clarke, T. E., Braun, V., Winkelmann, G., Tari, L. W., & Vogel, H. J. (2002). X-ray crystallographic structures of the *Escherichia coli* periplasmic protein FhuD bound to hydroxamate-type siderophores and the antibiotic albomycin. *Journal of Biological Chemistry, 277*(16), 13966–13972. https://doi.org/10.1074/jbc.M109385200.

Duvaud, S., Gabella, C., Lisacek, F., Stockinger, H., Ioannidis, V., & Durinx, C. (2021). Expasy, the Swiss bioinformatics resource portal, as designed by its users. *Nucleic Acids Research, 49*(W1), W216–W227. https://doi.org/10.1093/nar/gkab225.

Endicott, N. P., Lee, E., & Wencewicz, T. A. (2017). Structural basis for xenosiderophore utilization by the human pathogen *Staphylococcus aureus*. *ACS Infectious Diseases, 3*(7), 542–553. https://doi.org/10.1021/acsinfecdis.7b00036.

Endicott, N. P., Rivera, G. S. M., Yang, J., & Wencewicz, T. A. (2020). Emergence of ferrichelatase activity in a siderophore-binding protein supports an iron shuttle in bacteria. *ACS Central Science, 6*(4), 493–506. https://doi.org/10.1021/acscentsci.9b01257.

FDA. (2022). Drugs@FDA (*016267 Deferoxamine Mesylate*). FDAgov.

Fukushima, T., Allred, B. E., & Raymond, K. N. (2014). Direct evidence of iron uptake by the Gram-positive siderophore-shuttle mechanism without iron reduction. *ACS Chemical Biology, 9*(9), 2092–2100. https://doi.org/10.1021/cb500319n.

Fukushima, T., Allred, B. E., Sia, A. K., Nichiporuk, R., Andersen, U. N., & Raymond, K. N. (2013). Gram-positive siderophore-shuttle with iron-exchange from Fe-siderophore to apo-siderophore by *Bacillus cereus* YxeB. *Proceedings of the National Academy of Sciences of the United States of America, 110*(34), 13821–13826. https://doi.org/10.1073/pnas.1304235110.

Grim, K. P., San Francisco, B., Radin, J. N., Brazel, E. B., Kelliher, J. L., Párraga Solórzano, P. K., et al. (2017). The metallophore staphylopine enables *Staphylococcus aureus* to compete with the host for zinc and overcome nutritional immunity. *MBio, 8*(5), https://doi.org/10.1128/mBio.01281-17.

Gwinner, T. (2008). Das siderophore-antibiotikum salmycin. In *Biologie*. Tübingen: Eberhard Karls Universitat Tübingen.

Hider, R. C., & Kong, X. (2010). Chemistry and biology of siderophores. *Natural Product Reports, 27*(5), 637–657. https://doi.org/10.1039/b906679a.

Holden, V. I., & Bachman, M. A. (2015). Diverging roles of bacterial siderophores during infection. *Metallomics, 7*(6), 986–995. https://doi.org/10.1039/c4mt00333k.

Jumper, J., Evans, R., Pritzel, A., Green, T., Figurnov, M., Ronneberger, O., et al. (2021). Highly accurate protein structure prediction with AlphaFold. *Nature, 596*(7873), 583–589. https://doi.org/10.1038/s41586-021-03819-2.

Koblan, L. W., Doman, J. L., Wilson, C., Levy, J. M., Tay, T., Newby, G. A., et al. (2018). Improving cytidine and adenine base editors by expression optimization and ancestral reconstruction. *Nature Biotechnology, 36*(9), 843–848. https://doi.org/10.1038/nbt.4172.

Kontoghiorghes, G. J., Kolnagou, A., Skiada, A., & Petrikkos, G. (2010). The role of iron and chelators on infections in iron overload and non iron loaded conditions: Prospects for the design of new antimicrobial therapies. *Hemoglobin, 34*(3), 227–239. https://doi.org/10.3109/03630269.2010.483662.

Lecerof, D., Fodje, M., Hansson, A., Hansson, M., & Al-Karadaghi, S. (2000). Structural and mechanistic basis of porphyrin 27etalation by ferrochelatase. *Journal of Molecular Biology, 297*(1), 221–232. https://doi.org/10.1006/jmbi.2000.3569.

Madeira, F., Pearce, M., Tivey, A. R. N., Basutkar, P., Lee, J., Edbali, O., et al. (2022). Search and sequence analysis tools services from EMBL-EBI in 2022. *Nucleic Acids Research, 50*(W1), W276–W279. https://doi.org/10.1093/nar/gkac240.

Madsen, J. L. H., Johnstone, T. C., & Nolan, E. M. (2015). Chemical synthesis of staphyloferrin B affords insight into the molecular structure, iron chelation, and biological activity of a polycarboxylate siderophore deployed by the human pathogen *Staphylococcus aureus. Journal of the American Chemical Society, 137*(28), 9117–9127. https://doi.org/10.1021/jacs.5b04557.

Mariotti, P., Malito, E., Biancucci, M., Lo Surdo, P., Mishra, R. P. N., Nardi-Dei, V., et al. (2013). Structural and functional characterization of the *Staphylococcus aureus* virulence factor and vaccine candidate FhuD2. *Biochemical Journal, 449*(3), 683–693. https://doi.org/10.1042/BJ20121426.

Martell, A. E., & Smith, R. M. (1974). *Critical stability constants, Vol. 1.* New York, NY: Plenum Press.

Mishra, R. P. N., Mariotti, P., Fiaschi, L., Nosari, S., MacCari, S., Liberatori, S., et al. (2012). *Staphylococcus aureus* FhuD2 is involved in the early phase of staphylococcal dissemination and generates protective immunity in mice. *The Journal of Infectious Diseases, 206*(7), 1041–1049. https://doi.org/10.1093/INFDIS/JIS463.

Murray, P. R., Baron, E. J., Pfaller, M. A., Tenover, F. C., & Yolken, R. H. (1999). *Manual of clinical microbiology* (7th ed.). Washington, DC: American Society for Microbiology.

Podkowa, K. J., Briere, L. A. K., Heinrichs, D. E., & Shilton, B. H. (2014). Crystal and solution structure analysis of FhuD2 from *Staphylococcus aureus* in multiple unliganded conformations and bound to ferrioxamine-B. *Biochemistry, 53*(12), 2017–2031. https://doi.org/10.1021/bi401349d.

Rahav, G., Volach, V., Shapiro, M., Rund, D., Rachmilewitz, E. A., & Goldfarb, A. (2006). Severe infections in thalassaemic patients: Prevalence and predisposing factors. *British Journal of Haematology, 133*(6), 667–674. https://doi.org/10.1111/j.1365-2141.2006.06082.x.

Rivera, G. S. M., Beamish, C. R., & Wencewicz, T. A. (2018). Immobilized FhuD2 siderophore-binding protein enables purification of salmycin sideromycins from *Streptomyces violaceus* DSM 8286. *ACS Infectious Diseases, 4*(5), 845–859. https://doi.org/10.1021/acsinfecdis.8b00015.

Ronan, J. L., Kadi, N., McMahon, S. A., Naismith, J. H., Alkhalaf, L. M., & Challis, G. L. (2018). Desferrioxamine biosynthesis: Diverse hydroxamate assembly by substrate-tolerant acyl transferase DesC. *Philosophical Transactions of the Royal Society B: Biological Sciences, 373*(1748), https://doi.org/10.1098/rstb.2017.0068.

Sebulsky, M. T., & Heinrichs, D. E. (2001). Identification and characterization of fhuD1 and fhuD2, two genes involved in iron-hydroxamate uptake in *Staphylococcus aureus. Journal of Bacteriology, 183*(17), 4994–5000. https://doi.org/10.1128/JB.183.17.4994-5000.2001.

Sebulsky, M. T., Shilton, B. H., Speziali, C. D., & Heinrichs, D. E. (2003). The role of FhuD2 in iron(III)-hydroxamate transport in *Staphylococcus aureus*: Demonstration that FhuD2 binds iron(III)-hydroxamates but with minimal conformational change and implication of mutations on transport. *Journal of Biological Chemistry, 278*(50), 49890–49900. https://doi.org/10.1074/jbc.M305073200.

Shen, Y., & Ryde, U. (2005). Reaction mechanism of porphyrin metallation studied by theoretical methods. *Chemistry – A European Journal, 11*(5), 1549–1564. https://doi.org/10.1002/chem.200400298.

Song, L., Zhang, Y., Chen, W., Gu, T., Zhang, S. Y., & Ji, Q. (2018). Mechanistic insights into staphylopine-mediated metal acquisition. *Proceedings of the National Academy of Sciences of the United States of America, 115*(15), 3942–3947. https://doi.org/10.1073/pnas.1718382115.

Speziali, C. D., Dale, S. E., Henderson, J. A., Vinés, E. D., & Heinrichs, D. E. (2006). Requirement of *Staphylococcus aureus* ATP-binding cassette-ATPase FhuC for iron-restricted growth and evidence that it functions with more than one iron transporter. *Journal of Bacteriology, 188*(6), 2048–2055. https://doi.org/10.1128/JB.188.6.2048-2055.2006.

Sriyosachati, S., & Cox, C. D. (1986). Siderophore-mediated iron acquisition from transferrin by *Pseudomonas aeruginosa*. *Infection and Immunity, 52*(3), 885–891. https://journals.asm.org/journal/iai.

Stintzi, A., Barnes, C., Xu, J., & Raymond, K. N. (2000). Microbial iron transport via a siderophore shuttle: A membrane ion transport paradigm. *Proceedings of the National Academy of Sciences of the United States of America, 97*(20), 10691–10696. www.pnas.orgcgidoi10.1073pnas.200318797.

Teufel, F., Almagro Armenteros, J. J., Johansen, A. R., Gíslason, M. H., Pihl, S. I., Tsirigos, K. D., et al. (2022). SignalP 6.0 predicts all five types of signal peptides using protein language models. *Nature Biotechnology, 40*(7), 1023–1025. https://doi.org/10.1038/s41587-021-01156-3.

Torres, V. J., Attia, A. S., Mason, W. J., Hood, M. I., Corbin, B. D., Beasley, F. C., et al. (2010). *Staphylococcus aureus* fur regulates the expression of virulence factors that contribute to the pathogenesis of pneumonia. *Infection and Immunity, 78*(4), 1618–1628. https://doi.org/10.1128/IAI.01423-09.

Varadi, M., Anyango, S., Deshpande, M., Nair, S., Natassia, C., Yordanova, G., et al. (2022). AlphaFold protein structure database: Massively expanding the structural coverage of protein-sequence space with high-accuracy models. *Nucleic Acids Research, 50*(D1), D439–D444. https://doi.org/10.1093/nar/gkab1061.

Wang, M., Lai, T. P., Wang, L., Zhang, H., Yang, N., Sadler, P. J., et al. (2015). "Anion clamp" allows flexible protein to impose coordination geometry on metal ions. *Chemical Communications, 51*(37), 7867–7870. https://doi.org/10.1039/c4cc09642h.

Waterhouse, A. M., Procter, J. B., Martin, D. M. A., Clamp, M., & Barton, G. J. (2009). Jalview Version 2-A multiple sequence alignment editor and analysis workbench. *Bioinformatics (Oxford, England), 25*(9), 1189–1191. https://doi.org/10.1093/bioinformatics/btp033.

CHAPTER FOURTEEN

Native metabolomics for mass spectrometry-based siderophore discovery

Marquis T. Yazzie[a] ⓘ **, Zachary L. Reitz[b]** ⓘ **, Robin Schmid[c]** ⓘ **, Daniel Petras[d,e]** ⓘ **, and Allegra T. Aron[a,*]** ⓘ

[a]Department of Chemistry and Biochemistry, University of Denver, Denver, CO, United States
[b]Department of Ecology, Evolution and Marine Biology, University of California, Santa Barbara, CA, United States
[c]Institute of Organic Chemistry and Biochemistry, Czech Academy of Sciences, Prague, Czechia
[d]Department of Biochemistry, University of California Riverside, Riverside, CA, United States
[e]Interfaculty of Microbiology and Infection Medicine, University of Tübingen, Tübingen, Germany
[*]Corresponding author. e-mail address: allegra.aron@du.edu

Contents

1. Introduction	318
2. General methods and assembly of instrumentation	322
3. Sample preparation for native metabolomics	325
3.1 Materials and equipment	326
3.2 Step-by-step method details	326
3.3 Alternative methods/procedures	327
4. Mass spectrometry acquisition parameters	328
4.1 Materials and equipment	328
4.2 LC-MS^2 acquisition parameters	329
4.3 Post-column pH adjustment	330
4.4 Post-column metal infusion	332
5. Data analysis using ion identity molecular networking	333
5.1 Materials and equipment	334
5.2 Feature finding in mzmine: step-by-step method details	334
5.3 Metal adduct specific modules: step-by-step method details	336
5.4 Mzmine compound annotation, molecular networking, data export for GNPS2	342
5.5 Visualization in GNPS, GNPS2 Cytoscape	343
5.6 Layering with other sources of info	345
6. Limitations and considerations	345
7. Summary and conclusions	347
Acknowledgments	347
References	347

Methods in Enzymology, Volume 702
ISSN 0076-6879, https://doi.org/10.1016/bs.mie.2024.07.001
Copyright © 2024 Elsevier Inc. All rights are reserved, including those for text and data mining, AI training, and similar technologies.

Abstract

Microorganisms, plants, and animals alike have specialized acquisition pathways for obtaining metals, with microorganisms and plants biosynthesizing and secreting small molecule natural products called siderophores and metallophores with high affinities and specificities for iron or other non-iron metals, respectively. This chapter details a novel approach to discovering metal-binding molecules, including siderophores and metallophores, from complex samples ranging from microbial supernatants to biological tissue to environmental samples. This approach, called *Native Metabolomics*, is a mass spectrometry method in which pH adjustment and metal infusion post-liquid chromatography are interfaced with ion identity molecular networking (IIMN). This rule-based data analysis workflow that enables the identification of metal-binding species based on defined mass (m/z) offsets with the same chromatographic profiles and retention times. Ion identity molecular networking connects compounds that are structurally similar by their fragmentation pattern and species that are ion adducts of the same compound by chromatographic shape correlations. This approach has previously revealed new insights into metal binding metabolites, including that yersiniabactin can act as a biological zincophore (in addition to its known role as a siderophore), that the recently elucidated lepotchelin natural products are cyanobacterial metallophores, and that antioxidants in traditional medicine bind iron. Native metabolomics can be conducted on any liquid chromatography-mass spectrometry system to explore the binding of any metal or multiple metals simultaneously, underscoring the potential for this method to become an essential strategy for elucidating biological metal-binding molecules.

1. Introduction

Metals are essential for life, serving as cofactors in biochemical reactions such as DNA replication and repair (iron and manganese), respiration (iron and copper), photosynthesis (iron and manganese), and biosynthesis of primary and secondary metabolites (for example, iron, zinc, vanadium, molybdenum, magnesium, and calcium) (Lippard & Berg, 1994). One common strategy for metal acquisition in microorganisms is the production, secretion, and processing of small-molecule metal chelators that form non–covalent complexes with metals (Hider & Kong, 2010). These complexes typically have a very high affinity for a given metal (Raymond, Dertz, & Kim, 2003; Timofeeva, Galyamova, & Sedykh, 2022), but may also bind metals promiscuously (Behnsen et al., 2021; Chaturvedi, Hung, Crowley, Stapleton, & Henderson, 2012; Koh et al., 2015). Siderophores are high–affinity chelators of ferric iron (Fe^{3+}), and these structures include ferrioxamines (deferoxamines B and E), catecholates (enterobactin), and carboxylates (rhizoferrin and aerobactin) (Hider & Kong, 2010). Like siderophores, metallophores are secondary metabolites secreted by microorganisms that enable the sequestration of metal ions from metal-deficient environments. There has been recent interest in discovering metallophores that bind non-iron metals or numerous metals promiscuously. For example, chalkophores such as methanobactins bind cuprous copper (Cu^+) (Kenney & Rosenzweig, 2018; Wang et al., 2017), zincophores like staphylopine, bacillopaline, pseudopaline, and yersinopine bind zinc (Zn^{2+}) (Behnsen et al., 2021; Łoboda & Rowińska-Żyrek, 2017; Morey & Kehl-Fie, 2020), and recently, the lanthanophore methylolanthanin has been observed to bind lanthanides (Zytnick et al., 2023).

Metallophores have various biological functions along with potential biomedical, energy, and ecological applications. Siderophores can modulate interactions between microorganisms (Endicott, Lee, & Wencewicz, 2017; Galdino et al., 2024; Santus et al., 2022; Strange, Zola, & Cornelissen, 2011; Zhu et al., 2020), shape interactions between microorganisms and their multicellular hosts (Pahari, Pradhan, Nayak, & Mishra, 2017; Qi & Han, 2018; Saha et al., 2020; Wang et al., 2024), and impact how organisms survive in different environments. Siderophores can provide a competitive advantage to pathogens but can also be protective against colonization by pathogens (Kramer, Özkaya, & Kümmerli, 2020). Given this, siderophores have been harnessed for medical applications, such as targeting pathogens through siderophore-antibiotic (Sassone-Corsi et al., 2016) and siderophore-antibody conjugates (Motz et al., 2024; Negash, Norris, & Hodgkinson, 2019; Pinkert et al., 2021) and treating iron overload conditions. Siderophores, metallophores, and the microorganisms that produce them have also been used for bioremediation of contaminated soil and water (Roskova, Skarohlid, & McGachy, 2022; Vijayaraj, Mohandass, & Joshi, 2020). Finally, siderophores and siderophore-producing bacteria can promote plant growth by supplying iron and other metals to plants (Lurthy et al., 2020; Pahari et al., 2017).

Traditionally, siderophores have been discovered by culturing organisms in iron-deficient medium compared to iron-rich medium to find differentially produced compounds. Among the first discovered siderophores were enterobactin and aerobactin, which were discovered from *Salmonella typhimurium* and *Aerobacter aerogenes 62–1* grown in low iron media, respectively (Gibson & Magrath, 1969; Pollack & Neilands, 1970). Other early siderophores were isolated from large-scale fermentations based on observations of colored media, such as in the case of ferrichrome produced by *Ustilago sphaerogena* (Neilands, 1967). New genome mining-based strategies that predict the presence of metallophores have emerged in recent years and have revolutionized siderophore discovery (Medema et al., 2011; Reitz & Medema, 2022; Reitz, Butler, & Medema, 2022). For example, genome mining can be used to find siderophore biosynthesis pathways generally homologous to known biosynthesis pathways (Butler et al., 2023; Carroll & Moore, 2018). However, predicting the selectivity and affinity of metal coordination sites is still difficult even with these tools, and a key drawback of tools is that they only predict metallophores related to known biosynthetic machinery. This can preclude the discovery of totally novel metallophores.

It remains challenging to detect and characterize siderophores from complex samples, including microbial supernatants, biological tissue (i.e., fecal, gastrointestinal tract, brain) and fluids (i.e., urine and serum), and environmental samples (i.e., soil, dissolved organic matter). Moreover, the only way to find siderophores produced by organisms that cannot be cultured in the laboratory is to find them in their native environment (D'Onofrio et al., 2010). A critical bottleneck arises from the limited strategies for assessing metal-binding preferences directly from these mixtures. Metal binding must be established experimentally and utilizes methods ranging from inductively coupled plasma-mass spectrometry (ICP-MS) (Boiteau, Fitzsimmons, Repeta, & Boyle, 2013; Dewey et al., 2023; Pluháček et al., 2016), atomic absorption spectroscopy (AAS) (Piper & Higgins, 1967), X-ray fluorescence spectroscopy (XRF) (Aschner et al., 2017), UV–visible absorption spectroscopy (Atkin et al., 1970; Neilands, 1981), nuclear magnetic resonance (NMR) spectroscopy (Atkinson et al., 1998), and colorimetric assays such as the chrome azurol S (CAS) assay (Himpsl & Mobley, 2019), one of the most widely used techniques for assessing whether siderophores are present. While typically low throughput, these strategies have been essential in discovering and characterizing numerous siderophores. However, these strategies are most accurate when carried out on pure compounds. Mass spectrometry (MS) strategies have been some of the most successful approaches to discovering siderophores from complex samples. Strategies include using characteristic metal isotope fingerprints (Baars et al., 2014), mass defect/Kendrick analysis (Pluháček et al., 2016), and hyperfine splitting isotope patterns (Walker et al., 2017). The development of software packages such as ChelomEx (Baars et al., 2014, 2016) and ion identity networking modules in mzmine (Pluskal et al., 2010; Schmid et al., 2021, 2023) have more recently enabled siderophore screening strategies. Multimodal approaches have also been employed that combine matrix-assisted laser desorption/ionization–Fourier transform ion cyclotron resonance–MS imaging (MALDI-FT-ICR MSI) (Perry et al., 2020) with LC–electrospray ionization–MS (LC-ESI-MS)/LC-ICP-MS-based methods.

This chapter presents and details a novel approach to discovering metallophores from complex samples using mass spectrometry coupled to a post-column pH modulation and metal infusion, which we call *native metabolomics*. In this approach, experimental modifications are interfaced with ion identity molecular networking (IIMN), a rule-based data analysis workflow that identifies metal-binding species based on defined mass (m/z) offsets with the same chromatographic profiles and retention times (Fig. 1A). This method works by first establishing an optimal pH for metal binding (see Section 4.3) and then infusing in metal salt (see Section 4.4). While siderophores and metallophores bind a specific metal with a high affinity, this binding may not occur without metals in extraction and chromatography solvents. Moreover, even if these metal-binding metabolites were previously bound to metals in their native environment, the low pH and high organic solvent conditions of liquid chromatography can destroy metal complexation (Lopes, Stark, Hong, Gates, & Staunton, 2001; Ross, Ikonomou, & Orians, 2000; Waska, Koschinsky, & Dittmar, 2016). In fact, when seven siderophores were inspected in public data available in the GNPS-MassIVE repository, we found that more than 70% of files contained the proton-bound adduct and more than 50% of files contained exclusively the proton-bound adduct. Given this, a post-LC pH adjustment can be employed to control the analysis pH for optimal metal binding (Aron et al., 2022). This strategy was motivated by native mass spectrometry in proteomics, which uses neutral pH and a given ionic strength to preserve the structure of intact biomolecules in their native structural form (Karch, Snyder, Harvey, & Wysocki, 2022; Tamara, den Boer, & Heck, 2022). Adjusting pH after

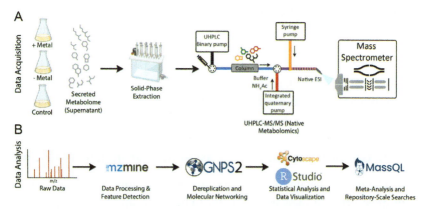

Fig. 1 Overview schematic. (A) Workflow begins with cultivation in high and low metal conditions followed by solid phase extraction of supernatant, UHPLC-MS2 with post-LC pH modulation and metal infusion. (B) A detailed schematic of our data analysis pipeline. Figure was created in part with Biorender.com.

chromatography instead of altering chromatography methods can preserve optimal peak shapes for data downstream analysis (detailed in Section 5). By re-establishing metal binding that may have been lost during processing and analysis, this method enables unbiased discovery by pairing these experimental steps with a computational workflow. IIMN (Section 5), performed in mzmine and GNPS2 and visualized in Cytoscape, connects correlated features that differ by a user-specified, metal-specific mass delta. These connected features are demarcated by an ion identity edge that is indicative of a metal-bound species (Fig. 1B).

This is a generalizable approach for finding metal-binding molecules. It can be performed on any sample type, ranging from a pure standard to complex microbial supernatants to increasingly complex matrices such as fecal or dissolved organic matter ocean samples. Any liquid chromatography-mass spectrometry system can be used in this workflow. Native electrospray metabolomics can be used to explore the binding of any metal or even mixtures of metals. This method has been used to discover biologically relevant binding of siderophores to non-iron metals, as in the case of yersiniabactin produced by *Escherichia coli* Nissle (Behnsen et al., 2021). In addition to yersiniabactin, applying this method to *E. coli* Nissle supernatant also found that several yersiniabactin truncations (such as HPTzTn-COOH) and derivatives bind zinc. This method has been used to prescribe function to novel natural products, including the recently elucidated leptochelins, which exhibit promiscuous metal-binding, copper detoxification properties, and enhanced production during iron deficiency (Avalon et al., 2024). Finally, this method has also been used to better understand complex extracts, such as the traditional Samoan traditional medicine "matalafi" by characterizing iron-binding compounds in the extracts (Molimau-Samasoni et al., 2021). We are pursuing this approach to find uncharacterized siderophores and other metal-binding molecules in human, phyllosphere, and rhizosphere microbiomes.

2. General methods and assembly of instrumentation

The native metabolomics workflow was developed on a Q-Exactive quadrupole orbitrap mass spectrometer (Thermo Fisher Scientific) with an integrated Vanquish binary ultra-high-performance liquid chromatography (UHPLC) system and Vanquish make-up pump system, but this method can easily be established on any other high-resolution LC-MS/MS system if an additional HPLC and syringe pump are available. The mass spectrometry acquisition parameters for this instrument are described in the text and in Tables 1 and 2. If the reader has access to a high-resolution (U)HPLC-MS2, the parameters described here can be closely followed. If the user can access another high-resolution UHPLC-MS2, such QTOF systems, the LC-MS2 parameters must be altered, but it is still possible to run this workflow. The workflow is described for a system in which a binary, make-up, and syringe pump have been fully integrated into the LC-MS2 platform, where the binary pump delivers the LC gradient, and the make-up pump delivers a consistent flow of ammonium acetate buffer, and the syringe pump delivers the metal solution (Fig. 2). We recognize that the readers may not have access to a fully integrated system, so we provide alternative workarounds. Readers with only one pump system can connect their mass spectrometer to an external HPLC system using PEEK™ tubing (0.005-in ID × 1/16-in. OD, red) or Viper tubing (0.13 mm ID × 750 mm L) through a T-joint connection (PEEK™ Tees and Crosses for High-Pressure HPLC Connections). This external HPLC system can be controlled through a separate computer. Readers without access to an integrated syringe pump can connect an external syringe pump for metal infusion through a T-joint connection, and flow rate can be controlled directly using the syringe pump.

This workflow should also be suitable to run on an HPLC-MS2 (instead of a UHPLC). If the user only has MS1 capability (but no ability to acquire MS2 data), experimental data can be collected and ion identity correlation can be assessed in mzmine. Since molecular networking relies

Table 1 Liquid chromatography gradient for deferoxamine B example, supplied by a binary pump.

Time (min)	Flow (mL/min)	%B
0	0.200	5
6	0.200	50
10	0.200	99
13	0.200	99
13.1	0.200	5
16	0.200	5

This example was run on a 2.1 mm column diameter, as this is a routine diameter used in many laboratories. The quaternary pump delivers a consistent 0.200 mL/min flow of ammonium acetate buffer solution.

Table 2 MS1 and MS2 settings on the Thermo Q Exactive Orbitrap for deferoxamine B example.

Parameter: Full MS settings	Value
Injection volume	2–5 μL
Full MS	m/z 150–1500
Resolution at m/z 200	120,000
Automatic gain control (AGC)	1e6
Maximum ion injection time	100 ms
Parameter: HESI	**Value**
Sheath gas flow	50 AU
Auxiliary gas flow	15 AU
Spray voltage	3.5 kV
S-Lens RF	55%
Parameter: DDA MS2 Settings	**Value**
Top (N)	5
Resolution at m/z 200	15,000
AGC	5e5
Dynamic exclusion	5 s
Normalized collision energy (CE)	Stepped (25,35,45)
Maximum ion injection time	200 ms
Apex trigger	2–15 s

on MS2 spectral similarity, an experiment that only collects MS1 will lack the molecular networking component but can still reveal correlated features with metal-characteristic mass deltas of interest. This workflow is also possible with a low-resolution instrument, but low-resolution m/z would have to be used in the modules discussed in Section 5. The workflow is described using a C18 reverse phase column, which is frequently used in siderophore characterization, isolation, and separation (Aron et al., 2022; Baars et al., 2016; Boiteau et al., 2013), though is amenable to other chromatography types.

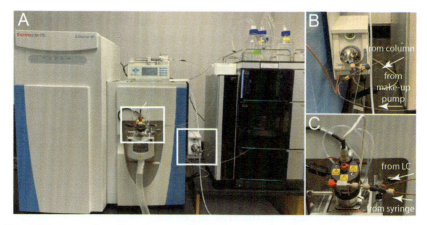

Fig. 2 Photographs of native metabolomics experimental setup. Photograph of the (A) Native metabolomics set-up. (B) T-joint connecting the make-up and binary pump. (C) T-joint connecting metal infusion by syringe pump into the LC flow.

Standard safety protocols for biosafety level 1 and analytical chemistry laboratories should be followed, including personal protective equipment and proper ventilation, waste disposal, and laboratory signage. It is recommended that all samples be analyzed in triplicate.

3. Sample preparation for native metabolomics

Sample preparation for native metabolomics experiments follows similar sample preparation protocols as those for MS-based metabolomics experiments. Chemical standards can be solubilized in methanol-water mixtures based on solubility, with most siderophores soluble in either 50% methanol-water mixture or 80% methanol-water mixture. One of the most common sample types run using the native metabolomics workflow is microbial supernatants because metallophores are secreted to scavenge metals from the extracellular environment. Microorganisms are cultured in iron-deficient medium. Iron-deficient minimal media can promote siderophore production, along with conditions that increase the cellular requirement for iron (Payne, 1994). These include using carbon sources like succinate and lactate as opposed to glucose. Modulating other parameters like increasing aeration during growth can also promote siderophore production, and are the subject of various reviews (Payne, 1994; Soares, 2022). Maximal siderophore production occurs in late log phase/early stationary phase, so microbial cultures should be grown to stationary phase. When growing microorganisms in metal-deficient media, it is important to address possible sources of metal contamination. This includes iron contamination in culture glassware, which should be washed with a 6M HCl acid solution before culturing. In the case of rare earth elements (including lanthanide metals), glassware can contain these elements from the manufacturing process (Hemmerle, Ochsner, Vonderach, Hattendorf, & Vorholt, 2021). When working with small molecules that bind rare earth elements, it is best to work with single-use plastic containers when possible.

For a single mass spectrometry sample, 35–60 mL of microbial supernatant (at optical density (OD) between 0.5 and 1) provides enough material when analyzed on a Q-Exactive. Native metabolomics can also work with significantly less material depending on production amount. Before following the steps outlined below, supernatant should be removed from cells either through filtration or centrifugation, then each sample should be lyophilized. The solid phase extraction protocol is outlined below.

3.1 Materials and equipment
- Optima LC-MS grade methanol (Thermo-Fisher Scientific)
- Optima LC-MS grade water (Thermo-Fisher Scientific)
- Solid-phase extraction manifold (Thermo-Fisher Scientific)
- Diaphram vacuum pump (Welch)
- HLB solid-phase extraction cartridges, 60 µm particle size, 3 mL volume, 200 mg sorbent (Chromabond or Oasis)
- Borosilicate glass disposable culture tubes (Fisher Scientific)
- Centrifugal vacuum concentrator (Eppendorf)
- Sulfamethazine and sulfadimethoxine internal standards (Sigma Aldrich)
- Mass spectrometry vials and caps (Thermo-Fisher Scientific)
- Mass spectrometry inserts (Thermo-Fisher Scientific)

Alternatives: Any solid-phase extraction manifold, diaphram vacuum pump, borosilicate culture tubes, and centrifugal vacuum concentrators can be used. Divinylbenzene stationary phases (such as HLB) have shown excellent recovery of siderophores (McCormack et al., 2003), but other stationary phases can also be used. Additionally, other compounds can be used as internal standards for extraction efficiency and analysis.

3.2 Step-by-step method details
1. Prepare wash and elution solutions for the solid-phase extraction. The wash solution is 1–3% LC-MS grade methanol in LC-MS grade water.

The elution solutions are 20–80% LC-MS grade methanol in LC-MS grade water and 100% LC-MS grade methanol solution.

2. Prepare the reconstitution solution that contains the same methanol/water ratio as the wash solution and an internal standard to measure extraction efficiency. The extraction efficiency internal standard facilitates assessment of solid-phase extraction efficiency. Our reconstitution solution contains 50 nM sulfamethazine as our extraction efficiency standard, which results in a final 1.25 µM concentration after sample concentration.

3. 5 mL of reconstitution solution can be added to each lyophilized sample.

4. Activate solid-phase extraction cartridges with LC-MS grade methanol (3×3 mL) then wash with a wash solution (3×3 mL).

5. Load samples dropwise onto washed cartridges, then add the wash solution to remove any salts (3×3 mL).

6. The sample-containing cartridges are dried under vacuum before finally eluting samples into borosilicate glass culture tubes. For maximal resolution, samples can be eluted using a stepwise gradient. Specifically, samples can be eluted in 2 mL 20% methanol, 2 mL of 50% methanol, and 2 mL of 80% MeOH before finally washing cartridges with 2 mL of 100% MeOH.

7. The eluted samples are transferred into mass spectrometry vials and dried using a centrifugal vacuum concentrator.

8. After concentration, sample residue is weighed, reconstituted with 80% methanol/water containing an internal standard, and added to mass spectrometry inserts.

3.3 Alternative methods/procedures

Samples that contain even more complexity than microbial extracts can also be analyzed using the native metabolomics method and previously published sample processing, including fecal samples (Hosseinkhani et al., 2021), dissolved organic matter samples (Cancelada et al., 2022), and soil samples (Hallberg et al., 2024). Optimization of sample extraction methods for metabolomics is the subject of a number of reviews (Eshawu & Ghalsasi, 2024; Martias et al., 2021) with new strategies for selective enrichment (i.e., through the use of TiO_2 nanoparticles) (Egbers, Harder, Koch, & Tebben, 2020) or Fe-IMAC chromatography (Li et al., 2018) are emerging.

4. Mass spectrometry acquisition parameters

This section details the liquid chromatography and tandem mass spectrometry acquisition parameters for the native metabolomics method originally described in (Aron et al., 2022). Chromatography separation parameters are discussed for the Vanquish binary pump (Thermo Fisher Scientific). MS1 and MS2 parameters are tabulated for the Q-Exactive HF (Thermo Fisher Scientific) with an additional integrated make-up (quaternary) pump (Fig. 2A, Table 1). These guidelines serve as a representative example, and parameters must be adjusted if a different LC or MS are used. While this section cannot provide an exhaustive tutorial for every LC-MS2, it aims to highlight the most important parameters for native metabolomic analysis. Additionally, this section illustrates how the parameters discussed below can affect proton- and iron-adduct formation using a 10 μM deferoxamine B (Sigma Aldrich D9533-1G) standard prepared in 50% LCMS grade MeOH with 1 μM sulfadimethoxine internal standard. Prior to running the experimental samples in question, the user should run experimental and computational analysis on a control siderophore standard, such as the deferoxamine B highlighted in this section. Raw and centroided MS files (along with mzmine feature finding outputs, mzmine batch file, and metadata files) are available at MSV000094749 (ftp://massive. ucsd.edu/v07/MSV000094749/) to enable the user to compare experimental results to those obtained in our laboratory. If the user is obtaining confusing results, it is suggested that the user troubleshoot their experimental workflow using a deferroxamine B standard (Sigma-Aldrich) until they can reproduce the results in Fig. 4.

4.1 Materials and equipment
- Optima LC-MS grade methanol (Thermo Fisher Scientific)
- Optima LC-MS grade water with 0.1% formic acid (Thermo Fisher Scientific)
- Optima LC-MS grade acetonitrile with 0.1% formic acid (Thermo-Fisher Scientific)
- Ammonium acetate for LC-MS LiChropur (Sigma-Aldrich)
- Iron chloride (Sigma Aldrich, ≥99.9% trace metal basis purity)
- Deferoxamine B mesylate salt standard (Sigma Aldrich, ≥92.5% purity)
- Tubing (PEEK), 0.005-in ID × 1/16-in. OD, red and/or (Viper, 0.13 mm ID × 750 mm L)
- T-joints (PEEK)
- Syringe pump (Chemyx Inc.)

Native metabolomics for mass spectrometry-based siderophore discovery

- 3.27 mm diameter 500-mL syringe (Hamilton)
- High-pressure Vanquish binary gradient system (Thermo–Fisher Scientific)
- High-pressure Vanquish quaternary gradient system (Thermo–Fisher Scientific)
- 1.7 µM pore size, 100 Å particle size, 150 mm length, 1.0–2.1 mm diameter Kinetex C-18 chromatography column (Phenomenex)
- Q-Exactive quadrupole orbitrap system (Thermo Fisher Scientific)

Alternatives: The protocol below specifically outlines LC-MS2 parameters for the vanquish binary UHPLC system coupled to a Q-Exactive mass spectrometer and fully integrated make-up pump (quaternary vanquish UHPLC pump) and syringe pump. As stated in Section 2, native metabolomics is suitable to any mass spectrometry system, with necessary modifications made to the parameters described below. Iron chloride is included as the metal salt for the deferoxamine example, but other metal salts (i.e., copper chloride or sulfate, zinc chloride or sulfate, etc.) can be used if searching for non-iron metallophores.

4.2 LC-MS^2 acquisition parameters

Instead of providing a step-by-step method guide, we have provided Tables 1 and 2 detailing the acquisition settings used on the Q Exactive (Thermo Fisher) with integrated high-pressure binary gradient (Thermo Fisher Scientific Vanquish) and make-up pump (Thermo Fisher) systems. This is because native metabolomics can be optimized for use on most mass spectrometry platforms (as described in Section 2). Chromatographic separation uses a high-pressure binary gradient system. The mobile phase consists of solvent A (water with 0.1% formic acid) and solvent B (acetonitrile with 0.1% formic acid). A Phenomenex Kinetex C-18 chromatography column (1.7 µM pore size, 100 Å particle size, 150 mm length, 1.0–2.1 mm diameter) is routinely used for analysis. While a 200 µL/min flow rate is used for the larger, 2.1 mm column diameter, we suggest using a smaller diameter column to accommodate a lower flow rate (100–150 µL/min with an equivalent make-up flow of ammonium acetate buffer) to result in sharper peak shapes, which enable better performance of peak shape correlation modules (Section 5) . The LC gradient and injection volume used can be tailored to a specific sample or analysis type. We provide the LC and MS parameters used for the deferoxamine B standard in Tables 1 and 2, respectively.

MS1 and MS2 information are required for ion identity molecular networking analysis (Section 5). Full MS can be collected in positive or negative ionization modes, with different compounds and different ion adducts exhibiting different preferences for one analysis mode versus the other. Positive mode was used for the deferoxamine B example (and is tabulated in Table 2) because this compound ionizes well in positive mode due to the presence of a protonatable amine; siderophores that contain primarily carboxylate or catechol moieties may be suited for negative ionization mode. Notably, the analysis mode will dictate which adducts the user searches for in Section 5.3. The source parameters, including S-lens RF and spray voltage, can influence the intensity of the proton-bound and metal-bound siderophore complex observed in MS1. The S-lens RF describes the radio frequency applied to the S-lens, which is a transmission device of the ion source interface that consists of the electrodes providing a path for the passage of ions. ESI spray voltage describes the voltage applied to the electrospray needle to produce the spray current. Previous literature has investigated these parameters in non-targeted metabolomics in complex samples (Assress, Ferruzzi, & Lan, 2023), metal-ligand binding mass spectrometry (Tsednee, Huang, Chen, & Yeh, 2016), and small molecule bound to membrane proteins spectrometry (Gault et al., 2016). For example, a non-targeted metabolomics study found that the number of features increased as the RF level was increased from 10% to 60% but then saturated at around 70% (Assress et al., 2023). Spray voltages of 3.5 kV are optimal for the highest intensity for a mix of compounds (Skogvold et al., 2021). While there is currently limited optimization of mass spectrometry parameters for metal-siderophore complex formation, we found that S-lens RF and spray voltages influence the intensity of the iron and proton adducts (Fig. 3C and D).

Best practice includes analyzing media blanks (or another other matrix blank), extraction blanks, instrument blanks, and a quality control mix alongside samples of interest. Extraction blanks can highlight any sources of contamination, and media blanks are used for background subtraction, particularly when metal salts (which can contain various background signals) are directly infused. Instrument blanks are used to monitor any column carryover between samples. A quality control mixture is used to monitor retention time and feature intensity across the run.

4.3 Post-column pH adjustment

Post-pH adjustment uses a post-LC infusion of 10 mM ammonium acetate at a 200 µL/min flow rate for the full duration of the gradient. The

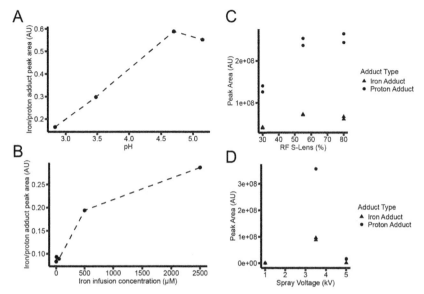

Fig. 3 LC and MS parameters affect the iron/proton-adduct ratio of the deferoxamine B standard sample. (A) The iron/proton adduct peak area ratio for deferoxamine B (10 μM) at varying ammonium acetate concentrations: 0 mM, 10 mM, 50 mM and 100 mM. Ammonium acetate concentrations correspond to pH values 2.8, 3.5, 4.7, and 5.2, respectively. Sample size at each ammonium acetate concentration is N = 1. (B) The iron/proton adduct peak area ratio for deferoxamine B (10 μM) at varying FeCl₃ infusion concentrations: 0 μM, 5 μM, 50 μM, 500 μM and 100 μM. Sample size at each FeCl₃ infusion concentration is N = 1. (C) The iron and proton adduct peak areas for deferoxamine B (10 μM) were evaluated at 30, 55, and 80 (%) RF S-Lens. Sample size at each RF S-lens was N = 2. FeCl₃ was infused at 500 μM concentration. (D) The iron and proton adduct peak areas for deferoxamine B (10 μM) were evaluated at 1, 3.5, 5 (kV) spray voltage. Sample size at each spray voltage was N = 2. FeCl₃ was infused at 500 μM concentration. For all experiments, deferoxamine-iron adduct ([M+Fe^{3+}-2H^{+}]$^{+}$) was identified as 614.2722 *m/z* and deferoxamine-proton adduct ([M+H]$^{+}$) was identified as 561.3606 *m/z*.

ammonium acetate solution is prepared using LCMS grade ammonium acetate salt (Sigma-Aldrich) added to LCMS grade water to buffer pH between 4 and 5 or around 9 (Konermann, 2017). This solution is infused using an integrated make-up (quaternary pump, Thermo Fisher Scientific Vanquish) through a T-connection. The T-connection connects the flow from the binary pump with the flow from the make-up pump just before the divert switch value of the Q Exactive (Fig. 2B). PEEK™ tubing (0.005-in ID × 1/16-in. OD, red) or Viper tubing (0.13 mm ID × 750 mm L) connects the make-up pump to the T-joint. We note that the reader may not have access to a second integrated pump; therefore, alternative strategies for post-LC infusion are described in (Section 2).

Fig. 3A highlights the effect of pH on the abundance of proton-bound, iron-bound, and the ratio of iron-bound to proton-bound peak abundance. In the case of the deferoxamine standard used in this experiment, raising the pH can increase the abundance of the iron adduct. However, this observation is compound-dependent, given that iron chelation exhibits pH dependence based on the structural moieties in siderophores. For example, ferric iron chelation requires the deprotonation of a phenolic proton for catecholate (pK_{a1} of 9.2) or a hydroxyl proton for hydroxamates (pK_a 8–9) and carboxylates (pK_a 3.5–5) (McCleverty, 2003; Neilands, 1995). Different siderophores have different behaviors, which may explain why some organisms produce multiple siderophores. For example, at higher pH values (>6), the catecholate siderophore enterobactin has a higher iron-binding affinity than the mixed-type (hydroxamate and carboxylate) siderophore aerobactin. In contrast, aerobactin exhibits an iron-binding affinity slightly greater than enterobactin (Valdebenito, Crumbliss, Winkelmann, & Hantke, 2006) at lower pH values (<4.3). This demonstrates that there is not one optimal pH for siderophore discovery (i.e., no "one-size-fits-all" approach), but rather iterating through multiple pH levels using the modular infusion setup described below can facilitate discovery. This concept also extends to metals beyond iron. For example, Behnsen et al. (2021) report the ratio of yersiniabactin-iron binding versus yersiniabactin-zinc binding exhibits pH-dependence (Behnsen et al., 2021). The bis-catecholate siderophore azotochelin (produced by *Azotobacter vinelandii*) also exhibits preferential binding of molybdate at pH 6–7 over other pH values (Bellenger et al., 2007).

4.4 Post-column metal infusion

Aron et al. (2022) report introducing transition metals (iron, copper, zinc, manganese, cobalt, and nickel) after LC and pH adjustment by direct infusion. While the protocol describes iron infusion for siderophore discovery, the same principles can be applied to other metal salts if the goal is to find chalkophores, zincophores (Behnsen et al., 2021), lanthanophores (Zytnick et al., 2023), or other metallophores. Additionally, mixtures of metal salts can be added as one mixed metal solution, though high salt levels may result in ion suppression. For this reason, samples should be run with and without metal infusion to assess for any ion suppression that could occur due to infusion.

After the flow from binary and make-up pumps is mixed through a T-joint, a solution of iron salt can be infused through an external syringe pump and connected through another T-joint just before the source (Fig. 2C). A ferric iron chloride (FeCl$_3$) solution is prepared in LCMS grade water, then a Hamilton syringe (3.27 mm diameter, 500 μL volume) is used to infuse this solution through a T-joint connection before the ion source. The suitable flow rate of infusion ranges from 2 to 5 μL/min depending on the concentration of the iron salt solution. Suitable iron concentrations range from 5 μM to 5 mM with the abundance of iron-bound versus proton-bound species dependent on iron-concentration (Fig. 3B). The iron solution can be infused for the analysis phase of the LC gradient, specifically from 0.25 min to 11 min of run time. Iron infusion is not necessary when the flow is diverted to the waste. Additionally, when deciding between various counter-ions for iron salts (i.e., FeCl$_3$, FeSO$_4$, etc.), aqueous solubility of metal is essential to consider (Haynes, 2011; Seidell & Linke, 1919). The distance from the T-connection to the source is 6.5 cm in the set-up used for experiments in this paper, though it was experimentally determined that this distance has little effect on the iron-adduct/proton-adduct ratio of the tested compounds (data not shown).

5. Data analysis using ion identity molecular networking

The advent of specialized data processing software alongside the modifications to standard UHPLC-MS/MS experimental set-up described in Section 4 facilitates the identification of putative metal-binding molecules. This section details the post-acquisition data analysis pipeline to find metal-binding molecules from the complex datasets generated using the experimental workflow detailed in Section 4, beginning with the conversion of vendor-specific raw MS data to the open.mzML format. Data is then processed in the open-source software mzmine to detect "ion" features, defined by their m/z and retention time (RT). This is the foundation for ion identity molecular networking, the strategy that groups the unbound and bound ion species based on similar MS2 fragmentation spectra. This section highlights the key innovation facilitating this workflow—the development of the ion identity molecular networking module in mzmine (Schmid et al., 2021) and integration in the GNPS2 web-ecosystem. Raw and centroided MS files along with mzmine feature finding outputs, mzmine batch file, and metadata files are available at

MSV000094749 (ftp://massive.ucsd.edu/v07/MSV000094749/) to enable the user to perform data analysis steps alongside this tutorial workflow.

5.1 Materials and equipment

- Computer (Windows, Mac, or Linux)
- Internet connection
- mzmine (can be downloaded at http://github.com/mzmine/mzmine/releases)
- MSConvert (can be downloaded at https://proteowizard.sourceforge.io/download.html)
- Cytoscape (can be downloaded at https://cytoscape.org/download.html)

5.2 Feature finding in mzmine: step-by-step method details

Metal-specific mass deltas are observed in MS1 data. Given this, feature detection must be performed before finding features with correlated retention times, peak shapes, and metal-specific mass deltas. A feature refers to a chromatographic peak described by accurate mass, retention time, and signal intensity (Minkus, Bieber, & Letzel, 2022). Data processing steps are outlined to convert raw MS spectra to a list of detected ion features, abundances, and structural annotations (Damiani et al., 2023). Parameters for feature detection, feature grouping, and feature alignment are described briefly below and are also subject of the following articles (Damiani et al., 2023; Schmid et al., 2021, 2023). For siderophore discovery, feature grouping tools such as ion identity networking and metaCorrelate are essential modules in mzmine to pair with the experimental workflow described in Section 4. Before applying these modules, the following steps are taken to prepare raw data acquired on a Thermo MS instrument:

1. Mass spectrometry files are converted to centroid format.mzML format using MSConvert GUI (Chambers et al., 2012) in the following steps (for the MSConvert command-line interface (CLI) see their documentation):
   a. Select raw file(s) for conversion and add these files to the MSConvert GUI.
   b. Choose an Output Directory as a destination for the mzML files.
   c. Select the mzML file format for the output format then uncheck "Use zlib compression".
   d. To centroid the profile data, choose "Peak Picking" (with Vendor checked), indicate MS-Levels 1- without specifying an upper limit to apply to this to all scans, then add this filter by clicking Add. Be sure to

update msconvert (older versions came with a default title maker step that required another step of pushing the peak picking step to the top to be performed first, but this was fixed in new msconvert versions).

2. We recommend the deposition of raw data files and converted mzML files to a data repository like MassIVE/GNPS or MetaboLights (Haug et al., 2013) to follow FAIR data guidelines (Wilkinson et al., 2016).

3. The centroided mzML files are now ready to be imported and processed in mzmine (Damiani, 2023; Minkus, Bieber, & Letzel, 2022; Schmid, 2021, 2023). The LC-MS/MS feature detection workflow in mzmine is described below, and in more detail in the following resources (Damiani et al., 2023; Schmid et al., 2021, 2023). The following mzmine modules are performed sequentially to produce an aligned feature table that can be queried for metal adducts of interest (following the procedure detailed in Section 5.3). Follow the mzwizard in mzmine for an easy workflow setup, choosing the hyphenated instruments and the DDA workflow, i.e., data-dependent acquisition.

   a. Import MS data. This module imports.mzML MS data files, a sample metadata table (optional), and spectral reference libraries (optional). The mzmine documentation details the supported spectral libraries.

   b. Mass detection. This module processes all mass scans and generates a centroided mass list (i.e., a list of m/z values and corresponding signal intensities) for each scan based on a user-defined noise threshold. The factor of lowest signal mass detector handles both centroid and profile scans by simply applying a noise level as the factor of the lowest intensity in a scan, which usually corresponds to noise. Absolute noise levels can be set in the 'exact' and 'centroid' mass detectors for profile and centroid spectra, respectively. Make sure to perform mass detection on all scans, optionally by applying different noise levels to MS1 and MS2 scans by setting the MS level in scan filters.

   c. Chromatogram builder. This module builds extracted ion chromatograms (EICs) by finding similar m/z values across all scans. EICs require a minimum height and a minimum number of consecutive data points above an intensity threshold; this number depends on the scan rate and the feature width. The scan-to-scan m/z tolerance is typically higher than the instrument's advertised accuracy as this relates to averaged values across multiple scans.

   d. Feature resolving using the Local minimum feature resolver (formerly Chromatogram deconvolution). EICs may contain multiple chromatographically separated or partially co-eluting features. The resolvers split chromatograms into separate features by applying peak-shape constraints. User parameters help to balance the number of detected noise and 'real' features. Activate the MS/MS scan pairing to group fragmentation scans with their corresponding features. The resulting feature list can be filtered and analyzed in subsequent steps.

e. ^{13}C isotope filtering (optional). Feature detection steps of chromatogram building and resolving consider all of the signals stored in the mass lists, which means that signals generated by isotopologues of the same compound are detected as distinct features. This information is redundant for downstream data analysis, so this algorithm checks for the presence of potential ^{13}C-related features in the feature lists and removes all but the main signal.

f. Isotopic peaks finder (optional). This module searches for all possible isotope signals for each feature, including isotopes with multiple charge states. Given that siderophore-metal complexes often form multiple charge states (especially for siderophores with higher molecular weights), increasing the maximum charge of isotopes to be equal to or greater than two was observed to result in more accurate molecular networks.

g. Feature alignment using the Join aligner. Detected features are aligned by their m/z and retention time across all samples.

h. Feature list rows filter (optional). Setting the minimum aligned features to an absolute or relative threshold removes features that were detected in less than X or X% samples.

i. Gap filling using the Peak finder (multithreaded) (optional). Missing features in the feature table may result from missed resolving or alignment among many error sources. Gap-filling is a secondary informed feature-finding step that uses the m/z and retention time from other samples to extract features from the raw data. This may improve the power of downstream statistical tests.

j. Duplicate peak filter (optional). After gap-filling, duplicate features tend to be more similar in their average m/z and retention time. The duplicate peak filter then merges this information into a single row, discarding the duplicates.

5.3 Metal adduct specific modules: step-by-step method details

The workflow above results in a feature list that was aligned across samples. With this input, the correlation grouping and ion identity networking modules can be used to find features that exhibit metal-binding behavior.

The experimental set-up described in Section 4, in which metal salt is infused after chromatography, was developed to observe proton- and infused metal-adducts at the same retention time for integration with the metaCorrelate feature correlation grouping module. It identifies features originating from the same molecule based on shared retention time, chromatographic peak shape Pearson correlation, and an optional feature height correlation across all samples. The second step is the ion identity networking module that uses m/z deltas of grouped features to annotate ion species of the same molecule as adducts (e.g., $[M+Na]^+$), in-source fragments (e.g., $[M-H_2O+H]^+$), and multimers (e.g., $[2M+H]^+$) (Schmid et al., 2021). The user-defined list of ion species facilitates tailoring the computational method to the specific experimental setup described in Section 4 (Schmid et al., 2021). For the iron-infusion experimental set-up run in positive ionization mode, we suggest querying the following iron-adducts $[M+Fe^{3+}-2H^+]^+$ and $[M+Fe^{3+}-H^+]^{2+}$, which have m/z values of 53.91815 and 27.4627, respectively. The $[M+Fe^{2+}]^{2+}$ or $[M+Fe^{2+}-H^+]^+$ adducts can also be observed given that ESI is a reductive process (McIndoe & Vikse, 2019; Stocks & Melanson, 2018) and that siderophores can bind Fe^{2+} with a lower affinity (Khasheii, Mahmoodi, & Mohammadzadeh, 2021). The m/z associated with these ions is 27.96663 and 54.92598, respectively. A unique feature of ion identity networking in mzmine is that this program can find all combinations of adducts specified by the user. The ion identity library (Fig. 4B) contains all selected adducts, which are then combined with neutral loss modifications like water losses and a multimer multiplier to form potential in-source clusters and fragments like $[2M+H]^+$ and $[M-H_2O+H]^+$, respectively. Grouped feature pairs are then annotated by ion identity networking by pairwise comparison of ions from the user-defined library. As another result, features are connected into networks if they originate from the same neutral molecule (Fig. 4C and D). Searching for more than one molecule per cluster is advantageous as many side-rophores form dimers (two ligands) or complexes with other ligand-to-metal ratios (Khan, Singh, & Srivastava, 2018). Additionally, larger side-rophores may form doubly charged ion species when complexed with iron, and siderophores containing hydroxyl moieties can exhibit water losses. Given all the potential adducts and neutral losses that can form, this strategy that interrogates all adduct combinations in an unbiased manner is a comprehensive strategy to detect metal-binding adducts. This is in part because we often observe unexpected adducts (McIndoe & Vikse, 2019). For example, besides the Fe adducts described above, other Fe adducts may

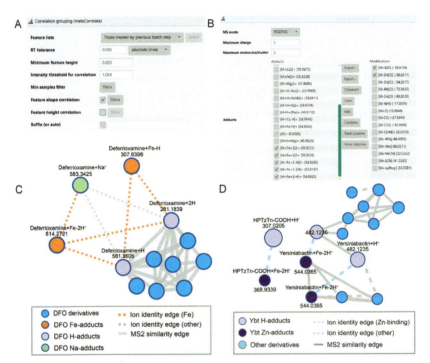

Fig. 4 Ion identity molecular network of the molecular family of deferoxamine (DFO) visualized in Cytoscape from a GNPS2 output. Deferoxamine (10 μM) adducts can be connected through ion identity molecular networking when analyzed with a 500 μM FeCl$_3$ infusion, as described in the protocol outlined in Section 4. mzmine is used to identify the [M+H$^+$]$^+$ and [M+2H$^+$]$^{2+}$ adducts and connect these to the [M+Fe^{3+}-2H$^+$]$^+$, [M+Fe^{3+}-H$^+$]$^{2+}$, and the [M+Na$^+$]$^+$ adduct using the modules illustrated in A and B. (C) The deferrioxamine molecular family visualized using ion identity molecular networking. Each node is labeled with its mass-to-charge (*m/z*) values. (D) Ion identity networking reveals that yersiniabactin and HPTzTn-COOH bind zinc. *This figure was modified from Behnsen, J., et al. (2021). Siderophore-mediated zinc acquisition enhances enterobacterial colonization of the inflamed gut. Nature Communications, 12, 7016.*

appear when FeCl$_3$ is infused and samples are analyzed in positive ionization mode. These include but are not limited to FeCl^{2+} and FeCl^{2+}. These adducts can be indicated as potential adducts in mzmine along with the more common adducts described above. It is important to note that ESI source parameters including desolvation gas, temperature, and source voltages influence observed adducts (McIndoe & Vikse, 2019). A detailed discussion of how to set up the parameters for both modules is described in the following steps below. Table 3 outlines ion–adducts of interest when various metals are infused in positive and negative analysis modes.

Table 3 Common adducts of interest found in positive and negative ionization modes for various biologically relevant transition metals (Aron et al., 2022).

Adduct	Monoisotopic mass	Ionization mode
Iron		
$[M+Fe^{3+}-2H^+]^+$	53.91815	Positive
$[M+Fe^{3+}-H^+]^{2+}$	27.4627	Positive
$[M+Fe^{3+}-4H^+]^-$	51.9036	Negative
$[M+FeCl^{2+}-3H^+]^-$	87.8803	Negative
Copper		
$[M+Cu^{2+}-H^+]^+$	61.9212	Positive
$[M+Cu^{2+}]^{2+}$	31.4643	Positive
$[M+Cu^+]^+$	62.9290	Positive
$[M+Cu^{2+}-3H^+]^-$	59.9067	Negative
Zinc		
$[M+Zn^{2+}-H^+]^+$	62.9207	Positive
$[M+Zn^{2+}]^{2+}$	31.964	Positive
$[M+Zn^{2+}-3H^+]^-$	60.9062	Negative
Cobalt		
$[M+Co^{2+}-H^+]^+$	57.9248	Positive
$[M+Co^{2+}]^{2+}$	29.466	Positive
$[M+Co^{2+}-3H^+]^-$	55.9103	Negative
Manganese		
$[M+Mn^{2+}-H^+]^+$	53.9296	Positive
$[M+Mn^{2+}]^{2+}$	27.4685	Positive

It is important to note that the counterion can influence ionization (e.g., adducts containing $FeCl^{2+}$ and $FeCl^{2+}$ are observed when $FeCl_3$ is infused). Additionally, this list is not exhaustive of all the metal adducts that may be present in an infusion experiment.

1. The first step in correlation grouping uses the metaCorrelate module (Fig. 4A). The following parameters are described and depend upon data acquisition parameters.
    a. RT tolerance. This pre-filter ensures that potential correlated features lie within the same retention time window. The stringency of this filter depends on whether feature shape correlation (see below) is active. The RT tolerance can be wider, i.e., less stringent, when combined with feature shape correlation, otherwise choose a narrow RT tolerance closer to the feature's full-width at half maximum (FWHM)/3.
    b. Min height. This parameter sets the minimum height for features considered in the correlation analysis. A zero value is recommended so that all detected features will be included. It should be noted that other feature constraints from the previous steps already filtered the final feature list.
    c. Intensity correlation threshold. This defines the minimum intensity of the data points in a feature to be considered during feature shape Pearson correlation. All data points below this value are disregarded. Given this, we recommend setting this value to zero so that the noise levels set in the mass detection steps are used.
    d. Min samples filter. This parameter only groups features that were detected in a minimum number of samples. Leaving the default values works well in most cases. This excludes gap-filled features and requires a minimum intensity overlap of 60%.
    e. Correlation grouping. This optional parameter applies a feature shape correlation filter in the retention time dimension. This should only be used when enough data points are present (i.e., 5 data points total with 2 on each side of the chromatographic peak apex). Apply the default minimum feature shape correlation of 85%.
    f. Feature height correlation. This option needs to be deactivated for the described metal-infusion method, due to varying ionization and sample matrix effects. This filter takes the Pearson correlation over all the feature heights across samples for feature pairs and therefore relies on stable ionization efficiencies throughout the sample set.
2. After completing correlation grouping, the user can define adducts of interest using the ion identity module (Fig. 4B). All paired features found using the metaCorrelate module in step **5.2.1** will be searched against an ion library for every possible combination of these ions. Given this, it is essential to perform correlation grouping before ion identity

networking. The ion identity networking parameters are described, and additional ion types are defined for the iron infusion method.

a. m/z tolerance. This refers to the acceptable m/z difference between two ions found in the same sample at the same retention time. Therefore, the expected m/z deviation is typically smaller than the instrument accuracy in relation to the exact mass. Still, this parameter depends upon instrument resolution. For example, this can be set to 3 ppm on a calibrated Q Exactive (Thermo Fisher). This parameter allows checking one feature, all features, or the average.

b. Min height. This optional parameter sets the minimum height features considered in the analysis. It is recommended to set this to zero.

c. Ion identity library. This module defines the set of searched ions. In addition to commonly occurring adducts (Na^+, K^+, etc.), the user must include the ion types for the metal infused in the experimental workflow.

   i. The user first sets the MS polarity (positive or negative mode) and then specifies the maximum charge and number of molecules per cluster. The maximum molecules per cluster defines how many molecules may form non-covalent clusters, i.e., $[2M+H]^+$.

   ii. For the infusion of Fe, it is necessary to define the Fe^{3+} ion adduct with mass 55.9333. Select the "Add" button and specify Fe^{3+} as the formula and click ok. The new ion is added to the bottom of the adduct list. Alternatively, specify a name, mass, and charge to define new ions (Fig. 4B), resulting in the output shown in Fig. 4C.

   iii. Next, the user can define a combination of ions by selecting "Combine". For example, the $[M+Fe^{3+}-2\,H^+]^+$ and $[M+Fe^{3+}-H^+]^{2+}$ adducts are commonly observed for siderophores. The user must manually add these ion–adducts in this step (Fig. 4B). Open the "Combine" dialog, double-click on $[M+Fe]^{3+}$ and on $[M-H]^-$; they should appear on the right. Click on the add button to add the combined ion type to the list and finally click finish to close the dialog.

   iv. An important note for the user is that this step should focus on the main ions of interest. Any exotic ions can be added in a later module using the "Add ion identities to networks" module. This module uses similar input parameters but only adds ion identities to existing ion identity networks.

d. An optional annotation refinement can be performed in which the user species what size networks are retained, whether to delete small networks without a major ion or a monomer, etc.

5.4 Mzmine compound annotation, molecular networking, data export for GNPS2

Compound annotation in mzmine is enabled through open spectral libraries that may be downloaded from GNPS, MassBank EU, MassBank of North America, and other sources (Schmid et al., 2023). Open spectral libraries are imported during the "Import MS data" step as mgf, json, or msp format. The spectral library search module is then applied to the final feature list, annotating features by matching their MS2 fragmentation scans against the libraries using the weighted cosine similarity, a minimum number of matches signals, and m/z tolerances for the precursors and signals. For small molecules, 4–6 minimum matched signals are recommended. The default minimum cosine similarity of 0.7 usually performs well for library matching and molecular networking but remains an arbitrary threshold without a real false-discovery rate. The next optional step in mzmine is the "MS/MS spectral networking (molecular networking)" module to compute the pairwise spectral similarity between all detected features. The parameters' default values for this module include an m/z tolerance of 10 ppm to match spectral signals, 4 minimum matched signals, and 0.7 minimum cosine similarity. Visualizing the ion identity molecular networking results in mzmine involves opening the final feature list with a double-click, selecting a row, opening the right-click menu, and selecting feature overview & IIMN networks. The ion identities can be visualized by selecting "Adduct" as a node label under visuals. The dashed edges correspond to the ion identity networks while the regular edges connect features by spectral similarity. The search bar in the filter menu allows quickly finding features by their compound annotation. Start typing and click the focus nodes button to see the selected feature nodes corresponding to this annotation. By default, mzmine only shows the n-nearest neighbors of the focused rows to limit the complexity of the networks and allow for easy propagation of annotations through network adjacency. Finally, the results are exported for the subsequent analysis in GNPS2, which allows for molecular networking and integration with other tools. The "Export molecular networking files" module exports a feature quantification table, an mgf MS2 spectra file, and the additional edges file from ion identity networking. GNPS2 requires all these files for the final IIMN steps.

Before visualization in GNPS2 and statistical analysis, blank removal can remove signals from contaminants present in metal salt solution, ammonium acetate buffer solution, and extraction blanks. This step is especially

Native metabolomics for mass spectrometry-based siderophore discovery 343

important in this workflow where direct infusion of metal salt and ammonium acetate buffer solution will introduce noise into the analysis (described in more detail in Section 5). We suggest using a media (or extraction) blank plus metal infusion for blank removal. Blank removal can be performed using the workflow described in Pakkir Shah et al. (2023).

5.5 Visualization in GNPS, GNPS2 & Cytoscape

The mzmine outputs can be visualized using ion identity molecular networking through the feature based molecular networking (FBMN) (Nothias et al., 2020) workflows in GNPS (Wang et al., 2016) and GNPS2. FBMN allows for the visualization and annotation of features in untargeted mass spectrometry experiments. It is important to note that ion identity networking edges cannot be viewed in classical molecular networking because it does not contain MS1 information. Using GNPS and GNPS2 to visualize the analysis results in Sections 4 and 5.1–5.3 can streamline the discovery of novel metal-binding molecules. In the protocol below, we first describe how to run a FBMN job in GNPS2, then we describe key visualization strategies in Cytoscape (www.cytoscape.org) (Shannon et al., 2003).

GNPS2 can be used to generate an FBMN and requires a feature table quant file with the intensities of LC-MS ion features (.csv), an MS2 spectral summary file with a list of MS2 spectra associated with the LC-MS ion features (.mgf), and a metadata (.txt) file. For metallophore discovery, an ion edges annotation file (.csv) is also required which provides the "Supplementary Pairs" of additional edges. A FBMN containing these additional ion identity edges can be run in GNPS2 through the following steps.

1. Navigate to GNPS2 (gnps2.org)
2. Click on the Launch workflows button then navigate to "feature_based_molecular_networking_workflow" under the workflow name. Click on Launch workflow.
3. Provide a description of the job, then select the relevant files generated in your feature finding job and select mzmine as feature finding tool.
    a. Add the following input files.
        i. Input inputfeatures refers to the .csv feature quant table.
        ii. Input inputspectra refers to the .mgf MS2 spectra file.
        iii. Input Metadata File refers to the .txt file containing sample information.
        iv. Input Spectral Library Folder
        v. Input Additional Edges (optional) refers to the .csv file with ion edges

4. Search parameters are set in the Basic Options and Advanced Molecular Networking Options windows. These parameters have been reviewed previously (Aron et al., 2020).
5. The .graphml file is downloaded and opened in Cytoscape, an open-source software for network visualization (Shannon et al., 2003). Changing node colors, edge colors, and edge types in Cytoscape can facilitate visualization of metal-binding molecules. This .graphml file can be found in MSV000094749 to enable the user to follow along with the steps below and reproduce the network found in Fig. 4C.
   a. Node colors can be changed by selecting the Style button on the side panel, then selecting the Node button in the bottom panel.
      i. Once the Node button has been selected, the user can click the arrow next to Fill Color. This opens up two criteria for selection—Column and Mapping Type.
      ii. "Best ion" can be selected in the row next to Column.
      iii. "Discrete Mapping" can be selected for mapping type, and the adducts of interest can be colored by clicking on the three dots in the discrete mapping column.
   b. In the Style window, the Edge button in the bottom panel can be selected to change edge colors and edge type. Edge color can be altered by clicking the arrow next to Stroke Color. As with coloring nodes, this opens up two criteria for selection—Column and Mapping Type.
      i. "EdgeAnnotation" can be selected in the row next to Column.
      ii. "Discrete Mapping" can be selected for mapping type, and the edge annotations of interest can be colored by clicking on the three dots in the discrete mapping column.
   c. Edge type can be altered by clicking the arrow next to Line Type.
      i. "EdgeType" can be selected in the row next to Column.
      ii. "Discrete Mapping" can be selected for mapping type, and the cosine edges can be differentiated from the MS1 annotations by clicking on the three dots in the discrete mapping column.

An example network for deferoxamine visualized in Cytoscape using the strategies described above is illustrated in Fig. 4C. Fig. 4D highlights how this strategy was used to discover that yersiniabactin, the truncation HPTzTn-COOH, and other derivatives bind zinc, which was reported in Behnsen et al. (2021).

5.6 Layering with other sources of info

This native metabolomics approach can also be combined with other emerging *in silico* and computational strategies, including genome mining and orthogonal mass querying techniques. Genome mining strategies for siderophore discovery can be used to prioritized strains for culturing or to suggest moieties that are likely present in siderophores (i.e., ornithine, citrate, etc.) (Reitz et al.., 2022). The native metabolomics approach can also be paired with orthogonal mass spectrometry tools, including the Mass Query Language (MassQL) (Jarmusch et al., 2022). MassQL is a mass spectrometry language that facilitates pattern searching in MS data. Along with using IIMN to find metal-specific mass deltas, MassQL can also be used to find metal-specific mass deltas in unprocessed data (no feature-finding required) or metal-specific isotopic signatures. For example, a query can be written using MassQL that searches for the three distinct iron isotopes (^{54}Fe, ^{56}Fe and ^{57}Fe), as described in detail in the original paper describing this technique (Jarmusch et al., 2022). This can aid in the discovery of compounds that remain metal-bound after chromatography. Importantly, compounds that remain fully metal-bound (with no proton- or other ion-adduct peaks) cannot be found using the native metabolomics and ion identity strategy described here but will be found using a MassQL-based strategy.

6. Limitations and considerations

This protocol works well for identifying putative metal-binding molecules; however, routine infusion of high salt solutions into the mass spectrometer must be approached with caution and routine instrument cleaning. This method can introduce contaminants in both positive and negative electrospray ionization, including a contaminant with m/z 537.8790 in the positive electrospray ionization mode that is likely a Fe-O-acetate complex (Keller, Sui, Young, & Whittal, 2008). In negative ionization mode, interferences are seen at m/z 197.8074 and 160.8413, which may originate from $[^{56}Fe^{35}Cl_3^{37}Cl]^-$ and $[^{56}Fe^{35}Cl_3]^-$ clusters, respectively and can be circumvented by collecting $m/z > 250$ Da. Additional uncharacterized contaminants have been observed that likely originate from the direct introduction of metal salts. Given the persistence of these contaminants as background signal during metal infusion, we suggest using a blank run at the same pH with metal infusion for blank removal, as this will remove background signal originating from ammonium acetate buffer and metal salt.

Additionally, high salt concentrations can also result in ion suppression, so we caution infusion of salt concentrations that are in excess of what is discussed here.

When comparing different ion abundances using this method, it is important to note that different compounds and ion adducts have different ionization efficiencies (the number of gas-phase ions per mol of compound by electrospray ionization) (Liigand et al., 2021). For this reason and others, it is not straightforward to determine a dissociation constant using this approach. Additionally, one must exhibit caution when comparing iron-adduct/proton-adduct ratios across compounds, as this ratio is likely compound dependent.

The reader is finally offered a word of caution. While this is a discovery approach, the presence of a metal-adduct does not prove that the binding compound is a metallophore, it only strongly suggests this. There are interesting examples of compounds that bind metals but do not facilitate metal uptake. Metal-binding by these compounds can have other physiological roles. For example, ferroverdins are produced by some *Streptomyces* species in response to iron overload (Martinet, Baiwir, Mazzucchelli, & Rigali, 2022). Pulcherriminic acid is produced by yeast and many bacteria including *B. subtilis*. In a recent study, this compound was shown to locally precipitate ferric iron into a pulcherrimin-iron complex, from which the strong chelator bacillibactin could sequester iron (Charron-Lamoureux et al., 2023). In addition to mitigating oxidative stress, this compound also serves to provide a local iron source that is only accessible to some organisms. While the compound obafluorin contains structural characteristics of siderophores (Batey et al., 2023)—specifically a catechol moiety—this compound neither acts as a siderophore nor a Trojan horse antibiotic (Tillotson, 2016). Rather, obafluroin-Fe^{3+} binding specifically prevents the hydrolytic breakdown of this antibiotic. Given these diverse physiological roles associated with metal-binding, follow-up studies must be done to confirm metal-binding, metal-selectivity, modes of binding, and biological metal accumulation. These strategies include those described above, namely ICP-MS, UV–vis and/or nuclear magnetic resonance (NMR) spectroscopy, and colorimetric assays to confirm metal-binding, often on a purified compound. ICP-MS is also a powerful strategy for assessing bioaccumulation of a given metal of interest by a compound of interest. Finally, genetic mutants are an important tool for further interrogating biological roles of these compounds.

7. Summary and conclusions

In this chapter, we walk the reader through how to perform native metabolomics analysis to study siderophores from a microbial extract grown in the laboratory. We also provide an example case in which a deferoxamine B standard is experimentally and computationally analyzed using this workflow to encourage the reader to attempt this workflow alongside the tutorial steps. Along with new developments in hardware and software, we anticipate the future of native metabolomics lies in its further integration with other sources of information (i.e., genome mining, ICP-MS data, or metal-characteristic isotopic signatures). In addition to aiding in the discovery of novel metallophores, we hope this approach can also help elucidate biological roles for metallophores, such as shaping microbial interactions in complex systems. In summary, the methods described here can shine light on siderophore and metallophore dark matter to elucidate further the importance of this class of molecules across biology.

Acknowledgments

This work was supported by University of Denver start-up funds and a Dean's fellowship from the University of Denver (MTY). We also acknowledge Olivia Schwartz (Department of Chemistry and Biochemistry, University of Denver) for her helpful comments.

References

Aron, A. T., et al. (2020). Reproducible molecular networking of untargeted mass spectrometry data using GNPS. *Nature Protocols, 15*, 1954–1991.

Aron, A. T., et al. (2022). Native mass spectrometry-based metabolomics identifies metal-binding compounds. *Nature Chemistry, 14*, 100–109.

Aschner, M., Palinski, C., Sperling, M., Karst, U., Schwerdtle, T., & Bornhorst, J. (2017). Imaging metals in: Caenorhabditis elegans. *Metallomics, 9*(4), 357–364.

Assress, H. A., Ferruzzi, M. G., & Lan, R. S. (2023). Optimization of mass spectrometric parameters in data dependent acquisition for untargeted metabolomics on the basis of putative assignments. *Journal of the American Society for Mass Spectrometry, 34*, 1621–1631.

Atkin, C. L., Neilands, J. B., & Phaff, H. J. (1970). Rhodotorulic acid from species of Leucosporidium, Rhodosporidium, Rhodotorula, Sporidiobolus, and Sporobolomyces, and a new alanine-containing ferrichrome from Cryptococcus melibiosum. *Journal of Bacteriology, 103*(3), 722–733.

Atkinson, R. A., Salah El Din, Kieffer, B., Lefèvre, J. F., & Abdallah, M. A. (1998). Bacterial iron transport: 1H NMR determination of the three- dimensional structure of the gallium complex of pyoverdin G4R, the peptidic siderophore of *Pseudomonas putida* G4R. *Biochemistry, 37*(45), 15965–15973.

Avalon, N. E., Reis, M. A., Thornburg, C. C., Williamson, R. T., Petras, D., Aron, A. T., N ... Gerwick, W. H. (2024). Leptochelins A-C , Cytotoxic Metallophores Produced by Geographically Dispersed Leptothoe Strains of Marine Cyanobacteria. *Journal of the American Chemical Society*, 146, 18626–18638.

Baars, O., Morel, F. M. M., & Perlman, D. H. (2014). ChelomEx: Isotope-assisted discovery of metal chelates in complex media using high-resolution LC-MS. *Analytical Chemistry, 86*(22), 11298–11305.

Baars, O., Zhang, X., Morel, F. M. M., & Seyedsayamdost, M. R. (2016). The siderophore metabolome of Azotobacter vinelandii. *Applied and Environmental Microbiology, 82*(1), 27–39.

Batey, S. F. D., et al. (2023). The catechol moiety of obafluorin is essential for antibacterial activity. *RSC Chemical Biology, 4*, 926–941.

Behnsen, J., et al. (2021). Siderophore-mediated zinc acquisition enhances enterobacterial colonization of the inflamed gut. *Nature Communications, 12*, 7016.

Bellenger, J. P., Arnaud-Neu, F., Asfari, Z., Myneni, S. C. B., Stiefel, E. I., & Kraepiel, A. M. L. (2007). Complexation of oxoanions and cationic metals by the biscatecholate siderophore azotochelin. *Journal of Biological Inorganic Chemistry, 12*(3), 367–376.

Boiteau, R. M., Fitzsimmons, J. N., Repeta, D. J., & Boyle, E. A. (2013). Detection of iron ligands in seawater and marine cyanobacteria cultures by high-performance liquid chromatography-inductively coupled plasma-mass spectrometry. *Analytical Chemistry, 85*(9), 4357–4362.

Butler, A., et al. (2023). Mining elements of siderophore chirality encoded in microbial genomes. *FEBS Letters, 597*, 134–140.

Cancelada, L., et al. (2022). Assessment of styrene-divinylbenzene polymer (PPL) solid-phase extraction and non-targeted tandem mass spectrometry for the analysis of xenobiotics in seawater. *Limnology and Oceanography: Methods, 20*, 89–101.

Carroll, C. S., & Moore, M. M. (2018). Ironing out siderophore biosynthesis: A review of non-ribosomal peptide synthetase (NRPS)-independent siderophore synthetases. *Critical Reviews in Biochemistry and Molecular Biology, 53*, 356–381.

Chambers, M. C., et al. (2012). A cross-platform toolkit for mass spectrometry and proteomics. *Nature Biotechnology, 30*, 918–920.

Charron-Lamoureux, V., et al. (2023). Pulcherriminic acid modulates iron availability and protects against oxidative stress during microbial interactions. *Nature Communications, 14*, 2536.

Chaturvedi, K. S., Hung, C. S., Crowley, J. R., Stapleton, A. E., & Henderson, J. P. (2012). The siderophore yersiniabactin binds copper to protect pathogens during infection. *Nature Chemical Biology, 8*, 731–736.

D'Onofrio, A., Crawford, J. M., Stewart, E. J., Witt, K., Gavrish, E., Epstein, S., ... Lewis, K. (2010). Siderophores from Neighboring Organisms Promote the Growth of Uncultured Bacteria. *Chemistry and Biology, 17*(3), 254–264.

Damiani, T., et al. (2023). Mass spectrometry data processing in MZmine 3: Feature detection and annotation. *Analytical Chemistry*. https://doi.org/10.26434/chemrxiv-2023-98n6q.

Dewey, C., Kaplan, D. I., Fendorf, S., & Boiteau, R. M. (2023). Quantitative Separation of Unknown Organic-Metal Complexes by Liquid Chromatography-Inductively Coupled Plasma-Mass Spectrometry. *Analytical Chemistry, 95*(20), 7960–7967.

Egbers, H., Harder, P., Koch, T. P. B., & Tebben, J. (2020). Siderophore purification with titanium dioxide nanoparticle solid phase extraction. *Analyst, 145*, 7303–7311.

Endicott, N. P., Lee, E., & Wencewicz, T. A. (2017). Structural basis for xenosiderophore utilization by the human pathogen *Staphylococcus aureus*. *ACS Infectious Diseases, 3*, 542–553.

Eshawu, A. B., & Ghalsasi, V. V. (2024). Metabolomics of natural samples: A tutorial review on the latest technologies. *Journal of Separation Science, 47*, 2300588.

Galdino, A. C. M., et al. (2024). Siderophores promote cooperative interspecies and intraspecies cross-protection against antibiotics in vitro. *Nature Microbiology, 9*, 631–646.

Gault, J., et al. (2016). High-resolution mass spectrometry of small molecules bound to membrane proteins. *Nature Methods, 13*, 333–336.

Gibson, F., & Magrath, D. I. (1969). The isolation and characterization of a hydroxamic acid (aerobactin) formed by *Aerobacter aerogenes* 62-I. *Biochimica et Biophysica Acta, 192*, 175–184.

Hallberg, Z. F., et al. (2024). Vitamin B12 variants structure soil microbial communities despite soil's vast reservoir of B12. 2024.02.12.580003. Preprint at https://doi.org/10.1101/2024.02.12.580003.

Haug, K., Salek, R. M., Conesa, P., Hastings, J., De Matos, P., Rijnbeek, M., … Steinbeck, C. (2013). MetaboLights - An open-access general-purpose repository for metabolomics studies and associated meta-data. *Nucleic Acids Research, 41* (D1), 781–786.

Haynes, W. M. (Ed.). (2011). *CRC handbook of chemistry and physics* (pp. 4.69)(92nd ed.). Boca Raton, FL: CRC Press.

Hemmerle, L., Ochsner, A. M., Vonderach, T., Hattendorf, B., & Vorholt, J. A. (2021). Chapter Nine—Mass spectrometry-based approaches to study lanthanides and lanthanide-dependent proteins in the phyllosphere. In J. A. Cotruvo (Vol. Ed.), *Methods in enzymology: Vol. 650*, (pp. 215–236). Academic Press.

Hider, R. C., & Kong, X. (2010). Chemistry and biology of siderophores. *Natural Product Reports, 27*, 637–657.

Himpsl, S. D., & Mobley, H. L. T. (2019). Siderophore Detection Using Chrome Azurol S and Cross-Feeding Assays. *Proteus mirabilis. Methods in Molecular Biology, 2021*.

Hosseinkhani, F., Heinken, A., Thiele, I., Lindenburg, P. W., Harms, A. C., & Hankemeier, T. (2021). The contribution of gut bacterial metabolites in the human immune signaling pathway of non-communicable diseases. *Gut Microbes, 13*(1), 1–22.

Jarmusch, A. K., et al. (2022). *A universal language for finding mass spectrometry data patterns* 2022.08.06.503000 Preprint at https://doi.org/10.1101/2022.08.06.503000.

Karch, K. R., Snyder, D. T., Harvey, S. R., & Wysocki, V. H. (2022). Native mass spectrometry: Recent progress and remaining challenges. *Annual Review of Biophysics, 51*, 157–179.

Keller, B. O., Sui, J., Young, A. B., & Whittal, R. M. (2008). Interferences and contaminants encountered in modern mass spectrometry. *Analytica Chimica Acta, 627*, 71–81.

Kenney, G. E., & Rosenzweig, A. C. (2018). Chalkophores. *Annual Review of Biochemistry, 87*, 645–676.

Khan, A., Singh, P., & Srivastava, A. (2018). Synthesis, nature and utility of universal iron chelator—Siderophore: A review. *Microbiological Research, 212–213*, 103–111.

Khasheii, B., Mahmoodi, P., & Mohammadzadeh, A. (2021). Siderophores: Importance in bacterial pathogenesis and applications in medicine and industry. *Microbiological Research, 250*, 126790.

Koh, E.-I., et al. (2015). Metal selectivity by the virulence-associated yersiniabactin metallophore system. *Metallomics, 7*, 1011–1022.

Konermann, L. (2017). Addressing a common misconception: Ammonium acetate as neutral pH "buffer" for native electrospray mass spectrometry. *Journal of the American Society for Mass Spectrometry, 28*, 1827–1835.

Kramer, J., Özkaya, Ö., & Kümmerli, R. (2020). Bacterial siderophores in community and host interactions. *Nature Reviews. Microbiology, 18*, 152–163.

Li, Y., et al. (2018). Fe(III)-based immobilized metal–affinity chromatography (IMAC) method for the separation of the catechol siderophore from Bacillus tequilensis CD36. *3 Biotech, 8*, 392.

Liigand, P., Liigand, J., Kaupmees, K., & Kruve, A. (2021). 30 Years of research on ESI/MS response: Trends, contradictions and applications. *Analytica Chimica Acta, 1152*, 238117.

Lippard, S. J., & Berg, J. M. (1994). *Principles of bioinorganic chemistry*. University Science Books.

Łoboda, D., & Rowińska-Żyrek, M. (2017). Zinc binding sites in Pra1, a zincophore from *Candida albicans. Dalton Transactions (Cambridge, England: 2003), 46*, 13695–13703.

Lopes, N. P., Stark, C. B. W., Hong, H., Gates, P. J., & Staunton, J. (2001). A study of the effect of pH, solvent system, cone potential and the addition of crown ethers on the formation of the monensin protonated parent ion in electrospray mass spectrometry. *Analyst, 126*, 1630–1632.

Lurthy, T., et al. (2020). Impact of bacterial siderophores on iron status and ionome in pea. *Frontiers in Plant Science, 11*.

Martias, C., et al. (2021). Optimization of sample preparation for metabolomics exploration of urine, feces, blood and saliva in humans using combined NMR and UHPLC-HRMS platforms. *Molecules (Basel, Switzerland), 26*, 4111.

Martinet, L., Baiwir, D., Mazzucchelli, G., & Rigali, S. (2022). Structure of new ferroverdins recruiting unconventional ferrous iron chelating agents. *Biomolecules, 12*, 752.

McCleverty, J. A. (2003). *Comprehensive Coordination Chemistry II*. Elsevier.

McCormack, P., Worsfold, P. J., & Gledhill, M. (2003). Separation and detection of siderophores produced by marine bacterioplankton using high-performance liquid chromatography with electrospray ionization mass spectrometry. *Analytical Chemistry, 75*(11), 2647–2652.

McIndoe, J. S., & Vikse, K. L. (2019). Assigning the ESI mass spectra of organometallic and coordination compounds. *Journal of Mass Spectrometry: JMS, 54*, 466–479.

Medema, M. H., et al. (2011). antiSMASH: Rapid identification, annotation and analysis of secondary metabolite biosynthesis gene clusters in bacterial and fungal genome sequences. *Nucleic Acids Research, 39*, W339–W346.

Minkus, S., Bieber, S., & Letzel, T. (2022). Spotlight on mass spectrometric non-target screening analysis: Advanced data processing methods recently communicated for extracting, prioritizing and quantifying features. *Analytical Science Advances, 3*, 103–112.

Molimau-Samasoni, S., Woolner, V. H., Foliga, S. T., Robichon, K., Patel, V., Andreassend S. K., ... Munkacsi, A. B. (2021). Functional genomics and metabolomics advance the ethnobotany of the Samoan traditional medicine "matalafi." *Proceedings of the National Academy of Sciences of the United States of America*, 118 (45).

Morey, J. R., & Kehl-Fie, T. E. (2020). Bioinformatic mapping of opine-like zincophore biosynthesis in bacteria. *mSystems, 5*, e00554–20.

Motz, R. N., et al. (2024). Conjugation to native and nonnative triscatecholate siderophores enhances delivery and antibacterial activity of a β-lactam to gram-negative bacterial pathogens. *Journal of the American Chemical Society, 146*, 7708–7722.

Negash, K. H., Norris, J. K. S., & Hodgkinson, J. T. (2019). Siderophore–antibiotic conjugate design: New drugs for bad bugs? *Molecules (Basel, Switzerland), 24*, 3314.

Neilands, J. B. (1967). Hydroxamic acids in nature. *Science (New York, N. Y.), 156*, 1443–1447.

Neilands, J. B. (1981). Microbial iron compounds. *Annual Review of Biochemistry, 50*, 715–731.

Neilands, J. B. (1995). Siderophores: Structure and function of microbial iron transport compounds. *The Journal of Biological Chemistry, 270*, 26723–26726.

Nothias, L.-F., et al. (2020). Feature-based molecular networking in the GNPS analysis environment. *Nature Methods, 17*, 905–908.

Pahari, A., Pradhan, A., Nayak, S. K., & Mishra, B. B. (2017). Bacterial siderophore as a plant growth promoter. In J. K. Patra, C. N. Vishnuprasad, & G. Das (Vol. Eds.), *Microbial biotechnology: Applications in agriculture and environment. Vol. 1. Microbial biotechnology: Applications in agriculture and environment* (pp. 163–180). Singapore: Springer. https://doi.org/10.1007/978-981-10-6847-8_7.

Pakkir Shah, A. K., et al. (2023). The Hitchhiker's guide to statistical analysis of feature-based molecular networks from non-targeted metabolomics data. *Analytical Chemistry.* https://doi.org/10.26434/chemrxiv-2023-wwbt0.

Payne, S. M. (1994). Detection, isolation, and characterization of siderophores. *Methods in Enzymology, 235*, 329–344.

Perry, W. J., Patterson, N. H., Prentice, B. M., Neumann, E. K., Caprioli, R. M., & Spraggins, J. M. (2020). Uncovering matrix effects on lipid analyses in MALDI imaging mass spectrometry experiments. *Journal of Mass Spectrometry, 55*(4).

Pinkert, L., et al. (2021). Antibiotic conjugates with an artificial MECAM-based siderophore are potent agents against gram-positive and gram-negative bacterial pathogens. *Journal of Medicinal Chemistry, 64*, 15440–15460.

Piper, K., & Higgins, G. (1967). Estimation of Trace Metals in Biological Material by Atomic Absorption Spectrophotometry. *Proceedings of the Association of Clinical Biochemists, 4*(7).

Pluháček, T., Lemr, K., Ghosh, D., Milde, D., Novák, J., & Havlíček, V. (2016). Characterization of Microbial Siderophores by Mass Spectrometry. *Mass Spectrometry Reviews, 35*, 35–47.

Pluskal, T., Castillo, S., Villar-Briones, A., & Orešič, M. (2010). MZmine 2: Modular framework for processing, visualizing, and analyzing mass spectrometry-based molecular profile data. *BMC Bioinformatics,* 11.

Pollack, J. R., & Neilands, J. B. (1970). Enterobactin, an iron transport compound from Salmonella typhimurium. *Biochemical and Biophysical Research Communications, 38*, 989–992.

Qi, B., & Han, M. (2018). Microbial siderophore enterobactin promotes mitochondrial iron uptake and development of the host via interaction with ATP synthase. *Cell, 175*, 571–582.e11.

Raymond, K. N., Dertz, E. A., & Kim, S. S. (2003). Enterobactin: An archetype for microbial iron transport. *Proceedings of the National Academy of Sciences of the United States of America, 100*, 3584–3588.

Reitz, Z. L., Butler, A., & Medema, M. H. (2022). Automated genome mining predicts combinatorial diversity and taxonomic distribution of peptide metallophore structures. 2022.12.14.519525 Preprint at https://doi.org/10.1101/2022.12.14.519525.

Reitz, Z. L., & Medema, M. H. (2022). Genome mining strategies for metallophore discovery. *Current Opinion in Biotechnology, 77*, 102757.

Roskova, Z., Skarohlid, R., & McGachy, L. (2022). Siderophores: An alternative bioremediation strategy? *The Science of the Total Environment, 819*, 153144.

Ross, A. R., Ikonomou, M. G., & Orians, K. J. (2000). Electrospray ionization of alkali and alkaline earth metal species. Electrochemical oxidation and pH effects. *Journal of Mass Spectrometry JMS, 35*, 981–989.

Saha, P., et al. (2020). Enterobactin induces the chemokine, interleukin-8, from intestinal epithelia by chelating intracellular iron. *Gut Microbes, 12*(1), 18.

Santus, W., et al. (2022). Mycobiota and diet-derived fungal xenosiderophores promote Salmonella gastrointestinal colonization. *Nature Microbiology, 7*, 2025–2038.

Sassone-Corsi, M., et al. (2016). Siderophore-based immunization strategy to inhibit growth of enteric pathogens. *Proceedings of the National Academy of Sciences, 113*, 13462–13467.

Schmid, R., et al. (2021). Ion identity molecular networking for mass spectrometry-based metabolomics in the GNPS environment. *Nature Communications, 12*, 3832.

Schmid, R., et al. (2023). Integrative analysis of multimodal mass spectrometry data in MZmine 3. *Nature Biotechnology, 41*, 447–449.

Seidell, A., & Linke, W. F. (1919). *Solubilities of inorganic and organic compounds* (2nd ed.). New York: D. Van Nostrand Company, 343.

Shannon, P., et al. (2003). Cytoscape: A software environment for integrated models of biomolecular interaction networks. *Genome Research, 13*, 2498–2504.

Skogvold, H. B., Sandås, E. M., Østeby, A., Løkken, C., Rootwelt, H., Rønning, P. O., ... Elgstøen, K. B. P. (2021). Bridging the Polar and Hydrophobic Metabolome in Single-Run Untargeted Liquid Chromatography-Mass Spectrometry Dried Blood Spot Metabolomics for Clinical Purposes. *Journal of Proteome Research, 20*(8), 4010–4021.

Soares, E. V. (2022). Perspective on the biotechnological production of bacterial siderophores and their use. *Applied Microbiology and Biotechnology, 106*(11), 3985–4004.

Stocks, B. B., & Melanson, J. E. (2018). In-source reduction of disulfide-bonded peptides monitored by ion mobility mass spectrometry. *Journal of the American Society for Mass Spectrometry, 29*, 742–751.

Strange, H. R., Zola, T. A., & Cornelissen, C. N. (2011). The fbpABC operon is required for ton-independent utilization of *Xenosiderophores* by *Neisseria gonorrhoeae* strain FA19. *Infection and Immunity, 79*, 267–278.

Tamara, S., den Boer, M. A., & Heck, A. J. R. (2022). High-resolution native mass spectrometry. *Chemical Reviews, 122*, 7269–7326.

Tillotson, G. S. (2016). Trojan horse antibiotics—A novel way to circumvent gram-negative bacterial resistance? *Infectious Diseases, 9*, 45–52.

Timofeeva, A. M., Galyamova, M. R., & Sedykh, S. E. (2022). Bacterial siderophores: Classification, biosynthesis, perspectives of use in agriculture. *Plants, 11*, 3065.

Tsednee, M., Huang, Y.-C., Chen, Y.-R., & Yeh, K.-C. (2016). Identification of metal species by ESI-MS/MS through release of free metals from the corresponding metal-ligand complexes. *Scientific Reports, 6*, 26785.

Valdebenito, M., Crumbliss, A. L., Winkelmann, G., & Hantke, K. (2006). Environmental factors influence the production of enterobactin, salmochelin, aerobactin, and yersiniabactin in *Escherichia coli* strain Nissle 1917. *International Journal of Medical Microbiology, 296*, 513–520.

Vijayaraj, A. S., Mohandass, C., & Joshi, D. (2020). Microremediation of tannery wastewater by siderophore producing marine bacteria. *Environmental Technology, 41*, 3619–3632.

Walker, L. R., Tfaily, M. M., Shaw, J. B., Hess, N. J., Paša-Tolić, L., & Koppenaal, D. W. (2017). Unambiguous identification and discovery of bacterial siderophores by direct injection 21 Tesla Fourier transform ion cyclotron resonance mass spectrometry. *Metallomics, 9*(1), 82–92.

Wang, M., et al. (2016). Sharing and community curation of mass spectrometry data with global natural products social molecular networking. *Nature Biotechnology, 34*, 828–837.

Wang, L., et al. (2017). Diisonitrile natural product SF2768 functions as a chalkophore that mediates copper acquisition in *Streptomyces thioluteus*. *ACS Chemical Biology, 12*, 3067–3075.

Wang, N., et al. (2024). Microbiome convergence enables siderophore-secreting-rhizobacteria to improve iron nutrition and yield of peanut intercropped with maize. *Nature Communications, 15*, 839.

Waska, H., Koschinsky, A., & Dittmar, T. (2016). Fe- and Cu-complex formation with artificial ligands investigated by ultra-high resolution Fourier-transform ion cyclotron resonance mass spectrometry (FT-ICR-MS): Implications for natural metal-organic complex studies. *Frontiers in Marine Science, 3*.

Wilkinson, M. D., Dumontier, M., Aalbersberg, Appleton, G., Axton, M., ... Mons, B. (2016). Comment: The FAIR Guiding Principles for scientific data management and stewardship. *Scientific Data, 3*, 1–9.

Zhu, W., et al. (2020). Xenosiderophore utilization promotes *Bacteroides thetaiotaomicron* resilience during colitis. e8 *Cell Host & Microbe, 27*, 376–388.e8.

Zytnick, A. M., et al. (2023). Discovery and characterization of the first known biological lanthanide chelator. 2022.01.19.476857 Preprint at https://doi.org/10.1101/2022.01.19.476857.

CHAPTER FIFTEEN

Anaerobic heme recycling by gut microbes: Important methods for monitoring porphyrin production

Ronivaldo Rodrigues da Silva, Arnab Kumar Nath, Victoria Adedoyin, Emmanuel Akpoto, and Jennifer L. DuBois*
Department of Chemistry and Biochemistry, Montana State University, Bozeman, MT, United States
*Corresponding author. e-mail address: jennifer.dubois1@montana.edu

Contents

1. Introduction		354
2. Methods		355
	2.1 Preparing heme and porphyrin-containing extracts from biological materials	355
3. Procedure		355
	3.1 Discontinuous quantification of hemes and porphyrins by HPLC	358
	3.2 Quantification of heme and porphyrins by UV–visible (UV–vis) and fluorescence emission spectroscopy	360
	3.3 Distinguishing porphyrins with catabolic and anabolic origins using stable isotopes and LC-MS	364
References		370

Abstract

Heme is the most abundant species of iron inside the human body and an essential cofactor for numerous electron/chemical group transfer reactions and catalyses, especially those involving O_2. Whole anaerobic biomes exist that also depend on heme but lack widespread, O_2-dependent pathways for heme synthesis and breakdown. The gastrointestinal tract is an anaerobic ecosystem where many microbes are auxotrophic for heme, and where the abundant members of the Bacteroidetes phylum convert heme into iron and porphyrins. Working with mixtures of these hydrophobic compounds presents challenges for analyses, especially when their source is biological. In this brief chapter, we detail a handful of important methods and point out caveats necessary for their concurrent detection, separation, and quantification.

Methods in Enzymology, Volume 702
ISSN 0076-6879, https://doi.org/10.1016/bs.mie.2024.07.002
Copyright © 2024 Elsevier Inc. All rights are reserved, including those for text and data mining, AI training, and similar technologies.

353

1. Introduction

Sometimes, biochemistry astounds. Every second, the human body generates 2.5 billion new red blood cells, each one packed with O_2-transporting hemoglobin (Muckenthaler, Rivella, Hentze, & Galy, 2017). The new cells replace old ones which have been damaged, destroyed, and then scavenged by macrophages. The heme thus recovered is scrapped so that its iron can be recovered, limiting our dietary iron requirements to manageable amounts. Heme oxygenases catalyze the critical reaction in the iron recovery process, using O_2 to convert the heme into biliverdin, CO, and Fe^{2+}.

Yet, heme also serves essential roles and heme homeostasis must be maintained in anaerobic niches where heme oxygenases are largely unavailable. The human gastrointestinal (GI) tract is anaerobic, and the major species of the GI microbiome are anaerobes from the Bacteroidetes and Firmicutes phyla. Most lack a complete biosynthetic pathway for heme, though many require heme as a nutrient for growth.

We recently showed that the *hmu* operon, encoding a pathway for heme uptake and metabolism, is largely confined to and ubiquitous within the Bacteroidetes phylum. We observed that a representative of this phylum, *Bacteroides thetaiotaomicron*, can use heme as a sole source of iron, provided the 6-gene *hmuYRSTUV* pathway is intact (Meslé, Gray, Dlakić, & DuBois, 2023). The products of heme catabolism by these species include Fe(II), protoporphyrin IX (PpIX), and possibly other, modified porphyrins. In addition, some pathogenic species of the GI tract, such as *Helicobacter pylori*, secrete porphyrins (Hamblin et al., 2005), while others like *Salmonella enterica* thrive on the iron derived from heme (Sebastiàn et al., 2022). The GI tract is consequently an environment populated by heme and porphyrins, compounds with significant influence over both microbial and human health.

Understanding anaerobic heme and porphyrin metabolism requires methods for concurrently quantifying heme, PpIX, and other porphyrins, over time and in the context of other biological molecules. Importantly, heme and porphyrins have different solubilities and absorbance/fluorescence properties that complicate their co-analysis. The following chapter describes methods for the extraction, detection, and quantification of heme, PpIX, and related porphyrins, from microbial cultures and other biochemical extracts. These methods may be useful for understanding heme and porphyrin metabolism in anaerobic microbes and microbiomes.

2. Methods
2.1 Preparing heme and porphyrin-containing extracts from biological materials

Heme and PpIX are not soluble in water, making it necessary to use other solvents to solubilize them, such as DMSO and alkaline water (1 M NaOH) (Meslé et al., 2023). This section presents solvents that can be used to extract heme and PpIX from cellular lysates and to solubilize concentrated heme and porphyrins for use in enzymatic assays. In the case of more dilute enzymatic assays that consume or produce heme or porphyrins, addition of methanol (followed by centrifugation) is typically sufficient to precipitate the proteins and solubilize the tetrapyrrole substrates/products for analyses.

2.1.1 Equipment
FastPrep Lysis B-matrix tubes (for bacterial cells)
FastPrep 24 5 g instrument (MP Biomedicals™) or sonicator

2.1.2 Reagents
Acetonitrile
Dimethyl sulfoxide (DMSO)
Ethyl acetate
Hemin chloride
Hydrochloric acid (HCl)
Protoporphyrin IX

3. Procedure

Different solvent combinations can be used to extract either heme or PpIX, separately, from bacterial cells. Ethyl acetate/12 M HCl (3:1, v/v) is appropriate for extracting heme, and acetonitrile/12 M HCl (82:18, v/v) for PpIX. For significant extraction of heme and PpIX together from the cell pellet, a solution composed of acetonitrile/12 M HCl/DMSO (41/9/50, v/v/v) may be used (Fig. 1). Details for each of these methods are shown below and in Table 1.

Ethyl acetate/12 M HCl protocol: To extract heme from cell pellets while minimizing extracted proteins, it is recommended to use acidified ethyl acetate in aqueous solution, which creates two layers, facilitating the recovery of heme.

Resuspend 0.1–0.2 g of cell pellet in 700 μL of 20 mM Tris-HCl (pH 7.4) and transfer it into a 2 mL tube containing silica beads to lyse the cells: e.g.

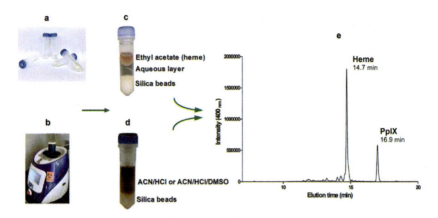

Fig. 1 Schematic illustrating heme/PpIX extraction steps. (A) Empty tubes containing 0.1 mm silica spheres; (B) FastPrep 24 5 g instrument, MP Biomedicals™ used for shaking the silica beads tubes for cell lysis; (C) Lysed cells (Ethyl acetate/HCl protocol) after centrifugation; (D) Lysed cells (ACN/12 M HCl, 82/18 v/v; or ACN/12 M HCl/DMSO, 41/9/50, v/v) after centrifugation; (E) Heme and PpIX separation via HPLC (C18 column).

FastPrep Lysis B-matrix tubes/FastPrep 24 5 g instrument, MP Biomedicals™, run for 2 cycles: 6.0 m/s for 40 s. Then, add 300 μL of ethyl acetate and 90 μL of 12 M HCl to the lysed cell mixture, mix gently for 20 min and centrifuge the tubes at 10,000 g for 10 min, room temperature. Next, collect the upper layer (ethyl acetate) containing the extracted heme, dry it in a vacuum centrifuge at room temperature and resuspend it in DMSO. This protocol creates two solvent layers, the upper ethyl acetate layer and the lower aqueous layer, through which the heme is easily collected from the upper (reddish) layer and separated from the other water-soluble cell components. The extraction can be repeated up to 3 times to ensure quantitative release of heme from the cellular material, and the extracts concentrated by drying on a rotary evaporator or vacuum centrifuge (speed-vac). For safety reasons, handle HCl in a fume hood.

Advantages and Limitations of Ethyl acetate/HCl protocol: With the formation of two layers (ethyl acetate and aqueous layers), this protocol allows for a faster extraction of heme in a reduced volume, free or with little protein content, but does not allow a considerable co-recovery of PpIX.

Acetonitrile/12 M HCl (82:18, v/v) and acetonitrile/12 M HCl/DMSO (41/9/50, v/v) protocols. Resuspend 0.1–0.2 g of cell pellet in 700 μL of acetonitrile/12 M HCl (82:18, v/v) for extracting PpIX or acetonitrile/12 M HCl/DMSO (41/9/50, v/v) for extracting heme and PpIX together. In either case, transfer the cell suspension into a 2 mL tube containing silica

Table 1 Methods for extracting heme and PpIX from bacterial cells.

Methods	Description
Ethyl acetate/HCl for extracting heme	
Step 1	Resuspend 0.1–0.2 g cell pellet in 700 μL of 20 mM Tris-HCl (pH 7.4)
Step 2	Cell lysis using a FastPrep Lysis B-matrix tubes (2 cycles: 6.0 m/second for 40 s)
Step 3	In a fume hood, add 300 μL of ethyl acetate and 90 μL of 12 M HCl
Step 4	Mix gently for 20 min and centrifuge the tubes at 10,000 g for 10 min at room temperature
Step 5	Collect heme from the upper layer (ethyl acetate)
Step 6	Finally, dry the sample in a vacuum centrifuge (room temperature), resuspend it in DMSO and load onto C18 column or store the dried sample at −20 °C, protected from light, until analysis
Acetonitrile/HCl for extracting PpIX; Acetonitrile/HCl/DMSO for PpIX and heme	
Step 1	For PpIX: Resuspend 0.1–0.2 g of cell pellet in 700 μL of acetonitrile/12 M HCl (82:18, v/v) For PpIX and heme: Resuspend cell pellet in acetonitrile/12 M HCl/DMSO (41/9/50, v/v)
Step 2	Cell lysis using a FastPrep Lysis B-matrix tubes (2 cycles: 6.0 m/s for 40 s)
Step 3	Centrifuge the tubes at 10,000 g for 10 min at room temperature and collect the supernatant
Step 4	Finally, for quantification, load the sample onto C18 column via HPLC and monitor the porphyrins at 400 nm

beads to lyse the cells (see above). After cell lysis, centrifuge the tubes at $10,000 \times g$ for 10 min and collect the supernatant, which can be loaded onto the C18 column for analysis by HPLC. If concentration is necessary, dry the sample in a vacuum centrifuge at room temperature and resuspend it in a smaller volume of the same extraction solvent. For safety reasons, handle HCl in a fume hood.

Advantages and Limitations of Acetonitrile/HCl protocol: This protocol allows better PpIX extraction and low heme content, with little to no protein content.

Advantages and Limitations of Acetonitrile/HCl/DMSO protocol: This allows considerable extraction of heme and PpIX, with little to no protein content.

When heme or PpIX is used as a substrate in individual enzymatic reactions, such as that catalyzed by the heme-degrading HmuS (Meslé et al., 2023), the ACN/12 M HCl/DMSO solution (41/9/50, v/v) is able to solubilize both the heme and the PpIX to be quantified by HPLC, and also to denature the enzyme, interrupting the catalysis. Dilute, denatured enzymes can be precipitated by brief centrifugation.

3.1 Discontinuous quantification of hemes and porphyrins by HPLC

Given the differences in polarity of the porphyrins, reversed-phase high performance liquid chromatography (HPLC) is suitable for separating many of them. Fig. 2 shows the separation of, for example, (1) deuteroheme, (2) heme, (3) mesoporphyrin IX, and (4) protoporphyrin IX using

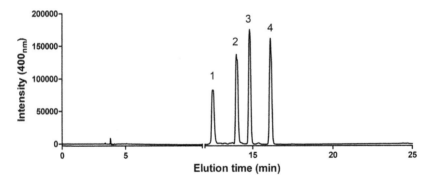

Fig. 2 Separation of the porphyrins by reversed-phase chromatography using a C18 column (Thermo Scientific, 250 × 4.6 mm) coupled to an HPLC instrument. The elution of the standards was monitored at 400 nm. The chromatography was carried out using a linear gradient between Solution A (ultrapure water + 0.1% trifluoroacetic acid, TFA) and Solution B (acetonitrile + 0.1% TFA), flow rate of 1 mL/min, oven temperature at 25 °C. The numbers from 1 to 4 represent the standard peaks and their elution time: (1) deuteroheme, eluted at 12.5 min/63% solution B; (2) heme, eluted at 14 min/71% solution B; (3) mesoporphyrin IX, eluted at 14.8 min/75% solution B; and (4) protoporphyrin IX, eluted at 16.1 min/81% solution B. Several factors can alter the elution time, including the size and type of column, column temperature and the mobile phase used.

a C18 column. A mixture composed of equimolar concentrations of these four porphyrins was loaded onto the C18 column to monitor their separation as the concentration of acetonitrile in water was increased (see method below).

3.1.1 Equipment
HPLC instrument (e.g., Prominence-i LC-2030 C 3D Plus, Shimadzu)
C18 column (e.g., Thermo Scientific, 250×4.6 mm)

3.1.2 Reagents
Acetonitrile
DMSO
HCl
Trifluoroacetic acid (TFA)
Protoporphyrin IX
Hemin chloride

3.1.3 Procedure
Reverse phase chromatography for separating and quantifying porphyrins. These porphyrins can be separated via reversed-phase chromatography using a linear gradient, i.e., where the stationary phase is nonpolar and the mobile phase is polar. Solution A (i.e. water) is considerably more polar than Solution B (i.e. acetonitrile or methanol). The more nonpolar the porphyrin, the stronger its interaction with the column. More weakly bound porphyrins elute first when the concentration of acetonitrile increases (Solution B). Regarding the porphyrins shown in Fig. 2, the sequence from most to least polar is: deuteroheme > heme > mesoporphyrin IX > protoporphyrin IX.

The method for separating the porphyrins shown here employed a C18 column coupled to an HPLC instrument, a linear gradient with solution A (ultrapure water + 0.1% trifluoroacetic acid, TFA) and solution B (acetonitrile + 0.1% TFA), flow rate of 1 mL/min, oven temperature at 25 °C, and monitoring at 400 nm. For absolute quantification, standards are used to prepare a standard curve (see note 2). Details for preparing a standard curve for hemin chloride and PpIX are described below:
- Prepare a 2 mM stock solution of hemin chloride dissolved in DMSO and a 2 mM solution of PpIX dissolved in acetonitrile/12 M HCl (82/18, v/v)
- Dilute each stock solution in the same solvents to get 200 µM concentration of each one (10 times dilution): 100 µL of 2 mM standards + 900 µL of DMSO for hemin chloride and acetonitrile/12 M HCl (82/18, v/v) for PpIX.

- Mix equal volumes of the 200 µM standard solutions to get 100 µM of each standard (2 times dilution): 500 µL of 200 µM hemin chloride + 500 µL of 200 µM PpIX.
- From the mixture containing both standards at 100 µM, perform dilutions using the solution acetonitrile/12 M HCl/DMSO (41/9/50, v/v) to get different concentrations from 0.1 to 100 µM standards.
- Inject these different standard mix solutions onto a C18 column coupled to an HPLC instrument. Monitor the standard elution at 400 nm.
- To construct the standard curve, plot the peak areas against the respective concentrations of hemin chloride and PpIX (from 0.1 to 100 µM). Linear regression is used to obtain the linear equation (note 3).

3.1.4 Notes

1. The elution time of these porphyrins can vary depending on a number of factors, including the size and type of column, the column temperature, and the mobile phase used. In addition, elution times tend to shift backwards (toward the more polar region) as the column undergoes prolonged use.
2. To prepare the samples for chromatography, it is recommended that they be protected from light and filtered with a 0.22 µm syringe filter before being injected onto the C18 column.
3. Calibration curves are used to determine the concentration of samples. Therefore, weigh and dilute the standards precisely. To be accurate, the experimental conditions for generating the standard curve need to be the same as for the samples, including the same chromatographic column, mobile phase, run temperature, volume of sample injected, monitoring absorbance, and solvents used to dissolve the samples.

3.2 Quantification of heme and porphyrins by UV–visible (UV–vis) and fluorescence emission spectroscopy

Extracted heme and porphyrins can be detected and quantified using UV–vis (heme) and fluorescence (PpIX). The focus here is on heme b (the complex of iron with protoporphyrin IX) and PpIX, though various functionalized forms of each can be treated similarly (Milgrom, L. R., 1997).

3.2.1 Equipment

Thermo Scientific NanoDrop
Thermo Scientific Micro 17 Microcentrifuge
Anaerobic Chambers, Coy Laboratory Products Inc.

Cary 60 UV-Vis spectrophotometer for absorbance measurement
BioTek Synergy H1 Multimode Reader for fluorescence measurement
pH meter, Thermo Scientific

3.2.2 Reagents
Tris base
NaCl
Hydrochloric acid (HCl)
Sodium hydroxide (NaOH)
Hemin chloride
Protoporphyrin IX (PpIX)
Polysorbate 80 or polyoxyethylene sorbitan monooleate (Tween 80)
Sodium dithionite ($Na_2S_2O_4$)

3.2.3 Procedure
Preparation and dilution of heme, PpIX, and related porphyrins. Limited aqueous
solubility of PpIX and heme demands modification of standard biological
buffers for the preparation of stock solutions, either by adding a detergent
or using alkaline solution. Here, PpIX is solubilized in an experimental
buffer (20 mM Tris-HCl, 250 mM NaCl (pH 7)) containing 4% Tween 80
by volume. Tween 80 is a nonionic surfactant and emulsifier often used to
dissolve hydrophobic solutes in aqueous solvent.

The low solubility of heme in water is addressed by dissolving solid hemin
chloride in 1 M NaOH solution and then diluting the stock to a desired
concentration using the experimental buffer. The alkaline water deprotonates
the propionic acid side chains, thus making the heme macrocycle ionic and
more readily dissolved. Stocks of PpIX and hemin can be conveniently
generated at 0.2 mM and 5 mM concentrations in these solvents.

Determining the concentrations of heme and PpIX. The concentrations of heme
and PpIX are measured on the Cary 60 UV-Vis spectrophotometer:

- The PpIX stock solution is prepared in 20 mM Tris-HCl, 250 mM NaCl
  (pH 7) containing 4% Tween 80. 50 μL of the stock PpIX is added to
  2 mL of the same buffer (2050 μL total volume) to record the spectrum.
- The concentration of PpIX is determined from Beer's Law (absorbance =
  ε × concentration × path length) using $\varepsilon = 262$ mM^{-1}cm^{-1} at 405 nm.
  The pH of the heme solution is adjusted to 8 before further use.
- The heme stock solution is prepared in 1 M NaOH and 10 μL of the
  stock heme is diluted to 2 mL using the experimental buffer above
  (2010 μL total volume).

- The concentration of heme is measured from the recorded spectrum using $\varepsilon = 58.44 \, mM^{-1}cm^{-1}$ at 385 nm. (See Mukherjee, Seal, & Dey, 2014; Pal, Roy, & Dey, 2021; Yi & Ragsdale, 2007).

Absorption properties of protoporphyrin IX (PpIX) and heme. Porphyrin macrocycles, both metal-free and metal-bound, exhibit characteristic absorption properties due to transitions between porphyrin π and π^* orbitals (Gouterman, 1978). The most intense absorption band, known as the Soret, is in the near UV region. Less intense bands in the visible region are known as the Q-bands. A major difference in the absorbance properties of metal-free and metal-bound porphyrins is in the number of Q-bands. The unbound porphyrins have four distinct Q-bands. Incorporation of a metal ion into the cavity of the macrocycle reduces the number of Q-bands to two (called α and β) because of the increased symmetry of the ring resulting from the deprotonation of the pyrrole nitrogen atoms (Milgrom, 1997). Additional metal-to-ligand or ligand-to-metal charge transfer bands may be present if the metal and porphyrin have available orbitals (occupied and unoccupied by electrons, respectively) with complementary symmetry and energies.

The reddish-purple solution of PpIX in pH 7 buffer containing Tween 80 has a Soret band at 405 nm and four Q-bands at 505, 540, 575, and 630 nm (Fig. 1, purple spectrum). However, absorption properties of the Fe(III) bound PpIX are completely different from that of PpIX with a broad Soret at 385 nm and α and β bands at 496 and 620 nm (green spectrum).

3.2.4 Notes

1. It is always recommended to use the protoporphyrin IX (PpIX) solution in the dark to avoid photoexcitation, which can lead to further reaction with O_2. Solutions under N_2 atmosphere will remain stable for longer times.
2. For absorption spectroscopy it is recommended to use the heme and protoporphyrin solutions within 3 h after preparation.

Reduction of ferric heme (Fe(III)) to ferrous heme (Fe(II)) by sodium dithionite $(Na_2S_2O_4)$ and measurement of the absorbance spectrum.

- Before the reduction process, the oxidized heme solution is degassed using a slow purge of inert N_2 gas over the surface of the solution in a septum-sealed vial or round bottom flask, taking care not to dry the solution.
- The reduction may be carried out inside the septum-sealed vessel using a gas-tight syringe to introduce the degassed reductant, or by bringing degassed solutions into a glove box, maintaining an atmosphere of 2.5% H_2/97.5% N_2.

- 450 μM Na$_2$S$_2$O$_4$ (prepared using degassed buffer inside the glove box) is added to the degassed 150 μM heme stock solution (three-fold excess compared to heme concentration).
- 20 mM Tris-HCl, 250 mM NaCl (pH 7) is measured as a blank in a screw cap sealed quartz cuvette with 1 cm path length.
- The spectrum of the reduced heme is measured in a screw cap or septum sealed quartz cuvette with 1 cm path length.

Absorption properties of oxidized and reduced heme. Reduction of the oxidized heme yields an absorbance spectrum with a sharpened Soret band at 385 nm, a prominent visible band at 580 nm, with a shoulder at 550 nm (Fig. 3B).

Fluorescence spectroscopy of Protoporphyrin (PpIX).
- The PpIX stock solution is prepared in 20 mM Tris-HCl, 250 mM NaCl (pH 7) containing 4% Tween 80. 50 μL of the stock PpIX is added in 2 mL of the buffer to record the spectrum.
- The concentration of PpIX in the stock is confirmed by measuring its UV–vis spectrum in the Cary 60 UV-Vis spectrophotometer using $\varepsilon = 262 \text{ mM}^{-1}\text{cm}^{-1}$ at 405 nm.
- The PpIX stock solution is serially diluted in the working buffer above to different final concentrations from 0.1 μM up to 20 μM.
- The fluorescence spectra of the diluted stocks are measured using a BioTek Synergy H1 Multimode Reader. The emission spectra are recorded with the excitation wavelength at 400 nm.

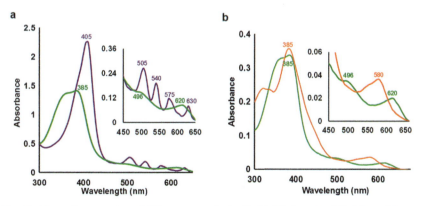

Fig. 3 Absorption spectra of (A) protoporphyrin IX (PpIX) (purple, 8.5 μM) and heme (green, 25 μM) and (B) oxidized (Fe(III), green) (6 μM) and reduced heme (Fe(II), red) (6 μM) measured in 20 mM Tris-HCl, 250 mM NaCl (pH 7) (4% Tween 80 added to PpIX).

Fluorescence properties of Protoporphyrin (PpIX). The extended conjugation of porphyrins makes them strongly fluorescent. Metallated porphyrins with available d-orbitals can re-absorb the emitted photons, thereby quenching the fluorescence spectrum. Heme is consequently not fluorescent.

Fluorescence emission from PpIX is recorded with the excitation wavelength at 400 nm. The emission spectrum has a maximum intensity band (λ_{max}) at 640 nm with another low-intensity band at 703 nm (not shown) (Fig. 4). The emission spectrum changes for natural precursors or variants of PpIX with different functional groups around the periphery of the macrocycle or with *N*-alkylation. The intensity of PpIX emission likewise increases with an increase in PpIX concentration up to 5 µM under the conditions used here (Fig. 4); however, the intensity is found to decrease at higher PpIX concentration (from 7 to 20 µM), possibly due to PpIX aggregation.

3.3 Distinguishing porphyrins with catabolic and anabolic origins using stable isotopes and LC-MS

Using radioactive labels is a classic way of tracing the metabolism of a compound from its initial introduction into a biological system to its ultimate fate. Stable isotopic labeling offers a safer and more ecological alternative that moreover opens the door to spectroscopic approaches that specifically detect the labeled atoms. Magnetically active nuclei can be specifically monitored using proton (^{1}H), carbon (^{13}C), and nitrogen (^{15}N) nuclear magnetic resonance (NMR). NMR can then be used to determine the structures of the endpoint metabolites. Similarly, mass-shifted metabolites can be identified in

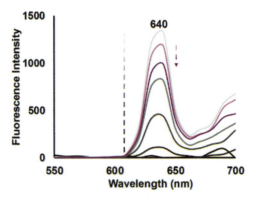

Fig. 4 Fluorescence emission spectra PpIX at concentrations from 0.1 to 20 µM (pH 7). The intensity of PpIX fluorescence emission increases from 0.1 µM to 7 µM as indicated by the ascending lines from black to gray. PpIX emission spectra at 10 and 20 µM concentrations are demonstrated by purple lines (light to dark gradient color).

liquid chromatography mass spectrometric (LC–MS) measurements and their structures predicted based on exact masses and fragmentation patterns.

To investigate heme metabolism by bacteria, stable-isotope labeled heme can be used as a component of a chemically defined culture medium, which allows the analysis of heme metabolites by NMR and LC–MS. Isotopically-labeled heme can be obtained from recombinant heme proteins, like the soluble fragment of cytochrome b5, expressed in *Escherichia coli* strains cultivated in a medium containing isotopically-labeled ^{15}N or ^{13}C components, such as ^{15}N ammonium chloride or ^{13}C glucose (used as sole sources of carbon and nitrogen, respectively). Alternatively, ^{13}C labeled 5-aminolevulinic acid (ALA), a heme biosynthetic precursor, can be added to the growth medium at the time of induction of protein expression (Rivera & Walker, 1995; Rodríguez-Marañón et al., 1996). This method is especially useful if it is desirable to label only specific locations on the tetrapyrrole. Labeled heme must be extracted from the cytochrome b5 or other protein carrier, and then introduced into the bacterial culture, mouse chow, enzymatic reaction, etc. Details of the production (based on the methods of Rivera and Walker) and extraction of labeled heme are given below.

3.3.1 Equipment
Autoclave
Shaker incubator
Electrophoresis system
Bruker 600 MHz Avance III solution NMR spectrometer
Agilent 1290 infinity UHPLC coupled to an Agilent 6538 high-resolution Q-TOF mass analyzer

3.3.2 Reagents
Ampicillin
Ethyl acetate
HCl
KH_2PO_4
$MgSO_4$
NaCl
Na_2HPO_4
$CaCl_2$
$FeCl_3$
Glucose or 13C-glucose
Ammonium chloride or 15N-ammonium chloride

Luria-Bertani medium (LB medium)
Isopropyl β-D-1-thiogalactopyranoside (IPTG)
DEAE resin

3.3.3 Procedure

Preparation of Labeled Minimal Medium and Generation of Labeled Heme. A minimal medium typically comprises salts such as Na_2HPO_4, KH_2PO_4, NH_4Cl, and NaCl, supplemented with a carbon source. M9 salts are a commonly used medium for bacterial growth in molecular biology experiments. The following is a recipe for preparing 1 liter of M9 medium and the general steps (not detailed) for obtaining isotopically-labeled heme.

- Weigh: 6 g of Na_2HPO_4, 3 g of KH_2PO_4, 1 g of NH_4Cl and 0.5 g of NaCl
- Mix these salts in distilled, deionized water and adjust the pH to 7.0. Final volume 1 liter.
- Sterilize the solution by autoclaving at 121 °C for 15 min, 1 atm.
- Add to the 1 liter autoclaved solution: 20 mL of 20% ^{13}C-labeled glucose (w/v), 2 mL of 1 M $MgSO_4$, 0.1 mL of 1 M $CaCl_2$, 0.5 mL of 100 mM ferric chloride, and 1 mL of 100 mg/mL ampicillin (used as a selective marker, see note 1). Prior to mixing with the 1 liter autoclaved M9 salts, these solutions should be filtered using a 0.22 μm pore filter. If ^{15}N-labeled ammonium chloride was used to produce ^{15}N-labeled heme, add non-labeled glucose. Alternatively, the given amounts of labeled carbon or nitrogen source can be added at the time of protein induction, in which case the medium should contain half of the amount of the unlabeled C/N source indicated above.
- From a growth on a selective LB agar (containing a selective antibiotic, like ampicillin), select a colony of the *E. coli* competent cells containing a plasmid that harbors the gene encoding the soluble portion of cytochrome b5. Inoculate a single colong into LB liquid medium and incubate overnight at 37 °C, 180 rpm; it is the starter culture. Next, transfer 1/100 (v/v) of the starter culture to the 1 liter of labeled minimal medium and incubate in a shaker incubator at 37 °C, 180 rpm.
- At the mid-log phase of the *E. coli* growth, induce the expression of cytochrome b5 by adding IPTG (Isopropyl β-D-1-thiogalactopyranoside), incubate the culture at 25 °C and determine the best time to collect the bacterial cells (see note 2). The labeled compound could be added at the point of protein expression induction, as stated above.
- Harvest the cells by centrifugation and lyse them (by sonication) to obtain the cell free soluble proteins.

- Partially purify the labeled cytochrome b5 using an anion exchange resin (DEAE-FF) and 20 mM Tris-HCl buffer, pH 8.0. This step ensures the isolation of labeled cytochrome b5 from other cellular components.
- Assess the purity of the labeled protein using Sodium Dodecyl Sulfate-Polyacrylamide Gel Electrophoresis (12% SDS-PAGE) and determine the concentration of the purified labeled cytochrome b5 using a quantitative assay such as the Bradford assay (Bradford, 1976).
- Extract the labeled heme from the partially purified cytochrome b5 using a suitable method, like the ethyl acetate protocol described in section A.
- Check the purity of the labeled-heme through reversed-phase chromatography, as described in section B.
- Now, the extracted labeled heme is used to feed the bacteria as part of the growth medium. After bacterial growth, the cell pellets are harvested by centrifugation, resuspended in acetonitrile/12 M HCl (82/18, v/v) and lysed to obtain heme metabolites for LC-MS (Fig. 5) or NMR analyses (Fig. 6).

Detection and quantification of heme and PpIX by LC-MS. The same procedure as described in sections A–C to generate both heme and PpIX standards of varying concentrations and bacterial cell extracts suitable for LC-MS analysis. Because HPLC detects only chromophoric compounds, it tends to underestimate the diversity of compounds retained in cellular extracts. By contrast, LC-MS only requires that the compounds can be successfully ionized and detected. The LC-MS total ion chromatogram (TIC) consequently reveals several additional compounds. Quantification is achieved by measuring standards of varying concentrations and integrating their peaks in the ion chromatograms. These standard curves are then compared with integrated peak intensities from the extracted ion chromatograms (EICs) for heme and PpIX (and additional porphyrins if required), respectively. EICs are extracted based on the expected exact theoretical masses of the metabolites of interest. The mass spectra measured for the heme and PpIX confirm the compounds' identities.

To measure LC-MS, porphyrins separation was carried out using the column Agilent Eclipse C18 (2.1 × 50 mm) coupled to an HPLC instrument, linear gradient with the mobile phase composed of water + 0.1% formic acid (solution A) and acetonitrile + 0.1% formic acid (Solution B) and a flow rate of 0.6 mL/min. Mass Spectrometer (detector) Agilent 6538 Q-TOF MS, (+) electrospray ionization, m/z 50–1700, 4 Hz scan speed, collision energy 25, top 2 and the Agilent MassHunter Acquisition 10.0 software (Fig. 5).

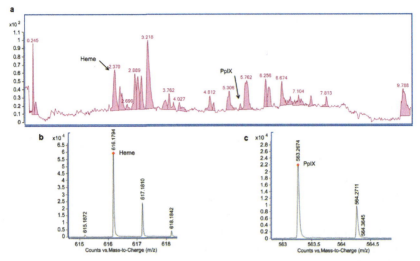

Fig. 5 Electrospray ionization (ESI) mass spectrometry of metabolites extracted from the human gut symbiont bacterium *Bacteroides thetaiotaomicron*. (A) Total Ion Chromatogram (TIC) from metabolites, arrows indicate heme and PpIX. (B and C) ESI spectrum of heme (616.1794 *m/z*) and PpIX (563.2674 *m/z*). The bacterium was cultivated in a minimal medium containing 15 μM hemin chloride (pH 7.1), under an anoxic atmosphere (2.5% H_2/97.5% N_2) at 37 °C. Cells were collected at mid-log phase growth and resuspended in ACN/12 M HCl to be lysed, see section A.

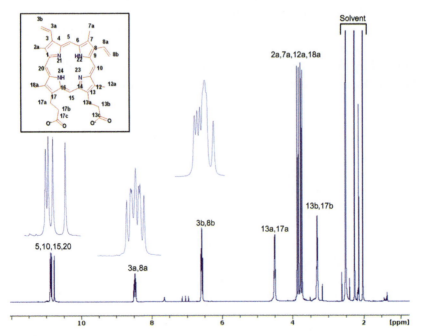

Fig. 6 ^{1}H NMR spectrum of standard PpIX dissolved in ACN/12 M HCl (82/18, v/v).

Detection of PpIX by NMR. Ferric heme and low-spin ferrous heme are paramagnetic and consequently difficult to characterize by NMR. However, PpIX and its metabolites can be readily identified by NMR. If the heme is ^{13}C or ^{15}N labeled, then these heteroatoms can also be identified among the downstream metabolites of PpIX and the compounds containing them can be structurally characterized using multinuclear/multidimensional NMR methods.

Depending on the scientific question being asked, cellular extracts may be prepared as described above. These extraction procedures are intended to isolate either heme and relatively polar compounds, or PpIX and less polar compounds. Alternatively, both heme and PpIX can be coextracted. In instances where PpIX, its more polar breakdown products, and potentially other cellular metabolites are required, an extraction procedure that retains a broad spectrum of most of the cell's small molecules is desirable. However, it may be essential first to remove paramagnetic heme, for example, by binding it to an affinity column L-histidine-immobilized sepharose (HIS) resin (Asher & Bren, 2010). Alternatively, the cellular lysates may be treated with a mixture of commercially available heme oxygenase and biliverdin reductase in order to convert the heme to bilirubin and free iron.

As an illustration, a PpIX spectrum (PpIX concentration $>10\,\mu M$) was recorded on a Bruker 600 MHz (^{1}H Larmor frequency) AVANCE III solution NMR spectrometer equipped with a 5 mm triple resonance (^{1}H, ^{13}C, ^{15}N), liquid helium-cooled cryoProbe, automatic sample loading system (SampleJet), and Topspin software (Billerica, MA, USA, Bruker version 3.2). The Bruker lc1pngpf2 (double solvent suppression, spoil gradient) pulse sequence was used for the acquisition of 1D ^{1}H NMR spectra, which was recorded at 300 K with 512 scans, 32 K data points, and a ^{1}H spectral window of 30 ppm (Fig. 6).

3.3.4 Notes

1. The antibiotic used in the media for *E. coli* growth depends on the selective marker gene in the plasmid used to express the recombinant protein.
2. The best time to collect *E. coli* cells after adding IPTG to the bacterial culture can be assessed on a small scale. Take culture samples at different time points, lyse the cells and check the protein expression by a 12% SDS-PAGE. Select the shortest time point that shows considerable protein expression.

References

Asher, W. B., & Bren, K. L. (2010). A heme fusion tag for protein affinity purification and quantification. *Protein Science, 19*, 1830–1839. https://doi.org/10.1002/pro.460.

Bradford, M. M. (1976). A rapid and sensitive for the quantification of microgram quantities of protein utilizing the principle of protein-dye binding. *Analytical Biochemistry, 72*, 248–254. https://doi.org/10.1006/abio.1976.9999.

Gouterman, M. (1978). *The porphyrins, Vol. III*, New York: Academic Press, Inc. 1–156.

Hamblin, M. R., Viveiros, J., Yang, C., Ahmadi, A., Ganz, R. A., & Tolkoff, M. J. (2005). *Helicobacter pylori* accumulates photoactive porphyrins and is killed by visible light. *Antimicrobial Agents and Chemotherapy, 49*(7), 2822–2827. https://doi.org/10.1128/AAC.49. 7.2822-2827.2005.

Meslé, M. M., Gray, C. R., Dlakić, M., & DuBois, J. L. (2023). *Bacteroides thetaiotaomicron*, a model gastrointestinal tract species, prefers heme as an iron source, yields protoporphyrin IX as a product, and acts as a heme reservoir. *Microbiology Spectrum, 11*, e0481522. https://doi.org/10.1128/spectrum.04815-22.

Milgrom, L. R. (1997). *The Colours of Life*. Oxford: Oxford University Press.

Muckenthaler, M. U., Rivella, S., Hentze, M. W., & Galy, B. (2017). A red carpet for iron metabolism. *Cell, 168*(3), 344–361. https://doi.org/10.1016/j.cell.2016.12.034.

Mukherjee, S., Seal, M., & Dey, S. G. (2014). Kinetics of serotonin oxidation by heme-Aβ relevant to Alzheimer's disease. *Journal of Biological Inorganic Chemistry, 19*, 1355–1365. https://doi.org/10.1007/s00775-014-1193-7.

Pal, I., Roy, M., & Dey, S. G. (2021). Interaction of apomyoglobin with heme-hIAPP complex. *Journal of Inorganic Biochemistry, 216*, 111348. https://doi.org/10.1016/j. jinorgbio.2020.111348.

Rivera, M., & Walker, F. A. (1995). Biosynthetic preparation of isotopically labeled heme. *Analytical Biochemistry, 230*, 2295–2302. https://doi.org/10.1006/abio.1995.1477.

Rodríguez-Marañón, M. J., Qiu, F., Stark, R. E., White, S. P., Zhang, X., Foundling, S., ... Rivera, M. (1996). ^{13}C NMR spectroscopic and X-ray crystallographic study of the role played by mitochondrial cytochrome b5 heme propionates in the electrostatic binding to cytochrome c. *Biochemistry, 35*, 16378–16390. https://doi.org/10.1021/bi961895o.

Sebastián, V. P., Moreno-Tapia, D., Melo-González, F., Hernández-Cáceres, M. P., Salazar, G. A., Pardo-Roa, C., ... Bueno, S. M. (2022). Limited heme oxygenase contribution to modulating the severity of *Salmonella enterica* serovar Typhimurium Infection. *Antioxidants (Basel), 11*, 1040. https://doi.org/10.3390/antiox11061040.

Yi, L., & Ragsdale, S. W. (2007). Evidence that the heme regulatory motifs in heme oxygenase-2 serve as a thiol/disulfide redox switch regulating heme binding. *Journal of Biological Chemistry, 282*, 21056–21067. https://doi.org/10.1074/jbc.M700664200.

CHAPTER SIXTEEN

Predicting metallophore structure and function through genome mining

Zachary L. Reitz* [iD]

Department of Ecology, Evolution and Marine Biology, University of California, Santa Barbara, CA, United States
*Corresponding author. e-mail address: zlreitz@gmail.com

Contents

1. Introduction		372
2. Materials and methods		374
	2.1 Retrieving records from the antiSMASH database	374
	2.2 Running antiSMASH jobs on the web server	374
	2.3 Retrieving records from MIBiG	375
	2.4 Running BLAST	376
3. Metallophore genomics: from sequence to molecule to function		376
	3.1 Biosynthesis of metal-chelating small molecules	377
	3.2 Active transport through the cell membrane and subsequent metal release	378
	3.3 Metal-responsive regulatory elements	379
4. Predicting metallophore pathways in antiSMASH		381
	4.1 Rule-based BGC detection	381
	4.2 Navigating antiSMASH results	382
	4.3 Identifying putative metallophore BGC regions among antiSMASH results	384
5. Predicting metallophore structure through comparative analyses		385
	5.1 Strategies for comparing BGCs	385
	5.2 Predicting the structure of a peptide metallophore from *Dickeya dadantii* 3937	387
	5.3 Predicting the structure of a putatively novel metallophore in *Acinetobacter pittii* PHEA-2	390
6. Next steps for the reader		396
Acknowledgments		398
References		398

Abstract

Metallophores are small molecule chelators that many microbes use to obtain trace metals from their environment. Through *genome mining*, where genomes are scanned for metallophore biosynthesis genes, one can not only identify which organisms are likely to produce a metallophore, but also predict the metallophore structure, thus preventing undesired reisolation of known compounds and

Methods in Enzymology, Volume 702
ISSN 0076-6879, https://doi.org/10.1016/bs.mie.2024.06.007
Copyright © 2024 Elsevier Inc. All rights are reserved, including those for text and data mining, AI training, and similar technologies.

371

accelerating characterization. Furthermore, the presence of accessory genes for the transport, utilization, and regulation can suggest the biological function and fate of a metallophore. Modern, user-friendly tools have made powerful genomic analyses accessible to scientists with no bioinformatics experience, but these tools are often not utilized to their full potential. This chapter provides an introduction to metallophore genomics and demonstrates how to use the free, publicly available antiSMASH platform to infer metallophore function and structure.

1. Introduction

Many microbes scavenge trace metals with *metallophores*, small molecules that chelate environmental metal ions and transport them back into the cell. Metallophores, particularly the Fe(III)-chelating siderophores, have drawn interdisciplinary attention for their natural roles in mediating a microbe's interactions with the environment, other microbes, and multi-cellular life (Kraemer, Duckworth, Harrington, & Schenkeveld, 2015; Kramer, Özkaya, & Kümmerli, 2020); as well as diverse applications to fields including agriculture, medicine, and bioremediation (Frei, Verderosa, Elliott, Zuegg, & Blaskovich, 2023; Gu et al., 2020; Soares, 2022). Hundreds of different metallophores have been discovered, each with unique physicochemical properties and biological activities (Hider & Kong, 2010; Kraemer et al., 2015). Most were found through traditional metallophore discovery techniques, where an organism of interest is grown under metal-starved conditions, and metal-chelating compounds are identified from the cell culture via assays then purified and characterized (Payne, 1994). By these methods, metallophore researchers often encounter the same problems facing other natural product discovery teams: many microbes do not produce detectable metallophores (at least under tested laboratory conditions), de novo structural elucidation by mass spectrometry and nuclear magnetic resonance can be difficult, and those immense efforts increasingly lead to rediscovery of known compounds (Cano-Prieto, Undabarrena, De Carvalho, Keasling, & Cruz-Morales, 2024).

Genome mining streamlines metallophore research by generating testable chemical and biological predictions in silico (Reitz & Medema, 2022). Genes encoding the production of a specialized metabolite are generally co-localized in the bacterial genome, forming *biosynthetic gene clusters* (BGCs). The genome miner can predict the biosynthetic capabilities of an organism of interest by using computational tools to scan its genome for putative BGCs. By inspecting the genes within a BGC, structural features of the corresponding metabolite can be predicted. These predicted features can inform isolation and characterization or allow for dereplication of an undesired compound. Furthermore, modern genome mining tools allow for the high-throughput analysis of many thousands of genomes, allowing for fast and cheap strain prioritization. In the past two decades, genome mining has revolutionized metallophore discovery (Reitz & Medema, 2022), as well as the field of natural products more broadly (Cano-Prieto et al., 2024).

Today, a number of automated tools for BGC prediction are available (Chavali & Rhee, 2018), the most popular of which is the antiSMASH platform (Blin et al., 2023; Medema et al., 2011). The latest version of antiSMASH (7.1) detects more than 80 different BGC families, including several classes of metallophores. Beside its scientific capabilities, its popularity is likely due in part to an easy-to-use graphical user interface and a free, publicly available web server, which has processed nearly 2 million jobs to date. Running antiSMASH has become a standard step in genome analysis pipelines, and the tool is routinely used by experts and non-experts alike. Unfortunately, most users do not know how to fully utilize the information provided in the antiSMASH output to predict gene functions and metabolite structures. Worse, many users treat antiSMASH (and other bioinformatics tools) as a black box and do not understand its limitations, leading to unsupported and incorrect conclusions.

Herein, I showcase genome mining as a powerful—albeit imperfect—tool for metallophore research. Rather than a comprehensive guide or review, this chapter is designed as a crash course to guide researchers unfamiliar with metallophore genomics towards generating useful hypotheses. I first describe how metallophore function can often be inferred from a gene cluster, then use several case studies to demonstrate the prediction of metallophore function and chemical structure, including the prediction of a novel siderophore from the emerging opportunistic pathogen *Acinetobacter pittii*. A significant portion of the text focuses on antiSMASH, which detects more metallophore BGCs than other automated tools and is suitable for beginners and experts alike. I outline specific features that are relevant for metallophore pathways and attempt to dispel some common misconceptions about antiSMASH that are relevant to researchers studying any class of natural products. Although genome mining does not require a lab coat, one must still develop research skills that become refined with practice. The methods demonstrated herein can all be performed in a web browser, and readers are encouraged to work through the exercises themselves.

2. Materials and methods

The following web-based tools and databases are used for the exercises in this chapter. Some additional genome mining tools of interest to the metallophore researcher are listed in Section 6. The usage, contents, and output formats of software and databases may differ across versions. Please read the official documentation and latest publication(s) for thorough and updated descriptions of their usage and capabilities.

2.1 Retrieving records from the antiSMASH database

Version 4 of the antiSMASH database contains precomputed antiSMASH v7.1 results for over 35,000 genomes from NCBI RefSeq (Blin, Shaw, Medema, & Weber, 2024). Accessing a database record, if available, can save both the user's time and the antiSMASH server's computational resources. Here, results for the *Dickeya dadantii* 3937 genome will be retrieved for use in Sections 4 and 5.

1. From a web browser, navigate to the antiSMASH database homepage at https://antismash-db.secondarymetabolites.org/ and click "Query".
2. Find the record of interest using one of several options under the "Select a category" dropdown menu. For *D. dadantii* 3937, any of the following will give identical results.
   - "Strain" = "3937"
   - "NCBI RefSeq Accession" = "NC_014500"
   - "NCBI assembly ID" = "GCF_000147055"
     Non-unique searches can also be used: "Species" = "dadantii" returns results for *D. dadantii* 3937 and two other strains.
3. Click "Search" to produce a list of matching BGC regions present in the database.
4. Click any of the rows with "*Dickeya dadantii* 3937" in the Species column to view antiSMASH results for the strain, then click "Overview" at the top of the page.

Interpretation of antiSMASH results towards predicting metallophore pathways and structures are discussed throughout the chapter.

2.2 Running antiSMASH jobs on the web server

The latest version of antiSMASH, 7.1.0, is available via the publicly available web server. Here, we will run antiSMASH on a single genomic contig from *Dickeya chrysanthemi* EC16. These results will be used for comparative genomics in Section 5.2.

1. From a web browser, navigate to the antiSMASH homepage at http://antismash.org.
2. Under "Data input", paste or type the relevant NCBI nucleotide accession: "JAFCAF010000006".
3. Under "Detection strictness", do nothing. All current metallophore detection rules (see Section 4.1) are under the "strict" category; however, the additional rules added in the "relaxed" category (see documentation) are generally not problematic and may be useful for catching unusual or fragmented BGCs.
4. "Under "Extra features", click "All off" for a slightly shortened runtime, or do nothing. No extra features will be needed for Section 5.2. See the documentation for a description of each feature. Entries in the antiSMASH database are equivalent to the results of running the same record on the antiSMASH web server with all Extra features.
5. Click Submit to be redirected to the status page. This small record should finish within minutes during normal server conditions.
6. When the green checkmark and "done" appear, click "results" to be directed to the overview page.

Note that a local file can also be uploaded to the web server. Genbank files are preferred to fasta files due to their gene annotation information. The file upload option allows for genomes with multiple records (ex: plasmids or contigs) to be run as a single job. Multi-record *genome assemblies* can be downloaded from NCBI and have accession numbers that start with "GCF_" (RefSeq) or "GCA_" (GenBank).

More options are available in the command line version. Full documentation is available at https://docs.antismash.secondarymetabolites.org/.

2.3 Retrieving records from MIBiG

Genome mining relies on gene homologs, and thus a database of known BGCs is an essential tool for genome miners. Currently, the most extensive collection of metallophore BGCs is found in the community-driven MIBiG repository (Terlouw et al., 2023). MIBiG is available at https://mibig.secondarymetabolites.org. Throughout the text, MIBiG v3.1 accessions (BGC000####) are given for BGCs when available. The corresponding entries can be retrieved from the "Search" or "Repository" pages of the website.

2.4 Running BLAST

NCBI BLAST can be used to find regions of similarity between sequences. Version 2.15.0 is available as a publicly available web server (Johnson et al., 2008). The web server has several subtypes that handle combinations of nucleotide versus protein sequences. Here tblastn is used, which accepts a protein sequence query and finds matches in a nucleotide sequence database. Using tblastn allows for straightforward examination of genes surrounding a match, which can help the researcher predict its function and relevance (see Section 5.3). In this example, we will search the genome of *Acinetobacter pittii* PHEA-2 for homologs of a decarboxylase involved in staphyloferrin B biosynthesis.

1. From a web browser, navigate to the NCBI BLAST web server at https://blast.ncbi.nlm.nih.gov and click the large "tblastn" button to be directed to the submission page. Note that the tblastn box at the top is highlighted blue.
2. Under "Enter Query Sequence", paste or type "SQF72452", the protein ID for *sbnH* (see Section 5.3).
3. Under "Choose Search Set", type "PHEA-2" into the "Organism" field, then click the dropdown entry "Acinetobacter pittii PHEA-2 (taxid:871585)".
4. Ignoring all other fields, click "BLAST" at the bottom of the screen and wait for results.

These results are interpreted in Section 5.3. Full documentation is available at https://blast.ncbi.nlm.nih.gov/doc/blast-help/.

3. Metallophore genomics: from sequence to molecule to function

Despite recent advances in metallophore genome mining techniques, there is no totally reliable method for automatically differentiating metallophore BGCs from other gene clusters (Reitz & Medema, 2022). Luckily, metallophore BGCs are among the easiest to identify manually, even for researchers unfamiliar with the intricacies of their biosynthesis. Metallophores are set apart from other natural products by their biological role in trace metal acquisition. The classic definition of a metallophore comprises three criteria: (1) metal ion chelation, (2) active transport of the metal-metallophore complex into the cell for use in metabolic functions, and (3) repression of biosynthesis by the transported metal (Drechsel & Winkelmann, 1997). Consequently, metallophore BGCs often show corresponding features rarely seen in other cluster types: (1) genes encoding chelator biosynthesis enzymes, (2) genes encoding machinery for metal-metallophore import and utilization, and (3) intergenic DNA sequences recognized by metal-sensing regulators (Reitz & Medema, 2022). This section provides a brief overview of metallophore genomics and practical considerations for their use in predicting whether a BGC encodes a metallophore pathway. Note that metallophores have been demonstrated to perform a variety of other biological functions, including toxic metal sequestration, signaling, antimicrobial activity, and oxidative stress protection (Johnstone & Nolan, 2015), and that even the presence of every expected metallophore feature cannot give an *in silico* guarantee that a molecule acts as a metallophore in vivo.

3.1 Biosynthesis of metal-chelating small molecules

The majority of metallophores are produced by one of two pathways: nonribosomal peptide synthetases (NRPSs) and the IucA/IucC-like enzymes commonly known as NRPS-independent siderophore (NIS) synthetases. NRPSs produce many families of nonribosomal peptides (NRPs) other than metallophores and are sometimes hybridized with polyketide synthase (PKS) genes. Structural predictions are facilitated by their assembly-line-like mechanism, details of which have been reviewed extensively (Miller & Gulick, 2016; Süssmuth & Mainz, 2017). NRP metallophore BGCs often contain one or more genes responsible for the biosynthesis of the chelating substructures, which can serve as markers to differentiate them from other NRPS systems (Reitz, Butler, & Medema, 2022). However, chelating moieties sometimes play non-metallophore roles. For example, the chelating catechol moiety of antibiotic obafluorin (BGC0001437) binds to Zn^{2+} in the active site of its target enzyme (Batey et al., 2023). NIS synthetases are purportedly specific to metallophore pathways, although a family member was recently found to play a role in the biosynthesis of a cytotoxic fungal compound (Yee et al., 2023). Distinct functional subtypes of NIS synthetases have been identified (Carroll & Moore, 2018); enhanced structural predictions can be made by subtyping a synthetase through phylogenetic methods reviewed previously (Adamek, Alanjary, & Ziemert, 2019; Schmitt & Barker, 2009). Several other biosynthetic classes of metallophores have been identified, as outlined in Table 1. Their detection in antiSMASH is discussed in Section 4.1.

Table 1 Biosynthetic classes of metallophores.

Class	Example	antiSMASH detection class
NRP metallophores	Amphibactins BGC0002465	NRP-metallophore (Added in v7.0) and/or NRPS(/PKS)
NRPS-independent siderophores	Aerobactin BGC0002682	NI-siderophore (Renamed from "siderophore" in v7.0)
Ribosomally synthesized and posttranslationally modified peptide (RiPP) chalkophores	Methanobactin LW3 BGC0002133	Methanobactin (Added in v7.0)
Opine-like metallophores	Staphylopine BGC0002487	Opine-like-metallophore (Added in v7.0)
Aminopolycarboxylic acid siderophores	EDHA BGC0002568	Aminopolycarboxylic-acid (Added in v7.0)
7-Hydroxytropolone siderophore	7-Hydroxytropolone (Yu, Chen, Jiang, Hu, & Xie, 2014)	Hydroxytropolone (Slated for next release)
Thiocarboxylic acid metallophores	Quinolobactin BGC0000925	*Not yet detectable*

3.2 Active transport through the cell membrane and subsequent metal release

Metallophore-metal complexes cross cell membranes through specific transporter proteins, in contrast to an ionophore that facilitates the passive membrane permeability of an ion (Frei et al., 2023). Most other classes of metabolites are not transported out of and into the cell, so genes encoding import machinery are among the clearest hallmarks of a metallophore BGC (Crits-Christoph, Bhattacharya, Olm, Song, & Banfield, 2020; Reitz & Medema, 2022). In Gram-negative bacteria, the canonical import pathway begins with import by a TonB-dependent outer membrane transporter (TBDT). The metallophore complex may then be escorted by a periplasmic binding protein (PBP) to an ATP-binding cassette (ABC) inner membrane transport complex. In Gram-positive bacteria, external complexes are recognized and transferred to the ABC importer by a homolog of the PBP called the siderophore binding protein (SBP), which is anchored at the cell surface. Crits-Christoph et al. found that genes encoding TBDTs, PBPs/SBPs, and type II ABC transporters—specifically using the Pfam entries TonB_dep_Rec, Peripla_BP_2, and FecCD, respectively—in a BGC were strong predictors of siderophore-like function (Crits-Christoph et al., 2020). However, many known metallophores do not follow this general pathway. For example, ferric legiobactin enters the periplasm through the TonB-independent transporter LbtU (Chatfield, Mulhern, Burnside, & Cianciotto, 2011); ferric pyoverdine (ex: BGC0002693) does not cross the inner membrane (Bonneau, Roche, & Schalk, 2020); and ferric yersiniabactin (ex: BGC0001055) is imported to the cytoplasm by YbtPQ, a repurposed member of the ABC exporter family (Wang, Hu, & Zheng, 2020). Transporter genes may also be located elsewhere in the chromosome; the zinc–kupyaphore transporter, currently unknown, is not encoded in the BGC (BGC0001627) (Mehdiratta et al., 2022). Finally, bacteria often encode transporters for xenosiderophores that they do not produce, and transporter genes may coincidentally be near an unrelated BGC (Crits-Christoph et al., 2020).

Once inside the cell, the metal must be released from the metallophore for use in metabolic functions. Siderophores generally have a far lower affinity for Fe^{2+} than Fe^{3+}, and thus reduction of the ferric complex is a common strategy (Bonneau et al., 2020; Matzanke, Anemüller, Schünemann, Trautwein, & Hantke, 2004). Genes encoding reductases may be annotated as such, or may be called a "siderophore interacting protein", "vibriobactin utilization protein", "PepSY domain-containing protein", or "FhuF". Note that NIS synthetases also contain a conserved FhuF-like domain that has not been reported to have any reductase activity. Ferric complexes of some catechol siderophores cannot be reduced without destruction of the siderophore; enterobactin-like BGCs often encode an "enterochelin esterase". For other metals, the mechanism of release is largely unknown. Methanobactins have a high affinity for Cu^{+} specifically, and MbnH and MbnP may be responsible for copper release by oxidation to Cu^{2+} (Dassama, Kenney, & Rosenzweig, 2017).

3.3 Metal-responsive regulatory elements

Bacterial metal homeostasis is generally regulated by metal-sensing transcription factors. When intracellular metal concentrations drop, the enzyme disassociates from its bound metal ion(s), changing its shape and ability to bind to DNA. Through this mechanism, a bacterium may respond to metal starvation by upregulating metallophore-dependent and -independent metal acquisition pathways. Several distinct families of global (genome-wide) metalloregulators have been identified; two of the most prevalent are the Fur (ferric uptake regulator) family and the DtxR family (Merchant & Spatafora, 2014; Sevilla, Bes, Peleato, & Fillat, 2021). Metalloregulators themselves are generally not encoded within a metallophore BGC; however, one can sometimes detect the transcription factor binding sites (TFBSs) *in silico* by searching the genome sequence for the binding motif, often called a "box" (e.g., Fur box or iron box). By identifying these TFBS motifs, the genome miner can predict which metal is transported by the metallophore and even find BGCs that are undetectable by homology-based techniques and software (Spohn, Wohlleben, & Stegmann, 2016).

Regulatory element detection was added to antiSMASH v7.0 as the TFBS Finder module (Blin et al., 2023). Metallophore researchers working with *Streptomyces* or other Actinobacteria will find use in the TFBS Finder results for ZuR (Zn), NuR (Ni), and DmdR (Fe). However, the underlying database of motifs, called LogoMotif, currently only contains regulator binding sites from *Streptomyces* (Augustijn et al., 2024). As of antiSMASH v7.1, TFBS Finder gives misleading results for non-Actinomycetes and further investigation is required. In particular, the Proteobacterial Fur binding sites are often marked as ZuR binding sites, if not missed entirely. For taxa and regulators not represented in LogoMotif/TFBS Finder, a standalone version of TFBS finder called Mini-Motif can be used (see Section 6).

Attempts to find regulatory elements may be stymied if the taxon of interest does not have a known TFBS motif, or if metallophores are regulated by more complicated networks. Rather than being directly controlled by a global metal-starvation regulator like Fur, some BGCs are indirectly controlled by a pathway-specific regulator. In Gram-negative bacteria, TBDT genes are sometimes found alongside genes encoding a sigma factor and antisigma factor (Pfam domains "Sigma-70" and "FecR", respectively). These form a "cell-surface signaling" pathway where the transport of a molecule through a TBDT triggers higher expression of that same TBDT (Llamas, Imperi, Visca, & Lamont, 2014). In *Pseudomonas*, pyoverdine production is controlled by the sigma factor PvdS, which is in turn controlled directly by Fur. Therefore, most operons in pyoverdine BGCs do not have Fur boxes. Note that a BGC can also be controlled by multiple regulators simultaneously, as exemplified by the cooperative

Predicting metallophore structure and function through genome mining 381

control of the metallophore staphylopine (BGC0002487) by Zur and Fur (Fojcik et al., 2018), and that metalloregulator pathways can themselves be modulated (Steingard & Helmann, 2023).

4. Predicting metallophore pathways in antiSMASH

The antiSMASH output includes a number of features that aid in the manual investigation of a detected BGC, including in-depth analyses and interactive tools. antiSMASH results are presented as a webpage that displays information in hierarchical detail: the most broadly useful data are prominently displayed, while advanced and specialized results may be found by opening tabs, expandable elements, and links (Blin et al., 2019). Sections 4 and 5 discuss specific antiSMASH features useful to the metallophore genome miner, including the interactive BGC viewer, KnownClusterBlast (KCB), and NRPS/ PKS analyses. Users may discover additional useful features by exploring the documentation (https://docs.antismash.secondarymetabolites.org) and simply by clicking everything.

4.1 Rule-based BGC detection

A biosynthetic class of natural products can often be defined by one or more enzyme domains that perform the key molecular transformation(s). For example, The IucA/IucC family of synthetases defines the NIS pathway (Carroll & Moore, 2018), while all methanobactin BGCs contain genes encoding maturation enzymes MbnB and MbnC (Dassama et al., 2017). antiSMASH and other "rule-based" detection platforms are built on a manually curated collection of hidden Markov model (HMM) profiles and associated logical rules that define a BGC class. For example, the antiSMASH v7 "methanobactin" rule is equivalent to the statement, "a gene matching the MbnB HMM profile must be within 5 kilobases of a gene matching the MbnC HMM profile". Table 1 gives the status of different metallophore detection rules in antiSMASH. Note that a biosynthetic pathway without a corresponding rule cannot not (yet) be detected, and thus rule-based techniques can only find homologs of known biosyntheses.

Automated tools currently cannot accurately predict the exact borders of a bacterial BGC purely from a submitted sequence. Instead, antiSMASH defines preset "neighborhood" distances around the core biosynthetic genes that are meant to be overly inclusive, as excluding a part of the real

382 — Zachary L. Reitz

BGC is more detrimental to the user than including unrelated genes that an expert eye can filter. For this reason, antiSMASH uses the term "BGC region" rather than "BGC" in its outputs. The antiSMASH algorithm is more fully described in the online documentation and has been reviewed in context with other BGC detection algorithms (Chavali & Rhee, 2018).

4.2 Navigating antiSMASH results

The antiSMASH v7.1.0 results for *Dickeya dadantii* 3937 were retrieved from the antiSMASH database in Section 2.1. The numbering in this section (ex: #1) refers to the annotations in Fig. 1.

The Overview page (Fig. 1A) lists all detected BGC regions, which are numbered by location on the underlying nucleotide sequence(s). The "Type" column indicates the detection rule(s) triggered by each region, with a link to their definition in the antiSMASH documentation. The "From" and "To" columns give the genomic location of the predicted region, which, as described above, will likely not match the true BGC. The "Most similar known cluster" and "Similarity" columns are discussed in Section 5. Results for individual regions can be accessed by clicking on the numbered buttons on the top bar or clicking any of the rows.

Fig. 1B–E shows the results page for Region 5, which can be divided into four main panels.

B. The top left panel features an interactive gene cluster. Genes in Region 5 triggered both the NRPS and NRP-metallophore rules (#1). The central bars above the BGC indicate the span of the core biosynthetic genes that triggered each rule, while the pale bars indicate the expanded "neighborhood" region that may or may not be part of the true BGC. Genes within the BGC region are colored according to their function (#2), which is predicted by the smCOG library (Medema et al., 2011). Note that the coloring may not be 100% accurate. This panel also contains buttons to download an svg file of the BGC for publication-quality figures or the region GenBank file for downstream analyses (#3).

C. The Gene details panel in the upper right is populated by clicking any gene in the interactive gene cluster (#4). Here, the gene details for DDA3937_RS14720 are shown. The panel lists information retrieved from the input sequence (#5), matches to antiSMASH's HMMs (#6), matches to third-party protein family HMM databases (#7), and links to additional tools or information (#8).

Predicting metallophore structure and function through genome mining   383

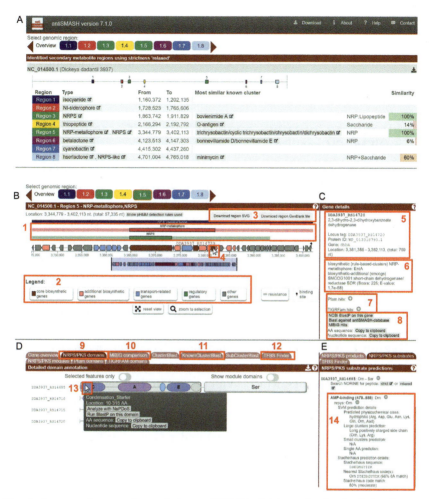

Fig. 1 Annotated screenshots of the antiSMASH v7.1.0 results for *D. dadantii* 3937 (GCF_000147055), retrieved from the antiSMASH database v4. (A) The results overview page. (B–E) The results page for Region 5. Annotated elements are described in Section 4.2.

D. The bottom left panel contains annotation details produced by analysis modules included in antiSMASH. Here, the NRPS/PKS domains results are shown (#9). This chapter also discusses the MIBiG comparison (#10), KnownClusterBlast (KCB, #11), and TFBS Finder (#12) modules. Most annotation detail tabs contain interactive elements with additional information (#13).

E. Some analysis modules give additional information in the bottom right panel. Details behind the adenylation domain substrate prediction of Orn are shown in #14 (see Section 5.2). Note that the structure predicted in the NRPS/PKS products tab is generally incorrect for metallophores.

4.3 Identifying putative metallophore BGC regions among antiSMASH results

This section provides an example of how a researcher can use antiSMASH to quickly determine if BGC regions putatively encode metallophore biosynthesis pathways. The antiSMASH v7.1.0 results for *Dickeya dadantii* 3937 were retrieved from the antiSMASH database in Section 2.1. From the Overview page (Fig. 1A), note that Region 2 is annotated as an "NI-siderophore", while Region 5 is an "NRP-metallophore". The automated class prediction may be enough to answer certain research questions. However, false positives and false negatives are possible (particularly for NRP metallophores), and a manual inspection of each BGC region is advisable.

Region 1: A few isocyanide chalkophores have been discovered (Chen et al., 2021). However, the reported examples (ex: SF2768, BGC0001574) also require NRPS genes, which aren't present in this region. Furthermore, there's no evidence of import machinery. Of the transport-related genes (colored light blue), DDA3937_RS05300 and DDA3937_RS05350 are annotated as MFS transporters (export) under Gene details, while clicking "MIBiG hits" for DDA3937_RS05380 (see Fig. 1C, #8) shows that homologs in known BGCs are generally annotated as exporters. **Not metallophore-like**.

Region 2: In addition to the IucA/IucC-like synthetases that triggered the NI-siderophore rule, a complete import pathway is encoded nearby: expanding Pfam hits (see Fig. 1C, #7) for the transport-related genes reveals a TBDT (DDA3937_RS07715), a PBP (DDA3937_RS07750), and a FecCD-family ABC transport complex (DDA3937_RS07755–65), as well as an MFS exporter. A literature search for "*Dickeya dadantii* NIS" reveals that this BGC region is responsible for production of the siderophore achromobactin (Franza, Mahé, & Expert, 2005). **Metallophore-like**.

Region 3: An NRPS region could encode a metallophore pathway that was not captured by the NRP-metallophore rule. However, there are no putative chelator biosynthesis genes near the NRPS, and clicking the "NRPS/PKS modules" tab (see Fig. 1D, #9) shows that the A domains are predicted to activate Pro and Gly, neither of which are known metallophore chelators or chelator precursors. The only putative transporter genes (DDA3937_RS08250–60) are likely unrelated: they are separated from the NRPS by many genes, do not contain FecCD domain according to Pfam (see Fig. 1C, #7), and are annotated as transporting amino acids. **Not metallophore-like**.

Region 4: No thiopeptide metallophores are known, and no importer genes are present. **Not metallophore-like**.

Region 5 (Fig. 1B): This region was annotated as an NRP-metallophore BGC region due to the presence of matches to EntA and EntC, two genes in the synthesis of the chelating group 2,3-dihydroxybenzoate. Clicking the dark red "core biosynthesis genes" reveals which HMM profiles were matched by each gene (Fig. 1, #4 and #6). Furthermore, a complete import pathway (TBDT, PBP, and ABC) is encoded in close proximity, as well as an esterase (DDA3937_RS14695) commonly found in catechol siderophore BGCs. Note that gene annotations are generally produced by automated HMM-based tools and may be inaccurate; while the protein family is called "enterochelin esterase", this BGC is not consistent with enterochelin specifically (see Section 5.2). **Metallophore-like**.

Regions 6–8: None of these regions show signs of metallophore production. Region 8 encodes a putative ABC importer complex (DDA3937_RS20920–30); however, they are annotated as importing xylose, and the neighboring xylose epimerase (DDA3937_RS20935) supports that annotation. **Not metallophore-like**.

5. Predicting metallophore structure through comparative analyses

5.1 Strategies for comparing BGCs

Genome mining relies on comparisons to known BGCs, which can allow for both BGC dereplication and the prediction of gene functions. The KnownClusterBlast (KCB) module, added in antiSMASH 3, automatically compares a detected BGC to every known BGC in the MIBiG database (Weber et al., 2015). The top KCB match is given on the Overview page as the "Most similar known cluster" (Fig. 1A), and full results are presented in the "KnownClusterBlast" tab of each BGC page (Fig. 1D, #11 and Fig. 2). Note that the related "ClusterBlast" tab contains matches from the antiSMASH database rather than the MIBiG database. In the KCB results, each matching

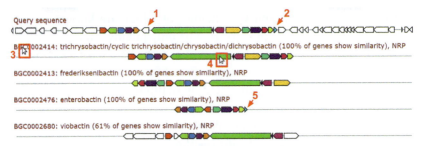

Fig. 2 Annotated screenshot of the antiSMASH v7.1.0 KnownClusterBlast results for *D. dadantii* 3937 (GCF_000147055) Region 5, retrieved from the antiSMASH database v4.

MIBiG BGC is presented with a similarity score such as "70% of genes show similarity". Specifically, this means that 70% of genes in the MIBiG entry have at least one homolog in the query BGC with at least 45% sequence identity. KCB is also used to generate the "MIBiG Hits" table found under "Gene details" for an individual gene (Fig. 1C, #8), which shows homologs meeting a more relaxed 20% sequence identity cutoff. ClusterCompare is a related module first added in antiSMASH 6 and available under the name "MIBiG comparison" (Fig. 1D, #10). The ClusterCompare algorithm was designed to perform better than KCB on multi-domain enzymes like NRPSs and PKSs, and the results can be interpreted similarly to KCB. Note that MIBiG entries are crowd-sourced and some contain errors, including truncated or overextended nucleotide ranges. Each entry contains citations that should be used to verify the information. Furthermore, many characterized metallophore BGCs are not yet in MIBiG; for example, the achromobactin locus in Region 2 of the *D. dadantii* 3937 genome (Franza et al., 2005). Comparative analyses can be done with tools such as BLAST or clinker in these cases (Johnson et al., 2008; Van Den Belt et al., 2023).

When comparing a metallophore BGC to a reference cluster, carefully consider the following questions. (1) Does the BGC of interest have missing or extra genes relative to the known cluster? If so, try to determine their relevance towards the research question by performing literature searches. Two gene clusters that differ only in the presence of transporter genes could reasonably produce the same compound, whereas a missing decarboxylase (see Section 5.3) or an extra methyltransferase suggest a modified structure. (2) Do NRPS/PKS genes, if present, have the same domains in the same order? Modular enzymes with different domain architectures are unlikely to produce the same compound, although exceptions are known (Jenner et al., 2023). Note that KCB and ClusterCompare often return matches with

Predicting metallophore structure and function through genome mining   **387**

completely different architectures, and these matches can still have a 100% KCB similarity score (for example, see Region 3 of the *D. dadantii* 3937 genome). (3) Do the NRPS adenylation (A) domains, if present, have the same predicted substrate? Substrate prediction is still an unsolved problem (Mongia et al., 2023), and modern tools frequently misannotate uncommon chelating amino acids and cationic amino acids such as Lys, Arg, ornithine (Orn), and 2,4-diaminobutyrate. Nevertheless, closely related A domains that activate the same substrate often have the same prediction (albeit an incorrect one). Note that each A domain substrate prediction tool and each version of antiSMASH gives different predictions due to differences in the underlying algorithms and training data, so only results from the same version and tool can be compared directly in this manner. In antiSMASH 7, a new prediction algorithm called NRPyS was introduced (Blin et al., 2023), which uses both a support vector machine (SVM) and comparisons of the active site residues (called the "Stachelhaus code") against a database of A domains with known selectivity (as detailed in Röttig et al., 2011).

5.2 Predicting the structure of a peptide metallophore from *Dickeya dadantii* 3937

Researchers sometimes ask me for the minimum KCB similarity score that can be "trusted" to state that a BGC produces a certain compound. Unfortunately, no such cutoff exists. Only experimental evidence can confirm that a BGC is responsible for producing a certain molecule. As demonstrated herein, even 100% of genes showing similarity and $> 90\%$ sequence identity does not guarantee that two BGCs produce the same compounds.

1. Retrieve the antiSMASH results for *D. dadantii* 3937 from the antiSMASH database (NCBI assembly ID: GCF_000147055.1) as described in Section 2.1. Note that the Overview page shows that Region 5's "Most similar known cluster" from KCB (BGC0002414, trichrysobactin family) has 100% similarity.

2. Click on Region 5, then the KnownClusterBlast results tab (Fig. 1D, #11 and Fig. 2). In addition to trichrysobactin, two other BGCs, each producing different siderophores, are a "perfect match" with 100% of genes showing similarity: BGC0002413 (frederiksenibactin) and BGC0002476 (enterobactin). The clusters are ordered by total BlastP bit score, which corresponds to amino acid sequence similarity. Clicking gene arrows in the KCB tab provides more information on the individual gene comparisons

(Fig. 2, #4). The *D. dadantii* 3937 NRPS DDA3937_RS14685 is 92% identical to the trichrysobactin NRPS and 50% identical to the frederiksenibactin NRPS, and thus trichrysobactin is still the best candidate. Note that the enterobactin record in MIBiG v3.1 has incorrect boundaries that do not include the NRPS.

3. Examine the differences in gene content between the *D. dadantii* 3937 region and the trichrysobactin BGC using KCB results. The uncolored flanking genes DDA3937_RS14610–14650 and DDA3937_RS14730–14820 have no clear relationship to metallophore pathways. Mobile genetic elements such as DDA3937_RS14680 (Fig. 2, #1) are not likely to be relevant unless they disrupt protein function by insertion within a coding region (Naka, López, & Crosa, 2008). The only relevant gene not present in trichrysobactin locus is the thioesterase DDA3937_RS14725 (Fig. 2, #2). KCB shows that this gene is homologous to b0597 in the enterobactin locus (Fig. 2, #5), a proofreading thioesterase which aids in efficient enterobactin production but does not affect the final molecular structure (Leduc, Battesti, & Bouveret, 2007).

4. Next, compare NRPS domain architectures between Region 5 and the reference BGCs.

   a. Click the "NRPS/PKS domains" tab, then click each domain in the NRPS gene DDA3937_RS14685 to see the predicted domains (Fig. 1D, #9 and #13).

   b. Note the domains in order: Condensation_Starter, AMP-binding, PCP, Epimerization, TIGR1720, Condensation_DCL, AMP-binding, PCP, and Thioesterase (a glossary of NRPS domains is available in the antiSMASH documentation).

   c. From the KCB results, click "BGC0002414" to open the MIBiG entry for trichrysobactin in a new tab (Fig. 2, #3). Note that MIBiG entries have nearly the same layout as antiSMASH results.

   d. Click the "NRPS/PKS domains" tab, then click each domain in the NRPS gene JJO56_07700. Note that the domains are identical to the Region 5 BGC.

   e. Repeat this process with BGC0002413 (frederiksenibactin), and note that two domains (Epimerization and TIGR1720) are missing. Therefore, frederiksenibactin can be eliminated as a probable siderophore of *D. dadantii* 3937.

5. Next, compare the predicted NRPS substrates to the reference compound. Under the NRPS/PKS domains tab (Fig. 1D, #9), the predicted amino acids are D-Orn and [L-]Ser. In contrast, the published structure

of trichrysobactin contains D-Lys and [L-]Ser (Sandy & Butler, 2011). This either could indicate that the *D.dadantii* 3937 NRPS truly activates Orn to produce a different compound, or could simply be a poor substrate prediction.

6. Investigate the A domain predictions in the *D.dadantii* 3937 antiSMASH results.

   a. In the bottom right panel, click "NRPS/PKS substrates".

   b. Fully expand the prediction details for DDA3937_RS14685 and Orn (Fig. 1E, #14). Note that the NRPyS SVM model is no more specific than "Long positively charged side chain (Orn, Lys, Arg)". Orn was predicted based on the Stachelhaus sequence, which is an "80% (moderate)" match to a known sequence in the NRPyS training set. Stachelhaus predictive capabilities drop sharply between 100% and 80%; therefore, the Orn prediction has a significant chance of being incorrect.

   c. In contrast, expand predictions for the second A domain of DDA3937_RS14685 to see a more trustworthy prediction: both algorithms predict Ser, and the Stachelhaus code match is "100% (strong)".

   d. Finally, note that the standalone A domain DDA3937_RS14710 is labeled as "X", indicating that the NRPyS predictive algorithms could not determine a consensus substrate. Fully expand the details for "X" to see that both underlying models predict "diOH-Bz" (2,3-dihydroxybenzoate, DHB), but "Sal" (salicylate) is also listed as a possible match. The former is more likely given that this A domain is part of a DHB biosynthesis subcluster (DDA3937_RS14705–20).

7. Retrieve the A domain predictions for the closest known BGC, trichrysobactin, for comparison.

   a. The MIBiG entry for BGC0002414 states "This entry is originally from NCBI GenBank JAFCAF010000006.1." Searching the antiSMASH database v4 for this accession yields no results.

   b. Generate antiSMASH results for the record according to Section 2.2.

   c. Expand the "NRPS/PKS substrates" results for JJO56_07700. Note that NRPyS incorrectly predicts Orn instead of the known substrate Lys, and the prediction details are identical to those of DDA3937_RS14685 (Fig. 1E, #14). Based on the fact that NRPyS misannotates the trichrysobactin NRPS and their high sequence identity, we can reject the Orn prediction for the *D.dadantii* 3937 NRPS in favor of Lys.

Based on these observations, the antiSMASH results strongly suggest that Region 5 contains a catechol siderophore BGC spanning from DDA3937_RS14655 to DDA3937_RS14725 that putatively produces siderophores in the (tri)chrysobactin family. We found no indication that Region 5 and the trichrysobactin BGC would produce different compounds, and the two BGCs share > 90% sequence identity. Nevertheless, stating that the genome "contains a trichrysobactin BGC" would be unsubstantiated. In fact, trichrysobactin is not produced by this strain under previously reported conditions: *D. dadantii* 3937 produces the chrysobactin monomer, but not the oligomers dichrysobactin, trichrysobactin, and cyclic trichrysobactin (Persmark, Expert, & Neilands, 1989).

5.3 Predicting the structure of a putatively novel metallophore in *Acinetobacter pittii* PHEA-2

1. Retrieve the antiSMASH results for *A. pittii* PHEA-2 from the antiSMASH database (NCBI assembly ID: GCF_000191145.1), by adapting the procedure in Section 2.1.
2. From the Overview page, note that Region 6 is predicted to contain an NI-siderophore BGC. The most similar cluster by KCB is the BGC for the *S. aureus* siderophore staphyloferrin B, with a low similarity of 16%.
3. View the results for Region 6. Examine each gene in the BGC region using the "Gene overview" tab at the bottom and the "Gene details" panel on the right, in addition to literature searches. Record observations, as exemplified in Table 2. Focus on genes that surround the core biosynthetic genes (colored dark red) and assign putative metallophore-related functions. As we move outward towards the edge of the region, genes stop having any clear association to metallophores. The BGC, which we have named *apm* (*Acinetobacter pittii* metallophore) for convenience, putatively ranges from BDGL_001497 to BDGL_001503.
4. Consider possible structural features based on the information gathered in Table 2. The three NIS synthetase domains of *apmC* and *apmE* suggest the presence of at least three amide or ester bonds between citrate or α-ketoglutarate and a nucleophile (alcohol or amine). The citrate synthase *apmF* further supports a citrate component. 2,3-Diaminopropionate (Dap) is likely based on the presence of *apmA* and *apmB*, and could serve as a substrate for the NIS synthetases. No other putative biosynthetic genes are present in the cluster.

Table 2 Example annotation of the NIS BGC Region 6 in A. pittii PHEA-2.

Locus tag	Notes
BDGL_001490–BDGL_001495	These two operons are likely involved in nitrate reduction and uptake.
BDGL_001496	Only 38 amino acids; unlikely to be relevant.
BDGL_001497 _apmA*_	First glance suggests that these two genes mark the beginning of the BGC. They are each annotated as encoding an "2,3-diaminopropionate biosynthesis protein" and appear to be co-operonic with the core genes. Furthermore, a literature search for "SbnA SbnB" reveals that the two proteins are involved in the biosynthesis of a known siderophore, staphyloferrin B (Kobylarz, Grigg, Takayama, et al., 2014). KCB results also show this match.
BDGL_001498 _apmB_	
BDGL_001499 _apmC_	Although annotated as a hypothetical protein by NCBI, this gene was identified as "IucA_IucC", a core biosynthetic gene by antiSMASH. Note that there are two Pfam domains: PF04183.15 (IucA/IucC family), and PF06276.15 (Ferric iron reductase FhuF-like transporter). This is typical of the family.
BDGL_001500 _apmD_	MFS transporters are commonly encoded in metallophore BGCs. Characterized examples export metallophores through the inner membrane.
BDGL_001501 _apmE_	Note that this _iucA/iucC_-like gene is about twice as long as the other _iucA/iucC_-like gene _apmC_. This suggests that more than one domain is present. Expanding the "Pfam hits" under Gene details reveals a fusion of two NIS synthetases, each of which putatively has a separate function. Gene fusions are easy to miss, but are otherwise unlikely to be relevant for structural prediction.

(continued)

Table 2 Example annotation of the NIS BGC Region 6 in A. pittii PHEA-2. (*cont'd*)

Locus tag	Notes
BDGL_001502 *apmF*	Click "MIBiG Hits" under Gene details. The top hit is "Siderophore_staphylobactin_biosynthesis_protein_SbnG", which is known to be a citrate synthase (Kobylarz, Grigg, Sheldon, Heinrichs, & Murphy, 2014).
BDGL_001503 *apmG*	Encodes a TonB-dependent outer membrane receptor.
BDGL_001504	DNA polymerase in a family associated with UV damage repair. No clear relation to metallophores.
BDGL_001505-BDGL_001507	Annotated as aroQ/accB/accC, an operon involved in fatty acid synthesis (Chacón et al., 2023). Note that KCB reports homologs in Polyketide type BGCs. There are no putative acyltransferases in our BGC that could attach a fatty acid to our molecule, so this operon is likely unrelated.
BDGL_001508-BDGL_001512	Given that the previous four genes putatively lie outside the cluster, these genes just require a cursory check to look for MIBiG hits or siderophore-related annotations. BDGL_001509 encodes an MFS transporter, but the more likely exporter for this pathway is BDGL_001500.

*Genes that putatively make up the metallophore BGC have been named for use in the text.

5. Compare the *apm* locus to related siderophore BGCs. Consider that the closest BGC may not (yet) be in the MIBiG database used by KCB. For each reference BGC, record any differences in gene content that could lead to structural variation.
   a. A literature search for "*Acinetobacter* citrate siderophore" in PubMed or Google Scholar finds acinetoferrin (BGC0000295) and baumannoferrins (BGC0002471). Both are hydroxamate siderophores, and accordingly, each locus contains genes encoding an N-monooxygenase and an fatty-acyltransferase. Neither gene is present in *apm*, and thus neither a hydroxmate residue nor a fatty acid are predicted. Furthermore, the acinetoferrin locus contains only two NIS synthetase domains, compared to three in *apm*.
   b. The most similar cluster by KCB is BGC0000943, the staphyloferrin B locus (*sbn*) from *Staphylococcus aureus* NCTC 8325, at 18% of genes showing similarity. Only *apmAB* and *sbnAB* are similar enough to meet the 45% sequence identity cutoff used for the KCB results tab. However, a manual examination of the gene-wise "MIBiG hits" tables, which use a 20% cutoff, shows that all five putative biosynthetic genes (*apmABCEF*) have homologs in *sbn* (Fig. 3). Conversely, *sbn* contains two biosynthetic genes, *sbnH* and *sbnI*, with no homologs in *apm*. A literature search reveals that SbnI generates O-phosphoserine, the substrate for Dap synthesis by SbnAB (Verstraete, Perez-Borrajero, Brown, Heinrichs, & Murphy, 2018). SbnI is redundant with primary metabolism, so the absence of an *sbnI* homolog in *apm* is unlikely to affect the structure. On the other hand, SbnH is a decarboxylase required for staphyloferrin B biosynthesis (Scheme 1) that converts citryl-Dap (**1**) to citryl-diaminoethane (**2**) (Cheung, Beasley, Liu, Lajoie, & Heinrichs, 2009). The differential transporter genes (*apmG* versus *sirABC*) can be ignored for structural prediction.

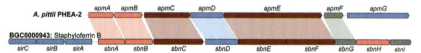

Fig. 3 Comparison of the putative NIS BGC of *A. pitti* PHEA-2 (BDGL_001497–001503) and the staphyloferrin B locus from *S. aureus* NCTC 8325 (BGC0000943). Homologous genes are connected. The loci images were retrieved from antiSMASH and MIBiG, respectively, using the SVG export buttons (Fig. 1B, #3). Genes are colored according to the antiSMASH scheme (Fig. 1B, #2).

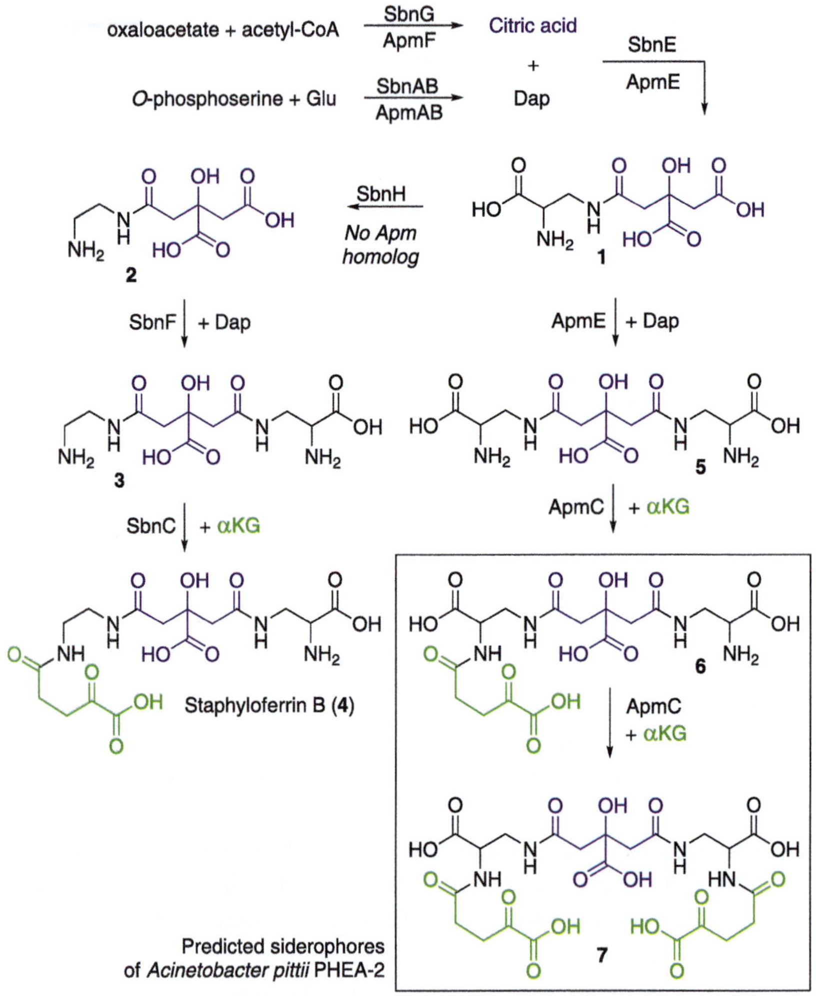

Scheme 1 Predicted biosynthesis and structures of *Acinetobacter pittii* PHEA-2 metallophores (**6** and/or **7**) by the *apm* locus, compared to the characterized biosynthetic pathway of staphyloferrin B (**4**) by the *sbn* locus. Dap, 2,3-diaminopropionate; αKG, α-ketoglutarate.

  c. Determine if a homolog of *sbnH* is located elsewhere in the *A. pittii* genome via NCBI BLAST.

    i. From the staphyloferrin B MIBiG entry (BGC0000943), click on *sbnH* (NCTC8325_00072). The Gene details give the NCBI Protein ID, SQF72452.1.

    ii. Follow the instructions in Section 2.4 to generate BLAST results.

iii. Click the "Alignments" tab to see that there is one homologous sequence in *A. pittii* PHEA-2 at 27% sequence identity and 45% sequence similarity (Fig. 4, #1).
iv. Click "Graphics" next to Range 1 then zoom out slightly to view the genomic context of the match (Fig. 4, #2 and #3).
v. Note that the matching gene is annotated as *lysA*, and the neighboring gene is *dapF* (Fig. 4, #4). A literature search shows that these genes are the final two steps of Lys biosynthesis. The most likely substrate for *lysA* is therefore the primary metabolite 2,6-diaminopimelate. Because there is no paralog in the genome, there is no candidate citryl-Dap decarboxylase in *A. pittii*.

6. Based on the above results, propose a biosynthetic pathway and product (s) consistent with the *apm* locus. Start with the reported biosynthesis of staphyloferrin B (Cheung et al., 2009), putatively assigning each role in *sbn* to the nearest homolog in *apm* (Scheme 1). ApmA and ApmB could synthesize Dap from O-phosphoserine and Glu, while ApmF could

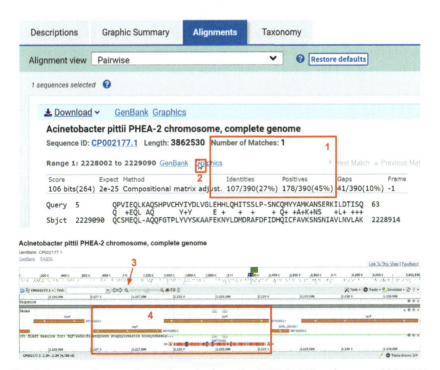

Fig. 4 Annotated screenshots of results from the NCBI BLAST web server. SQF72452 was used as a query against *A. pittii* PHEA-2 (taxid:871585) using tblastn.

synthesize citric acid from oxaloacetate and acetyl-CoA. ApmE could catalyze similar reactions to SbnE and SbnF, successively forming peptide bonds between citrate and two molecules of Dap to form compounds **1** and **5**. Decarboxylation to citryl-diaminoethane (**2**) cannot occur, as there is no *sbnH* homolog in *apm*. ApmC could then condense α-ketoglutarate (αKG) to one or both of the Dap residues, forming **6** and **7**, respectively. Based on comparisons to known αKG-containing siderophores, arguments could be made for either **6** or **7** as the final structure. In such cases of ambiguity, a full list of candidates is preferable for experimental detection and characterization. Finally, check that all biosynthetic genes in *apm* have been used in the proposed pathway.

In sum, a genome analysis predicts that the *apm* locus of *A. pittii* PHEA-2 encodes the biosynthesis of a novel metallophore that is structurally similar to staphyloferrin B and consists of the substructures citrate, Dap, and α-ketoglutarate. Specifically, products **6** and **7** are consistent with the BGC and known homologous BGCs. These results can help the experimentalist isolate and characterize the true compound(s) by predicting the molecular mass and expected fragmentation. If experimental results suggest that the predicted structure is inaccurate, consider cyclization (seen in other αKG-containing siderophores), gene inactivation, different enzyme substrates, or a secondary genomic locus, among other possibilities.

6. Next steps for the reader

The genome mining methods described in this chapter will enable the metallophore researcher to make testable structural and functional predictions in their system of interest. The careful, manual inspection of a BGC is an important foundational skill in metallophore genomics; however, the genome miner's toolkit extends far beyond antiSMASH and BLAST. Manual techniques are often not feasible for large-scale analyses such as prioritizing strains in a library, exploring the capabilities of a metagenome, or reconstructing a pathway's evolutionary history. A wide range of user-friendly bioinformatics tools have been developed to address and automate these and other problems. Some tools are available as publicly available web servers, but the majority will require learning basic command line skills. Researchers with programming experience may write their own custom workflows that process antiSMASH outputs to automate their analyses, or even locally modify the source code of existing software to meet their needs. Many programming languages have packages for parsing, converting, and manipulating biological files; a beginner might choose Biopython (Cock et al., 2009), which is used inside antiSMASH and many other Python-based tools. A small selection of free, open-source tools is presented here. While their usage is outside the scope of this chapter, many have tutorials in their documentation.

Web servers.

- **CAGECAT:** a platform for the comparative genomics tools **cblaster** and **clinker** (Van Den Belt et al., 2023). cblaster can be used to find homologs of a BGC in NCBI genome databases. clinker automatically generates BGC comparison figures similar to Fig. 3.
- **EFI:** a platform for the paired tools **EFI-EST** and **EFI-GNT** (Zallot, Oberg, & Gerlt, 2019). EFI-EST creates sequence similarity networks of protein-coding genes, and EFI-GNT analyzes and visualizes the genomic context of genes in those networks.
- **EMBL-EBI:** a huge collection of tools, including
  o **InterProScan**, which accepts a protein sequence and finds matches to protein family databases as well as signal peptides and other sequence features (Blum et al., 2021), and
  o **HMMER web server**, which allows for a protein sequence, sequence alignment, or HMM profile to be searched against sequence or domain databases. Results can be filtered based on domain architecture or taxonomy (Potter et al., 2018).
- **NGPhylogeny.fr:** a customizable "one-click" workflow for reconstructing phylogenetic trees from a collection of sequences, aimed at the non-specialist (Lemoine et al., 2019).

Command-line tools.

- **BGCFlow:** an extensive workflow for genome mining in bacterial pangenomes (Nuhamunada, Mohite, Phaneuf, Palsson, & Weber, 2024).
- **BiG-SCAPE:** a popular tool that performs BGC-to-BGC comparisons and clusters BGCs into related families, aiding in dereplication (Navarro-Muñoz et al., 2020). Packaged with **CORASON**, which constructs a phylogenetic tree of BGCs within a family.
- **lsaBGC:** a collection of tools for studying BGC diversity, evolution, and abundance within a taxon of interest (Salamzade et al., 2023).
- **MiniMotif:** the command-line version of antiSMASH's TFBS Finder (Augustijn, 2024). MiniMotif accepts a custom list of known TFBSs, constructs a probabilistic model called a position weight matrix (PWM), and scans a genome of interest for matches to the PWM motif.

- **multiSMASH:** a workflow that automates large-scale antiSMASH analyses (Reitz, 2023). Detected BGC regions are tabulated and can be sent to BiG-SCAPE/CORASON.

Acknowledgments

I appreciate the extensive testing and comments provided by Allegra Aron, Hannah Augustijn, Taruna Schuelke, Marquis Yazzie, and Alexa Zytnick. Funding for this work was provided by the Simons Foundation (Award 689265 to H.V. Moeller).

References

Adamek, M., Alanjary, M., & Ziemert, N. (2019). Applied evolution: Phylogeny-based approaches in natural products research. *Natural Product Reports, 36*(9), 1295–1312.

Augustijn, H. E. (2024). MiniMotif: Detect transcription factor binding sites in bacterial genomes. *Github*. https://github.com/HAugustijn/MiniMotif.

Augustijn, H. E., Karapliafis, D., Joosten, K. M. M., Rigali, S., Van Wezel, G. P., & Medema, M. H. (2024). LogoMotif: A comprehensive database of transcription factor binding site profiles in Actinobacteria. *Journal of Molecular Biology*, 168558.

Batey, S. F. D., Davie, M. J., Hems, E. S., Liston, J. D., Scott, T. A., Alt, S., ... Wilkinson, B. (2023). The catechol moiety of obafluorin is essential for antibacterial activity. *RSC Chemical Biology, 4*(11), 926–941.

Blin, K., Shaw, S., Augustijn, H. E., Reitz, Z. L., Biermann, F., Alanjary, M., ... Weber, T. (2023). antiSMASH 7.0: New and improved predictions for detection, regulation, chemical structures and visualisation. *Nucleic Acids Research, 51*(W1), W46–W50.

Blin, K., Shaw, S., Medema, M. H., & Weber, T. (2024). The antiSMASH database version 4: Additional genomes and BGCs, new sequence-based searches and more. *Nucleic Acids Research, 52*(D1), D586–D589.

Blin, K., Shaw, S., Steinke, K., Villebro, R., Ziemert, N., Lee, S. Y., ... Weber, T. (2019). antiSMASH 5.0: Updates to the secondary metabolite genome mining pipeline. *Nucleic Acids Research, 47*(W1), W81–W87.

Blum, M., Chang, H.-Y., Chuguransky, S., Grego, T., Kandasaamy, S., Mitchell, A., ... Finn, R. D. (2021). The InterPro protein families and domains database: 20 years on. *Nucleic Acids Research, 49*(D1), D344–D354.

Bonneau, A., Roche, B., & Schalk, I. J. (2020). Iron acquisition in *Pseudomonas aeruginosa* by the siderophore pyoverdine: An intricate interacting network including periplasmic and membrane proteins. *Scientific Reports, 10*(1), 120.

Cano-Prieto, C., Undabarrena, A., De Carvalho, A. C., Keasling, J. D., & Cruz-Morales, P. (2024). Triumphs and challenges of natural product discovery in the postgenomic era. *Annual Review of Biochemistry*. https://doi.org/10.1146/annurev-biochem-032620-104731.

Carroll, C. S., & Moore, M. M. (2018). Ironing out siderophore biosynthesis: A review of non-ribosomal peptide synthetase (NRPS)-independent siderophore synthetases. *Critical Reviews in Biochemistry and Molecular Biology, 53*(4), 356–381.

Chacón, L., Kuropka, B., González-Tortuero, E., Schreiber, F., Rojas-Jiménez, K., & Rodríguez-Rojas, A. (2023). Mechanisms of low susceptibility to the disinfectant benzalkonium chloride in a multidrug-resistant environmental isolate of *Aeromonas hydrophila*. *Frontiers in Microbiology, 14*, 1180128.

Chatfield, C. H., Mulhern, B. J., Burnside, D. M., & Cianciotto, N. P. (2011). *Legionella pneumophila* LbtU acts as a novel, TonB-independent receptor for the legiobactin siderophore. *Journal of Bacteriology, 193*(7), 1563–1575.

Chavali, A. K., & Rhee, S. Y. (2018). Bioinformatics tools for the identification of gene clusters that biosynthesize specialized metabolites. *Briefings in Bioinformatics, 19*(5), 1022–1034.

Chen, T.-Y., Chen, J., Tang, Y., Zhou, J., Guo, Y., & Chang, W.-C. (2021). Current understanding toward isonitrile group biosynthesis and mechanism. *Chinese Journal of Chemistry, 39*(2), 463–472.

Cheung, J., Beasley, F. C., Liu, S., Lajoie, G. A., & Heinrichs, D. E. (2009). Molecular characterization of staphyloferrin B biosynthesis in *Staphylococcus aureus*. *Molecular Microbiology, 74*(3), 594–608.

Cock, P. J. A., Antao, T., Chang, J. T., Chapman, B. A., Cox, C. J., Dalke, A., ... De Hoon, M. J. L. (2009). Biopython: Freely available Python tools for computational molecular biology and bioinformatics. *Bioinformatics (Oxford, England), 25*(11), 1422–1423.

Crits-Christoph, A., Bhattacharya, N., Olm, M. R., Song, Y. S., & Banfield, J. F. (2020). Transporter genes in biosynthetic gene clusters predict metabolite characteristics and siderophore activity. *Genome Research, 31*, 239–250.

Dassama, L. M. K., Kenney, G. E., & Rosenzweig, A. C. (2017). Methanobactins: From genome to function. *Metallomics: Integrated Biometal Science, 9*(1), 7–20.

Drechsel, H., & Winkelmann, G. (1997). *Iron Chelation and Siderophores. Transition Metals in Microbial Metabolism.* London: CRC Press, 1–49.

Fojcik, C., Arnoux, P., Ouerdane, L., Aigle, M., Alfonsi, L., & Borezée-Durant, E. (2018). Independent and cooperative regulation of staphylopine biosynthesis and trafficking by Fur and Zur. *Molecular Microbiology, 108*(2), 159–177.

Franza, T., Mahé, B., & Expert, D. (2005). *Erwinia chrysanthemi* requires a second iron transport route dependent of the siderophore achromobactin for extracellular growth and plant infection. *Molecular Microbiology, 55*(1), 261–275.

Frei, A., Verderosa, A. D., Elliott, A. G., Zuegg, J., & Blaskovich, M. A. T. (2023). Metals to combat antimicrobial resistance. *Nature Reviews Chemistry, 7*(3), 202–224.

Gu, S., Yang, T., Shao, Z., Wang, T., Cao, K., Jousset, A., ... Pommier, T. (2020). Siderophore-mediated interactions determine the disease suppressiveness of microbial consortia. *mSystems, 5*(3), e00811–e00819.

Hider, R. C., & Kong, X. (2010). Chemistry and biology of siderophores. *Natural Product Reports, 27*(5), 637–657.

Jenner, M., Hai, Y., Nguyen, H. H., Passmore, M., Skyrud, W., Kim, J., ... Tang, Y. (2023). Elucidating the molecular programming of a nonlinear non-ribosomal peptide synthetase responsible for fungal siderophore biosynthesis. *Nature Communications, 14*(1), 2832.

Johnson, M., Zaretskaya, I., Raytselis, Y., Merezhuk, Y., McGinnis, S., & Madden, T. L. (2008). NCBI BLAST: A better web interface. *Nucleic Acids Research, 36*(Web Server Issue), W5–W9.

Johnstone, T. C., & Nolan, E. M. (2015). Beyond iron: Non-classical biological functions of bacterial siderophores. *Dalton Transactions, 44*(14), 6320–6339.

Kobylarz, M. J., Grigg, J. C., Sheldon, J. R., Heinrichs, D. E., & Murphy, M. E. P. (2014). SbnG, a citrate synthase in *Staphylococcus aureus*: A new fold on an old enzyme. *The Journal of Biological Chemistry, 289*(49), 33797–33807.

Kobylarz, M. J., Grigg, J. C., Takayama, S.-I. J., Rai, D. K., Heinrichs, D. E., & Murphy, M. E. P. (2014). Synthesis of L-2,3-diaminopropionic acid, a siderophore and antibiotic precursor. *Chemistry & Biology, 21*(3), 379–388.

Kraemer, S. M., Duckworth, O. W., Harrington, J. M., & Schenkeveld, W. D. C. (2015). Metallophores and trace metal biogeochemistry. *Aquatic Geochemistry, 21*(2), 159–195.

Kramer, J., Özkaya, Ö., & Kümmerli, R. (2020). Bacterial siderophores in community and host interactions. *Nature Reviews. Microbiology, 18*(3), 152–163.

Leduc, D., Battesti, A., & Bouveret, E. (2007). The hotdog thioesterase EntH (YbdB) plays a role in vivo in optimal enterobactin biosynthesis by interacting with the ArCP domain of EntB. *Journal of Bacteriology, 189*(19), 7112–7126.

Lemoine, F., Correia, D., Lefort, V., Doppelt-Azeroual, O., Mareuil, F., Cohen-Boulakia, S., & Gascuel, O. (2019). NGPhylogeny.fr: New generation phylogenetic services for non-specialists. *Nucleic Acids Research, 47*(W1), W260–W265.

Llamas, M. A., Imperi, F., Visca, P., & Lamont, I. L. (2014). Cell-surface signaling in Pseudomonas: Stress responses, iron transport, and pathogenicity. *FEMS Microbiology Reviews, 38*(4), 569–597.

Matzanke, B. F., Anemüller, S., Schünemann, V., Trautwein, A. X., & Hantke, K. (2004). FhuF, part of a siderophore-reductase system. *Biochemistry, 43*(5), 1386–1392.

Medema, M. H., Blin, K., Cimermancic, P., de Jager, V., Zakrzewski, P., Fischbach, M. A., ... Breitling, R. (2011). antiSMASH: Rapid identification, annotation and analysis of secondary metabolite biosynthesis gene clusters in bacterial and fungal genome sequences. *Nucleic Acids Research, 39*(Web Server Issue), W339–W346.

Mehdiratta, K., Singh, S., Sharma, S., Bhosale, R. S., Choudhury, R., Masal, D. P., ... Gokhale, R. S. (2022). Kupyaphores are zinc homeostatic metallophores required for colonization of *Mycobacterium tuberculosis*. *Proceedings of the National Academy of Sciences of the United States of America, 119*(8), https://doi.org/10.1073/pnas.2110293119.

Merchant, A. T., & Spatafora, G. A. (2014). A role for the DtxR family of metalloregulators in gram-positive pathogenesis. *Molecular Oral Microbiology, 29*(1), 1–10.

Miller, B. R., & Gulick, A. M. (2016). Structural biology of nonribosomal peptide synthetases. *Methods in Molecular Biology, 1401*, 3–29.

Mongia, M., Baral, R., Adduri, A., Yan, D., Liu, Y., Bian, Y., ... Mohimani, H. (2023). AdenPredictor: Accurate prediction of the adenylation domain specificity of nonribosomal peptide biosynthetic gene clusters in microbial genomes. *Bioinformatics (Oxford, England), 39*(39 Suppl 1), i40–i46.

Naka, H., López, C. S., & Crosa, J. H. (2008). Reactivation of the vanchrobactin siderophore system of *Vibrio anguillarum* by removal of a chromosomal insertion sequence originated in plasmid pJM1 encoding the anguibactin siderophore system. *Environmental Microbiology, 10*(1), 265–277.

Navarro-Muñoz, J. C., Selem-Mojica, N., Mullowney, M. W., Kautsar, S. A., Tryon, J. H., Parkinson, E. I., ... Medema, M. H. (2020). A computational framework to explore large-scale biosynthetic diversity. *Nature Chemical Biology, 16*(1), 60–68.

Nuhamunada, M., Mohite, O. S., Phaneuf, P. V., Palsson, B. O., & Weber, T. (2024). BGCFlow: Systematic pangenome workflow for the analysis of biosynthetic gene clusters across large genomic datasets. *Nucleic Acids Research* gkae314.

Payne, S. M. (1994). *[25] Detection, isolation, and characterization of siderophores*. *Methods in enzymology, Vol. 235*, Elsevier, 329–344.

Persmark, M., Expert, D., & Neilands, J. B. (1989). Isolation, characterization, and synthesis of chrysobactin, a compound with siderophore activity from *Erwinia chrysanthemi*. *The Journal of Biological Chemistry, 264*(6), 3187–3193.

Potter, S. C., Luciani, A., Eddy, S. R., Park, Y., Lopez, R., & Finn, R. D. (2018). HMMER web server: 2018 update. *Nucleic Acids Research, 46*(W1), W200–W204.

Reitz, Z. L. (2023). multiSMASH: A workflow and scripts for large-scale antiSMASH analyses. *Zenodo*. https://doi.org/10.5281/zenodo.8276143.

Reitz, Z. L., Butler, A., & Medema, M. H. (2022). Automated genome mining predicts combinatorial diversity and taxonomic distribution of peptide metallophore structures (p. 2022.12.14.519525) *bioRxiv*. https://doi.org/10.1101/2022.12.14.519525.

Reitz, Z. L., & Medema, M. H. (2022). Genome mining strategies for metallophore discovery. *Current Opinion in Biotechnology, 77*, 102757.

Röttig, M., Medema, M. H., Blin, K., Weber, T., Rausch, C., & Kohlbacher, O. (2011). NRPSpredictor2—A web server for predicting NRPS adenylation domain specificity. *Nucleic Acids Research, 39*(suppl 2), W362–W367.

Salamzade, R., Cheong, J. Z. A., Sandstrom, S., Swaney, M. H., Stubbendieck, R. M., Starr, N. L., ... Kalan, L. R. (2023). Evolutionary investigations of the biosynthetic diversity in the skin microbiome using lsaBGC. *Microbial Genomics, 9*(4), 000988.

Sandy, M., & Butler, A. (2011). Chrysobactin siderophores produced by *Dickeya chrysanthemi* EC16. *Journal of Natural Products, 74*(5), 1207–1212.

Schmitt, I., & Barker, F. K. (2009). Phylogenetic methods in natural product research. *Natural Product Reports, 26*(12), 1585–1602.

Sevilla, E., Bes, M. T., Peleato, M. L., & Fillat, M. F. (2021). Fur-like proteins: Beyond the ferric uptake regulator (Fur) paralog. *Archives of Biochemistry and Biophysics, 701*, 108770.

Soares, E. V. (2022). Perspective on the biotechnological production of bacterial siderophores and their use. *Applied Microbiology and Biotechnology, 106*(11), 3985–4004.

Spohn, M., Wohlleben, W., & Stegmann, E. (2016). Elucidation of the zinc-dependent regulation in *Amycolatopsis japonicum* enabled the identification of the ethylenediaminedisuccinate ([S,S]-EDDS) genes. *Environmental Microbiology, 18*(4), 1249–1263.

Steingard, C. H., & Helmann, J. D. (2023). Meddling with metal sensors: Fur-family proteins as signaling hubs. *Journal of Bacteriology, 205*(4), e0002223.

Süssmuth, R. D., & Mainz, A. (2017). Nonribosomal peptide synthesis-principles and prospects. *Angewandte Chemie, 56*(14), 3770–3821.

Terlouw, B. R., Blin, K., Navarro-Muñoz, J. C., Avalon, N. E., Chevrette, M. G., Egbert, S., ... Medema, M. H. (2023). MIBiG 3.0: A community-driven effort to annotate experimentally validated biosynthetic gene clusters. *Nucleic Acids Research, 51*(D1), D603–D610.

Van Den Belt, M., Gilchrist, C., Booth, T. J., Chooi, Y.-H., Medema, M. H., & Alanjary, M. (2023). CAGECAT: The CompArative GEne Cluster Analysis Toolbox for rapid search and visualisation of homologous gene clusters. *BMC Bioinformatics, 24*(1), 181.

Verstraete, M. M., Perez-Borrajero, C., Brown, K. L., Heinrichs, D. E., & Murphy, M. E. P. (2018). SbnI Is a Free Serine Kinase That Generates O-Phospho-l-Serine for Staphyloferrin B Biosynthesis in Staphylococcus Aureus. *Journal of Biological Chemistry, 293*(16), 6147–6160.

Wang, Z., Hu, W., & Zheng, H. (2020). Pathogenic siderophore ABC importer YbtPQ adopts a surprising fold of exporter. *Science Advances, 6*(6), eaay7997.

Weber, T., Blin, K., Duddela, S., Krug, D., Kim, H. U., Bruccoleri, R., ... Medema, M. H. (2015). antiSMASH 3.0-a comprehensive resource for the genome mining of biosynthetic gene clusters. *Nucleic Acids Research, 43*(W1), W237–W243.

Yee, D. A., Niwa, K., Perlatti, B., Chen, M., Li, Y., & Tang, Y. (2023). Genome Mining for Unknown–Unknown Natural Products. *Nature Chemical Biology, 19*(5), 633–640.

Yu, X., Chen, M., Jiang, Z., Hu, Y., & Xie, Z. (2014). The two-component regulators GacS and GacA positively regulate a nonfluorescent siderophore through the Gac/Rsm signaling cascade in high-siderophore-yielding *Pseudomonas* sp. strain HYS. *Journal of Bacteriology, 196*(18), 3259–3270.

Zallot, R., Oberg, N., & Gerlt, J. A. (2019). The EFI web resource for genomic enzymology tools: Leveraging protein, genome, and metagenome databases to discover novel enzymes and metabolic pathways. *Biochemistry, 58*(41), 4169–4182.

Printed and bound by CPI Group (UK) Ltd, Croydon, CR0 4YY
02/12/2024
01798497-0008